PRACTICAL APPLICATIONS OF COMPUTATIONAL INTELLIGENCE TECHNIQUES

INTERNATIONAL SERIES IN
INTELLIGENT TECHNOLOGIES

Prof. Dr. Dr. h.c. Hans-Jürgen Zimmermann, Editor
European Laboratory for Intelligent
  Techniques Engineering
Aachen, Germany

Other books in the series:

Fuzzy Databases: Principles and Applications
by Frederick E. Petry with Patrick Bose

Distributed Fuzzy Control of Multivariable Systems
by Alexander Gegov

Fuzzy Modelling: Paradigms and Practices
by Witold Pedrycz

Fuzzy Logic Foundations and Industrial Applications
by Da Ruan

Fuzzy Sets in Engineering Design and Configuration
by Hans-Juergen Sebastian and Erik K. Antonsson

Consensus Under Fuzziness
by Mario Fedrizzi, Janusz Kacprzyk, and Hannu Nurmi

Uncertainty Analysis in Enginerring Sciences:
Fuzzy Logic, Statistics, and Neural Network Approach
by Bilal M. Ayyub and Madan M. Gupta

Fuzzy Modeling for Control
by Robert Babuška

Traffic Control and Transport Planning:
A Fuzzy Sets and Neural Networks Approach
by Dušan Teodorović and Katarina Vukadinović

Fuzzy Algorithms for Control
by H.B. Verbruggen, H.-J.Zimmermann. and R. Babŭska

Intelligent Systems and Interfaces
by Horia-Nicolai Teodorescu, Daniel Mlynek, Abraham Kandel
  and H.J. Zimmermann

PRACTICAL APPLICATIONS OF COMPUTATIONAL INTELLIGENCE TECHNIQUES

Edited by

Lakhmi Jain
University of South Australia, Adelaide

and

Philippe De Wilde
University of London

KLUWER ACADEMIC PUBLISHERS
Boston/Dordrecht/London

Distributors for North, Central and South America:
Kluwer Academic Publishers
101 Philip Drive
Assinippi Park
Norwell, Massachusetts 02061 USA
Telephone (781) 871-6600
Fax (781) 871-6528
E-Mail < kluwer@wkap.com >

Distributors for all other countries:
Kluwer Academic Publishers Group
Distribution Centre
Post Office Box 322
3300 AH Dordrecht, THE NETHERLANDS
Telephone 31 78 6392 392
Fax 31 78 6546 474
E-Mail < orderdept@wkap.nl >

 Electronic Services < http://www.wkap.nl >

Library of Congress Cataloging-in-Publication Data

Practical applications of computational intelligence techniques / edited by Lakhmi Jain
and Philippe De Wilde.
    p. cm. -- (International series in intelligent technologies ; 16)
  Includes bibliographical references and index.
  ISBN 0-7923-7320-0 (acid-free paper)
    1. Computational intelligence--Industrial applications. I. Jain, L. C. II. De Wilde,
Philippe, 1958- III. Series.

Q342 .P73 2001
006.3--dc21
                                                                        2001029127

Printed on acid-free paper.

Printed in the United States of America

Contents

Chapter 2.
Networked virtual park
N. Magnenat-Thalmann, C. Joslin, and U. Berner

Chapter 3.
Commercial coin recognisers using neural and fuzzy techniques
J.M. Moreno, J. Madrenas, and J. Cabestany

Chapter 4.
Fuzzy techniques in intelligent household appliances
M. Mraz, N. Zimic, I. Lapanja, J. Virant, and B. Skrt

Chapter 5.
Neural prediction in industry: increasing reliability through use of confidence measures and model combination
P.J. Edwards, G. Papadopoulos, and A.F. Murray

Chapter 6.
Handling the back calculation problem in aerial spray models using a genetic algorithm
W.D. Potter, W. Bi, D. Twardus, H. Thistle, M.J. Twery, J. Ghent, and M. Teske

Chapter 7.

Genetic algorithm optimization of a filament winding process modeled in WITNESS

E. Wilson, C.L. Karr, and S. Messimer

Chapter 8.
Genetic algorithm for optimizing the gust loads for predicting aircraft loads and dynamic response
R. Mehrotra, C.L. Karr, and T.A. Zeiler

Chapter 9.
A stochastic dynamic programming technique for property market timing
T.C. Chin and G.T Mills

Chapter 10.
A hybrid approach to breast cancer diagnosis
M. Sordo, H. Buxton, D. Watson

Chapter 11.
Artificial neural networks as a computer aid for lung disease detection and classification in ventilation-perfusion lung scans
G.D. Tourassi, E.D. Frederick, and R.E. Coleman

Chapter 12.
Neural network for classification of focal liver lesions in ultrasound images
H. Yoshida

Preface

Computational intelligence paradigms have attracted the growing interest of researchers, scientists, engineers and application engineers in a number of everyday applications. These applications are not limited to any particular field and include engineering, business, banking and consumer electronics.

Computational intelligence paradigms include artificial intelligence, artificial neural networks, fuzzy systems and evolutionary computing. Artificial neural networks can mimic the biological information processing mechanism in a very limited sense. Evolutionary computing algorithms are used for optimisation applications, and fuzzy logic provides a basis for representing uncertain and imprecise knowledge. It is not the question of using these techniques but people will wonder in the next century, how we lived without these techniques in the last century.

This book contains twelve chapters. The first chapter, by Konar and Jain, is an introduction to computational intelligence paradigms. In the second chapter, Magnenat-Thalmann, Joslin, and Berner present the networked virtual park. It is shown that the network virtual environments (NVEs) hold the key to interactivity in virtual worlds. This virtual park is more than just a park with trees and flowers as it contains virtual attractions. This chapter shows the construction of the scenes and attractions in the park. It also describes the software that permits the user to interact and watch the attraction along with other users.

The third chapter, by Moreno, Madrenas, and Cabestany, is on commercial coin recognisers using neural and fuzzy techniques. This chapter describes the implementation of a classification/decision engine for an automatic coin recogniser. It is shown that the use of artificial neural and fuzzy models overcome some of the problems encountered when using traditional techniques.

The fourth chapter, by Mraz, Zimic, Lapanja, Virant, and Skrt, is on fuzzy techniques in intelligent household applications. This chapter includes several concepts of the use of fuzzy techniques in the analysis and design of control systems for household appliances manufactured by Gorenje GA, Slovenia. It is demonstrated that the application of intelligent techniques has resulted in energy saving while the production cost remains unaltered.

The fifth chapter, by Edwards, Papadopoulos, and Murray is on the application of neural networks in the prediction of paper curl, an important quality metric in the papermaking industry. It is shown that paper curl can be predicted from parameters defining the current characteristics of a reel of paper and the plant machinery using neural network techniques.

The sixth chapter, by Potter, Bi, Twardus, Thistle, Twery, Ghent, and Teske, is on handling the back calculation problem in aerial spray models using the genetic algorithms. The detailed background discussions of aerial spray practice and simulation, and review of the fundamentals of genetic algorithms are presented in this chapter. The experimental results and future directions are included.

The seventh chapter, by Wilson, Karr, and Messimer, is on genetic algorithm optimization of assembly lines to model filament winding using the witness simulation environment. It is demonstrated that the genetic algorithm serves as an effective optimization method for this purpose, and that it is robust enough to be used with a popular simulation environment.

In the eight chapter, Mehrotra, Karr, and Zeiler present a genetic algorithm (GA) for optimizing the gust loads for predicting aircraft loads and dynamic response. The effectiveness of the GA-based search is demonstrated via its application to both linear and nonlinear aircraft models, and by considering several different types of loading in the objective function.

The ninth chapter, by Chin and Mills, is on a stochastic dynamic programming technique for property market timing. The development of a market timing strategy for a property investor who has to decide the allocation of investment funds between the risk-free savings deposit and the comparatively risky property investment is presented. It is demonstrated that the proposed market timing strategy is capable of achieving superior investment returns in the Singapore property market.

The tenth chapter, by Sordo, Buxton, and Watson, is on a hybrid approach to breast cancer diagnosis. A hybrid methodology is proposed that combines knowledge from a domain in the form of simple rules with connectionist learning. This combination allows the use of small sets of data to train the network. It is demonstrated that the proposed approach is capable of classifying complex and limited data in a medical domain.

The eleventh chapter, by Tourassi, Frederick, and Coleman, is on the application of artificial neural networks for the detection and classification of lung disease. The development of a new approach for the diagnostic interpretation of ventilation-perfusion lung scans for patients with clinical suspicion of acute pulmonary embolism.

The last chapter, by Yoshida, is on neural network for classification of focal liver lesions in ultrasound images. The method is unique in the sense that it integrates a process of selection of multiscale texture features and a process of classification by neural network for effective classification. It is demonstrated that the proposed technique has the potential to increase the accuracy of diagnosis of focal liver lesions in ultrasound images.

This book will be useful to researchers, practicing engineers/scientists, and students, who are interested to develop practical applications in computational intelligence environment.

We would like to express our sincere thanks to Berend Jan van der Zwaag, Shaheed Mehta, Ashlesha Jain, and Ajita Jain for their help in the preparation of the manuscript. We are grateful to the authors for their contributions. We also thank the reviewers and Dr Neil Allen for their expertise and time. Our thanks are due to Kluwer Academic Publishers for their excellent editorial assistance.

Chapter 1

An Introduction to Computational Intelligence Paradigms

A. Konar and L.C. Jain

In the past decade, Artificial Intelligence (AI) had a grand paradigm shift from its domain of symbolic to non-symbolic and numeric computation. Prior to the mid-eighties, symbolic logic was used as the unique tool in the development of algorithms for the classical AI problems like reasoning, planning, and machine learning. The incompleteness of the traditional AI was shortly realized, but unfortunately no handy solutions were readily available at the time. In the nineties the monumental developments in fuzzy logic, artificial neural nets, genetic algorithms and probabilistic reasoning models motivated the researchers around the world to explore the possibilities of building more humanlike machines using these new tools. Consequently, a large number of intelligent systems that can complement the behavior of the traditional symbol-processing machines were built by employing these tools. This chapter provides a brief overview on the fundamental AI tools and their synergism, which together is informally known as computational intelligence.

1 Computational Intelligence – a Formal Definition

The phrase 'Computational Intelligence' is a new coinage in the disciplines of machine intelligence [1]. There is no well-accepted definition on this subject and its domain till date. In fact the next generation of this discipline will come up with many newer computational models, whose performance may far exceed the performance of the currently used models. So, the definition of computational intelligence should not limit the scope of its subsequent generations. Keeping this in mind we can formally define *computational intelligence as the computational models and tools of intelligence that can improve the intellectual behavior of machines.* The currently available tools of computational intelligence include fuzzy sets [11]-[16], neural networks [19]-[45], genetic algorithms [46]-[65], probabilistic reasoning [2]-[10], and some aspects of chaos theory and computational learning theory [32], [68]. For understanding computational intelligence, we first need to learn the scope of these tools, so that their individual merits are not sacrificed and the limitation of one tool can be overcome by the judicious use of the others. Thus instead of competition we can use the collective and hence synergistic behavior of the computational tools in modeling complex systems. The next few sections discuss the scope of the individual tools of computational intelligence.

2   The Logic of Fuzzy Sets

Proposed by Zadeh in 1965 [13], this logic is an extension of the classical propositional and predicate logic that rests on the principles of the binary truth functionality. For instance, let us consider the well-known *modus ponens* rule in propositional logic, given by:

$$p$$
$$\underline{p \to q}$$
$$q$$

where p and q are binary propositions and $\to$ denotes an 'if-then' operator. The rule states that given p is true and 'if p then q' is true, it can be inferred that q is true. But what happens when the truth or falsehood of p cannot be guaranteed? There are three situations that may arise in this context. First, we may take the support or refute of other statements in the system, if any, to determine the truth or falsehood of p. If majority support for p is obtained, then p may be accepted to be true, else p is assumed to be false. This idea led to the foundation of a new class of logic, called **non-monotonic logic** [66]. But what happens when the exact binary truth functional value of p cannot be ascertained. This called for the other two situations. Suppose p is partially true. This can be represented by a truth functional value in between 0 and 1, and this idea was later formalized as the basis of

multi-valued logic. The second situation thus presumes a non-binary truth functional value of p and attempts to infer a truth functional value of q in the range [0,1]. Now, consider the third situation where a fact approximately same as p, called p' is available and the nearest rule whose antecedent part partially matches with p' is p→q. Thus, formally

$$p'$$
$$\frac{p \rightarrow q}{q'}$$

where q' is the inferred consequence. This partial matching of p' with p, and thus generating the inference q' comes under the scope of **fuzzy logic**. The truth functional value of the propositions here lies in the same interval of [0,1] like that of multi-valued logic. Thus in one sense fuzzy logic is a multi-valued logic. But the most pertinent feature of fuzzy logic for which it receives so much attention is its scope of **partial matching**, as illustrated in the last example. Another interesting situation occurs when a number of rules' antecedent parts match partially with the fact p'. The resulting inference, say q_{final}, under this circumstance is determined by the composite influence of all these rules. Thus, in any real time system, the inferences guided by a number of rules follow a middle decision trajectory over time. This particular behavior of **following a middle decision trajectory** [30] is humanlike and is a unique feature of fuzzy logic, which made it so attractive!

Very recently Prof. Zadeh highlighted another important characteristic [34] of fuzzy logic that can distinguish it from other multi-valued logics. He called it **f.g-generalization**. According to him any theory, method, technique or problem can be fuzzified (or **f-generalized**) by replacing the concept of a crisp set with a fuzzy set. Further, any theory, technique, method or problem can be granulated (or **g-generalized**) by partitioning its variables, functions and relations into granules (information cluster). Finally, we can combine f-generalization with g-generalization and call it f.g-generalization. Thus ungrouping an information system into components by some strategy and regrouping them into clusters by some other strategy can give rise to a new kind of information sub-systems. Determining the strategies for ungrouping and grouping, however, rests on the designer's choice. The philosophy of f.g-generalization undoubtedly will re-discover fuzzy logic in a new form.

The last few paragraphs just highlighted the significance of fuzzy logic and its special characteristics. A clear understanding of these issues, however, calls for a brief explanation of the theory of fuzzy sets. A **fuzzy set** S can be formally described by

$$S = \{x / \mu_S(x): x \in U\} \qquad (1)$$

where x is any variable of a universal set U and $\mu_S(x)$ is the grade of membership value of x to belong to the set S. It may be noted that unlike conventional set, here the fuzzy set S includes all members of the universal set U with a membership value $\mu_S(x)$ in the range [0, 1]. For example, let us consider a universal set U of AGE where the element u in U has a range of 0 to 120 years. We also consider 3 fuzzy sets, say, OLD, YOUNG and CHILD. Here all ages of [0..120] years will be the elements of each fuzzy set OLD, YOUNG and CHILD, but $\mu_{OLD}(x)$, $\mu_{YOUNG}(x)$ and $\mu_{CHILD}(x)$ for a given x (say 25) must be different in general. As the age 25 corresponds to a young person, $\mu_{YOUNG}(25)$ will be very high (say 0.98) in comparison to $\mu_{OLD}(25)$ and $\mu_{CHILD}(25)$. Membership functions such as $\mu_{OLD}(x)$, $\mu_{YOUNG}(x)$ and $\mu_{CHILD}(x)$ are generally represented by curves with x as independent variable. These curves are usually called **membership curves**. The membership curves for the fuzzy sets OLD, YOUNG and CHILD are given in Figure 1.

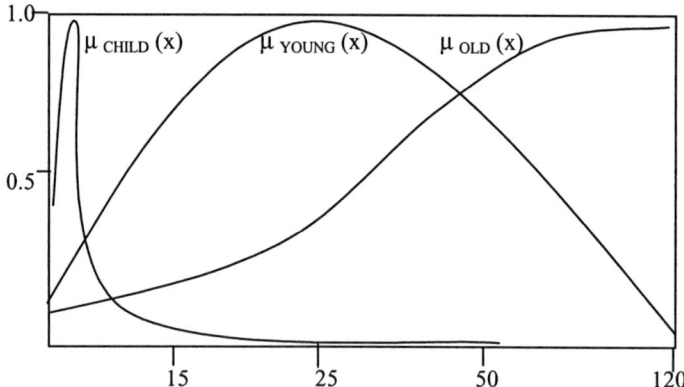

Figure 1. Membership curves for the fuzzy sets: OLD, YOUNG and CHILD.

In a **continuous fuzzy system**, membership functions are continuous over variable x. A **discrete fuzzy system**, however, employs discrete membership functions sampled over x at a regular or irregular interval.

For instance, the discrete version of $\mu_{YOUNG}(x)$ can take the following form:

$$\mu_{YOUNG}(x) = [2/0.2 \ 20/0.9 \ 22/1.0 \ 30/0.8 \ 40/0.6 \ 80/0.2 \ 100/0.01]$$

where the arguments in the last expression such as 22/1.0 denote that the membership of age = 22 to be young is 1.0.

For designing a fuzzy system we should be familiar with fuzzy relations. A **fuzzy relation** $R(x,y)$ informally denotes the relation between two fuzzy sets A and B under the universes $X = \{x\}$ and $Y = \{y\}$, respectively. The fuzzy relations are usually constructed based on the logical relationship of the sets A and B following a fuzzy rule. For example, consider the fuzzy If-Then rule:

   If *age* is **Young** Then *speed* is **High**.

Here, Young and High are fuzzy sets under the universal sets of AGE and SPEED respectively and italic words age and speed denote corresponding variables in the range [0..120] years and [5..9] m/s respectively. The fuzzy relation between sets Young and High guided by the above rule states that when membership of age to be Young is large, the membership of speed to be High should also be large. Further, when membership of age to be Young is small, the membership of speed to be High should also be small. One simple logical relation that supports it is the fuzzy AND relation. Thus we can formally construct the fuzzy implication relation between 'age is Young' and 'speed is High' denoted by R(age, speed) as follows.

$$R(age, speed) = Min[\mu_{YOUNG}(age), \mu_{HIGH}(speed)] \qquad (2)$$

where Min denotes a logical fuzzy AND operator. The novice readers may wonder as to how to compute the Min. A little understanding of the geometry will make it clear. Suppose we plot age, speed and R(age, speed) in three dimensions, where age and speed are independent variables. Now, for each value of the variable age and the variable speed, we can compute R by applying the Min operator following expression (2). Thus R can be obtained by considering all possible values of age and speed.

It needs mention at this stage that Min is not the only operator to represent fuzzy implication relation. In fact there exist as many as around 20 ways to represent fuzzy implication relation. For the sake of illustration, let us consider another important implication relation. Lukasiewicz, a great Polish logician, devised this long before fuzzy logic was invented and thus in honor of him it is called the **Lukasiewicz implication relation**. The Lukasiewicz implication relation [75] for the proposed rule can be stated as

$$R(\text{age, speed}) = \text{Min}[1, \{1 - \mu_{\text{YOUNG}}(\text{age}) + \mu_{\text{HIGH}}(\text{speed})\}] \qquad (3)$$

It may be noted that the above implication relation follows the propositional implication rule: $p \rightarrow q \Rightarrow \neg p \vee q$ for any two propositions p and q. Naturally the question arises: how? A look at the second braced part of expression (3) reveals the above analogy, where $\{\neg \mu_{\text{YOUNG}}(\text{age}) \vee \mu_{\text{HIGH}}(\text{speed})\}$ has been equivalently replaced by $\{1 - \mu_{\text{YOUNG}}(\text{age}) + \mu_{\text{HIGH}}(\text{speed})\}$. Unfortunately, $\{1 - \mu_{\text{YOUNG}}(\text{age}) + \mu_{\text{HIGH}}(\text{speed})\}$ can exceed 1, so a minimum of 1 and $\{1 - \mu_{\text{YOUNG}}(\text{age}) + \mu_{\text{HIGH}}(\text{speed})\}$ has been considered in the definition of the relation R(age, speed) in expression (3).

Another point to note is that implication is not the only fuzzy relation. There are as simple relations as conjunction (AND) or disjunction (OR) of fuzzy sets. On the other hand many complex relations such as implication of implications satisfying the following type of fuzzy rules:

If (If (*age* is **Young** Then *Speed* is **High**))
Then the *metabolic rate* of the person is **High**

are also prevalent in the current literature [16]. It is clear from the above rule that the outer implication constructs a relation $R'(R(\text{age, speed}), \text{metabolic rate})$, where R(age, speed) has been defined earlier.

Why are fuzzy relations so important in the logic of fuzzy sets? This is because fuzzy relations describe the input-output (antecedent–consequence) relationships of the fuzzy rules. Thus, if it can be constructed once for known membership distributions of the antecedent and consequent parts, it can be used for deriving the membership distribution of the consequences for a known membership distribution of a given similar antecedent clause. In other words, the fuzzy inferences (derived consequences) for rules with partially similar antecedent clauses can be evaluated with the implication relations. To illustrate the process of this computation, suppose the membership distribution of 'age is more-or-less Young', by notation $\mu_{MORE\_OR\_LESS\_YOUNG}(age)$ is given and the relation R(age, speed) for the rule 'If age is Young Then speed is High' is given. We are interested to determine the membership distribution of 'speed is more-or-less High' or $\mu_{MORE\_OR\_LESS\_HIGH}(speed)$. Here,

$$\mu_{MORE\_OR\_LESS\_HIGH}(speed)$$
$$= Max[Min\{\mu_{MORE\_OR\_LESS\_YOUNG}(age), R(age, speed)\}], \qquad (4)$$
$$age \in AGE$$

What does expression (4) mean? Here AGE is the universal set of ages: {0..120] and variable age can take any value of set AGE. Suppose, we are interested first to compute, say, $\mu_{MORE\_OR\_LESS\_HIGH}(speed = 7\ m/s)$. Let age be represented by X-axis, speed by Y-axis and μs and R by Z-axis. Note that we have 2 μs, $\mu_{MORE\_OR\_LESS\_HIGH}(speed)$ and $\mu_{MORE\_OR\_LESS\_YOUNG}(age)$ in the present context. First, we place the membership distribution of age, i.e., $\mu_{MORE\_OR\_LESS\_YOUNG}(age)$ versus age parallel to X-Z plane at Y = 7 m/s. Take a longitudinal section of the fuzzy surface R(age, speed) by a plane Y = 7 m/s, and then take minimum of the sectioned fuzzy surface and the distribution $\mu_{MORE\_OR\_LESS\_YOUNG}(age)$ for all possible ages in [0,120]. If we consider only integer ages we will have 121 minima, and finally we take the maxima of these 121 minimum memberships. The result corresponds to $\mu_{MORE\_OR\_LESS\_HIGH}(speed)$ at speed = 7 m/s. Thus $\mu_{MORE\_OR\_LESS\_HIGH}(speed)$ for other values of speed can be evaluated by shifting the $\mu_{MORE\_OR\_LESS\_YOUNG}(age)$ distribution along the Y-axis for different y.

In a discrete fuzzy system, we represent the membership distributions and the relations by vectors and matrices respectively. The Max-Min method for deriving fuzzy inferences, described by expression (4) thus takes the following form:

$$\mu_{MORE\_OR\_LESS\_HIGH}(speed)$$
$$= \mu_{MORE\_OR\_LESS\_YOUNG}(age) \circ \mathbf{R}(age, speed). \qquad (5)$$

The μs in expression (5) are the row vectors, $\mathbf{R}$ is the relational matrix and the operator "o" denotes the composition operation, which is similar with conventional matrix multiplication operation, with the replacement of product by Min and sum by Max operators. The elements R_{ij} of matrix R denotes the fuzzy implication relation from the i-th element of μ MORE_OR_LESS_YOUNG(age) to the j-th element of $\mu_{MORE\_OR\_LESS\_HIGH}$(speed). Example 1 below explains the computation involved in expression (5).

Example 1: Given the vector $\mu_{MORE\_OR\_LESS\_YOUNG}$(age) and the relational matrix R(age, speed) as follows, how to determine $\mu_{MORE\_OR\_LESS\_HIGH}$(speed)?

Let $\mu_{MORE\_OR\_LESS\_YOUNG}$(age) = [2/0.2 20/0.9 22/1.0 30/0.8] and

		speed (m/s)			
R =	age	5	6	7	8
	02	0.3	0.4	0.5	0.6
	20	0.4	0.5	0.6	0.7
	22	0.5	0.6	0.7	0.8
	30	0.6	0.7	0.8	0.9

Now, by expression (5), $\mu_{MORE\_OR\_LESS\_HIGH}$(speed)

$$= [0.2\ 0.9\ 1.0\ 0.8] \circ \begin{bmatrix} 0.3 & 0.4 & 0.5 & 0.6 \\ 0.4 & 0.5 & 0.6 & 0.7 \\ 0.5 & 0.6 & 0.7 & 0.8 \\ 0.6 & 0.7 & 0.8 & 0.9 \end{bmatrix}$$

$$= [0.6\ 0.7\ 0.8\ 0.8].$$

For the sake of interpretation, we will write the resulting vector as:

$\mu_{MORE\_OR\_LESS\_HIGH}$(speed)
= [5m/s / 0.6 6m/s / 0.7 7m/s / 0.8 8m/s / 0.8].

The linguistic meaning of the above vector is clear following its definition. For instance, the first component means that the membership of speed = 5 m/s to belong to the set MORE-OR-LESS-YOUNG is 0.6. The meaning of the other components of the vector can be derived similarly.

The last point without which the discussion on **fuzzy reasoning** [75] remains incomplete is the construction of the relational matrices. We already discussed about the formulation of fuzzy relations for continuous fuzzy system. For discrete systems, if the membership distribution of the antecedent and the consequent clauses are known, we can determine the fuzzy relational matrix for the system easily by using the composition operator. For example, consider the following rules:

If x is A1 and y is B1 then Z is C1

If x is A2 and y is B2 then Z is C2

where x, y and z are fuzzy variables and A1, A2, B1, B2, C1 and C2 are fuzzy sets such that A1, A2 $\subseteq$ X, B1, B2 $\subseteq$ Y and C1, C2 $\subseteq$ Z. X, Y and Z are three universal sets. The fuzzy relational matrices R1 and R2 for such rules can be stated as follows:

$$R1 = \{\mu_{A1}(x) \wedge \mu_{B1}(y)\}^{T} \ o \ \{\mu_{C1}(z)\} \tag{6}$$

$$R2 = \{\mu_{A2}(x) \wedge \mu_{B2}(y)\}^{T} \ o \ \{\mu_{C2}(z)\}] \tag{7}$$

In expressions (6) and (7) the $\wedge$ (Min) operation of the vectors are done component-wise, and the T above a vector denotes its transpose.

Thus for observed distributions of x is A$'$ and y is B$'$, where A$'$ and B$'$ are two fuzzy sets, such that A$'$ is close to both A1 and A2 and B$'$ is close to both B1 and B2 respectively, the composite inference vector z is C$'$ can be derived following expression (8).

$$\mu_{C1}'(z) = [\{\mu_{A}'(x) \wedge \mu_{B}'(y)\} \ o \ R1] \tag{8}$$

$$\mu_{C2}'(z) = [\{\mu_{A}'(x) \wedge \mu_{B}'(y)\} \ o \ R2] \tag{9}$$

$$\mu_{C}'(z) = \mu_{C1}'(z) \vee \mu_{C2}'(z) \tag{10}$$

where $\vee$ denotes the Max operator, which is also carried out over the vectors component-wise. The inferencing procedure discussed above is self-explanatory and thus no example is given to illustrate it.

The next important aspects of fuzzy logic, we will discuss here, is its capability of **unsupervised classification**. A question then obviously arises: what is an unsupervised classification and why fuzzy logic is useful for such classification? To answer this let us take an example. Suppose, in an animal kingdom, we want to classify the animals by their speed and height to weight ratio. We can represent the classification process here by two dimensions. Let the X- and Y-axes denote speed and height/weight of the animals respectively. Now, if we plot the coordinates for each animal in the kingdom we will find that the animals of the same species (say, dogs) will occupy very close coordinates and thus form clusters. Consequently animals of different species will form different clusters. The process of cluster formation is in general called an **unsupervised classification**. But how do these clusters help us? They help us in classifying an unknown animal from its **feature space**, here speed and height to weight ratio, once the clusters are known. In other words, if one knows the measurements of speed and height to weight ratio of an animal, s/he can determine the class to which the animal should belong to. So, classification has some justifications. In the next paragraph we will discuss the need for fuzzy classification.

The clustering process we discussed so far does not consider overlapping of classes. But overlapping of classes occur in most practical classification problems. For instance, the cluster of foxes and dogs should overlap in our last example as their speed and height to weight ratio are very close. So, from the measurements of these two features only, one is unable to say whether the unknown animal, which we want to classify is a fox or a dog. Such problems can be handled using fuzzy clustering. Here a given animal can belong to different classes with a membership value in the interval [0..1], but sum of its memberships of belonging to different classes must be 1. One such classical clustering algorithm proposed by Prof. Bezdek, is popularly known as **Fuzzy c-means clustering (FCM)** [11]. The next section will outline this algorithm.

The objective of the fuzzy c-means clustering algorithm is to classify a given set of p dimensional data points $X = [x_1 \ x_2 \ x_3 \ \dots \ x_n]$ into a set of c fuzzy classes or partitions Ai [16], represented by clusters, such that the sum of the memberships of any component of X, say x_k, in all the c classes is 1. Mathematically, we can represent this by:

$$\sum_{i=1}^{c} \mu_{Ai}(x_k) = 1, \quad \text{for all } k = 1 \text{ to } n. \quad (11)$$

Further, all elements of X should not belong to the same class with membership 1. This is so because otherwise there is no need of the other classes. Thus mathematically, we state this by:

$$0 < \sum_{k=1}^{n} \mu_{Ai}(x_k) < n. \tag{12}$$

For example, if we have 2 fuzzy partitions A1 and A2, then for a given X = [x_1 x_2 x_3] say, we can take

$\mu_{Ai} = [0.6/x_1\ \ 0.8/x_2\ \ 0/x_3]$ and
$\mu_{A2} = [0.4/x_1\ \ 0.2/x_2\ \ 1/x_3]$.

It is to be noted that the conditions described by expressions (11) and (12) are valid in the present classification.

Given c classes A_1, A_2, ..., A_c, we can determine their cluster centers V_i for i = 1 to c by using the following expression.

$$V_i = \left[\sum_{k=1}^{n} [\mu_{Ai}(x_k)]^m x_k \right] / \sum_{k=1}^{n} [\mu_{Ai}(x_k)]^m \tag{13}$$

Here, m (m>1) is any real number that influences the membership grade. It is to be noted from expression (13) that the cluster center V_I basically is the weighted average of the memberships $\mu_{Ai}(x_k)$. A common question now naturally arises: can we design the fuzzy clusters A_1, A_2, ..., A_c in a manner so that the data point (feature) vector x_k for any k is close to one or more cluster centers V_i? This can be formulated by a performance criterion given by:

Minimize J_m over V_i (for fixed partitions U) and μ_{Ai} (for fixed V_i)

$$J_m(U, V_i) = \sum_{k=1}^{n} \sum_{i=1}^{c} [\mu_{Ai}(x_k)]^m \|x_k - V_i\|^2 \tag{14}$$

subject to $\sum_{i=1}^{c} \mu_{Ai}(x_k) = 1$

where $\|\cdot\|$ is an inner product induced norm in p dimension. Differentiating the performance criteria with respect to V_i treating μ_{Ai} as constants and to μ_{Ai} treating V_i as constants and setting them to zero, we find following Bezdek [11]:

$$\mu_{Ai}(x_k) = \left[\sum_{j=1}^{c} \left(\|x_k - V_i\|^2 / \|x_k - V_j\|^2 \right)^{1/(m-1)} \right]^{-1} \qquad (15)$$

This is a great development which led to the foundation of the fuzzy c-means clustering algorithm. The algorithm is formally presented below.

Procedure Fuzzy c-means clustering;
Input: Initial pseudo-partitions $\mu_{Ai}(x_k)$ for i= 1 to c and k= 1 to n
Output: Final cluster centers
Begin
  Repeat
    For i:= 1 to c
      Evaluate V_i by expression (13);
    End For;
    For k:= 1 to n
      For i:= 1 to c
        Call $\mu_{Ai}(x_k)$ OLD_$\mu_{Ai}(x_k)$;
        If $\|x_k - V_i\|^2 > 0$
        Then evaluate $\mu_{Ai}(x_k)$ by (15) and call it CURRENT_$\mu_{Ai}(x_k)$
        Else set $\mu_{Ai}(x_k)$:=1 and call it CURRENT_$\mu_{Ai}(x_k)$
        and set $\mu_{Aj}(x_k)$:=0 for j≠i;
      End For;
    End For;
  Until | CURRENT_$\mu_{Ai}(x_k)$ – OLD_$\mu_{Ai}(x_k)$ | <= ∈ ; ∈ = 0.01 (say)
End.

Example 2: Consider a set of 15 2-dimensional data points x_1 to x_{15} (Table 1), we want to classify them into 2 pseudo-partitions A_1 and A_2, say. Let the components of each point x_k be x_{k1} and x_{k2}. Thus we can plot these 15 points with respect to x_{k1} and x_{k2} axes.

Table 1. A set of 15 2-dimensional data points.

k	1	2	3	4	5	6	7	8	9	10	11	12	13	14	15
x_{k1}	0	0	0	1	1	1	2	3	4	5	5	5	6	6	6
x_{k2}	0	2	4	1	2	3	2	2	2	2	1	2	3	0	4

Let the initial pseudo-partitions be A_1 and A_2 such that
    μ_{A1} = {0.854/x_1 ... 0.854/x_{14} 1/x_{15}} and
    μ_{A2} = {0.146/x_1 ... 0.146/x_{14} 0/x_{15}}.

We now evaluate the cluster centers for the 2-partitions. Here, the cluster center for partition A_1 will have 2 coordinates, one for x_{k1} and the other for x_{k2}. Let the x_{k1} coordinate for partition A_1 be V_{11} and the x_{k2} coordinate for the same partition be V_{12}. Then with m = 1.25 we find

$$V_{11} = \{(0.854)^{1.25}(0+0+0+1+1+1+2+3+4+5+5+5+6+6) + (1)^{1.25}(6)\}/$$
$$\{14\,(0.854)^{1.25} + (1)^{1.25}\} = 3.042, \text{ and}$$
$$V_{12} = \{(0.854)^{1.25}(0+2+4+1+2+3+2+2+2+2+1+2+3+0) + (1)^{1.25}(4)\}/$$
$$\{14\,(0.854)^{1.25} + (1)^{1.25}\} = 2.028.$$

The cluster centers V_{21} and V_{22} for partition A_2 can easily be evaluated by replacing (0.854) by (0.146) and (1) by (0) in the last 2 expressions.

The computation of the new membership of each point x_k can now also be evaluated with the computed values of V_{11}, V_{12}, V_{21} and V_{22} and last coordinate of the points. After 6 iterations of execution of the algorithm the cluster centers are found as:

$$V_1 = (V_{11}, V_{12}) = (0.88, 2.0) \text{ and}$$

$$V_2 = (V_{21}, V_{22}) = (5.14, 2.0).$$

The membership values of the points to belong to partition A_1 and A_2 after 6 iterations are found as:

$$\mu_{A1} = \{0.99/x_1 \ \ 1/x_2 \ \ 0.99/x_3 \ \ 1/x_4 \ \ 1/x_5 \ \ 1/x_6 \ \ 0.99/x_7 \ \ 0.47/x_8 \ \ 0.01/x_9$$
$$0/x_{10} \ \ 0/x_{11} \ \ 0/x_{12} \ \ 0.01/x_{13} \ \ 0/x_{14} \ \ 0.01/x_{15}\} \text{ and}$$

$$\mu_{A2} = \{0.01/x_1 \ \ 0/x_2 \ \ 0.01/x_3 \ \ 0/x_4 \ \ 0/x_5 \ \ 0/x_6 \ \ 0.01/x_7 \ \ 0.53/x_8 \ \ 0.99/x_9$$
$$1/x_{10} \ \ 1/x_{11} \ \ 1/x_{12} \ \ 0.99/x_{13} \ \ 1/x_{14} \ \ 0.99/x_{15}\}.$$

It may indeed be noted that sum of the memberships of any point x_k to belong to the 2 partitions A_1 and A_2 to be 1 is always maintained irrespective of the iterations.

3 Computational Models of Neural Nets

Fuzzy logic, which has proved itself successful especially in reasoning and unsupervised classification, however, is not directly amenable for automated machine learning [67]-[69]. Artificial neural nets (ANN), which are electrical analogue of the biological neural nets, on the other hand can learn facts (represented by patterns) and determine the inter-relationship among the patterns. The process of determining the interrelationships among patterns, in general, is informally called **encoding**. Some ANN parameters, usually the weights and the thresholds, are generally encoded to represent the interrelationship among the patterns. The process of determining the inter-related pattern(s) from one given pattern using the encoded network parameters is usually called **recall**. Depending on the type and characteristics of the encoding and recall process, machine learning techniques are categorized into three major heads: (i) supervised learning, (ii) unsupervised learning, and (iii) reinforcement learning.

In a **supervised learning** system, there is a trainer who provides the input and the corresponding target (output) patterns. A learning algorithm is employed to determine a unique set of network parameters that jointly satisfy the input-output interrelationship of each two patterns. After the encoding process, as discussed above is over, the network on excitation with an unknown input pattern can generate its corresponding output pattern. The ANN when trained by a supervised learning algorithm, thus, behaves as a multi-input-multi output function approximator.

A common question that naturally arises: how does a supervised learning system actually work? Generally, the network is initially assigned a random set of parameters such as weights and thresholds. Suppose, we want to train the network with a single input-output pattern. What should we do? We supply the network with our input pattern, and let the network generate its output pattern. The generated output pattern is compared with the target pattern. The difference of these two patterns results in an error vector. A supervised learning algorithm is then employed to adjust the network parameters using the error vector. For multiple input-output patterns, the error vector for each pattern set is determined and a function of these error vectors is used to adjust the network parameters. There exist quite a large number of supervised learning algorithms using neural nets. The most popular among them is the Back-propagation algorithm.

An **unsupervised learning system,** employs no teacher, and thus the inter-relation among the patterns is not known. Generally in an unsupervised learning process one or more input pattern sets are automatically mapped to one cluster pattern. The classification of patterns by a fuzzy c-means clustering algorithm is an unsupervised learning for example. The encoding process in an unsupervised learning system varies greatly from one system to another. Most systems, however, employ a recursive learning rule that autonomously adjusts the network parameters for attaining some criteria like minimization of the network energy states. Among the unsupervised learning systems most popular are Hopfield nets and Self-organizing Map (SOM).

The third category of the learning system that bridges the gap between the supervised and the unsupervised learning is popularly known as the **reinforcement learning**. This learning scheme employs an **internal critic** that examines the response of the environment in turn of the action of the learning system on the environment. If the response is in favor of the goal, then the action is **rewarded** otherwise it is **penalized**. Determination of the status of the action: reward or penalty may, however, require quite a long time, until the goal is reached. Q-learning is one of the most common reinforcement learning, and we will say a few words on Q-learning later.

3.1   The Back-Propagation Learning Algorithm

As already discussed, Back-propagation learning algorithm is one of the most popular supervised learning algorithms. It employs a feed-forward topology of neurons, denoted by circles, (see Figure 2), with a number of layers (here 3) and each layer comprising of a number of neurons. The bottommost layer of the figure is the input layer, the topmost layer is the output layer and the intermediate layers (here only 1) are called the hidden layers. The neurons in the hidden and the output layers of the back-propagation algorithm receives weighted output from the neurons of the previous layer, which are summed up and then passed through a sigmoid type non-linearity (Figure 3). The neurons in the input layer simply pass the output to the next layer following the connectivity between these two layers.

The back-propagation algorithm is based on the principle of the **gradient descent learning**. We will shortly outline this principle and use it for deriving the training procedure of the back-propagation algorithm.

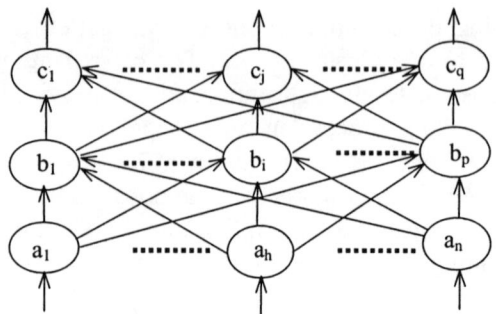

Figure 2. A feed-forward topology of a neural net used in the back-propagation learning.

Let $W_{p,q,k}$ (or simply W_{pq}) be the weight connected between neurons p in layer (k-1) with neuron q at the k-th (output) layer. Further, let E be the Euclidean norm of the error vector, for a given training pattern, produced at the output layer. Formally,

$$E = \tfrac{1}{2}\sum_{\forall r}(t_r - Out_r)^2 \qquad (16)$$

where t_r and Out_r denote the target (scaler) output and the computed output at node r in the output layer.

The gradient descent learning requires that for any weight W_{pq},

$$W_{pq} \leftarrow W_{pq} - \eta(\partial E/\partial W_{pq}) \qquad (17)$$

where η is the learning rate.

The computation of $\partial E/\partial W_{pq}$, however, is different in the last layer from the rest of the layers. We now consider two types of output layers: output layer with and without non-linearity of neurons.

When the neurons in the output layer contain no non-linearity,

$$Out_q = Net_q = \sum_{\forall p} W_{pq}Out_p \ , \qquad (18)$$

where p is the index of the neurons in the penultimate layer.

Now, $\qquad \partial E/\partial W_{pq} = (\partial E/\partial Out_q)\,(\partial Out_q/\partial W_{pq})$

$$= -\,(t_q - Out_q)\,Out_p \qquad (19)$$

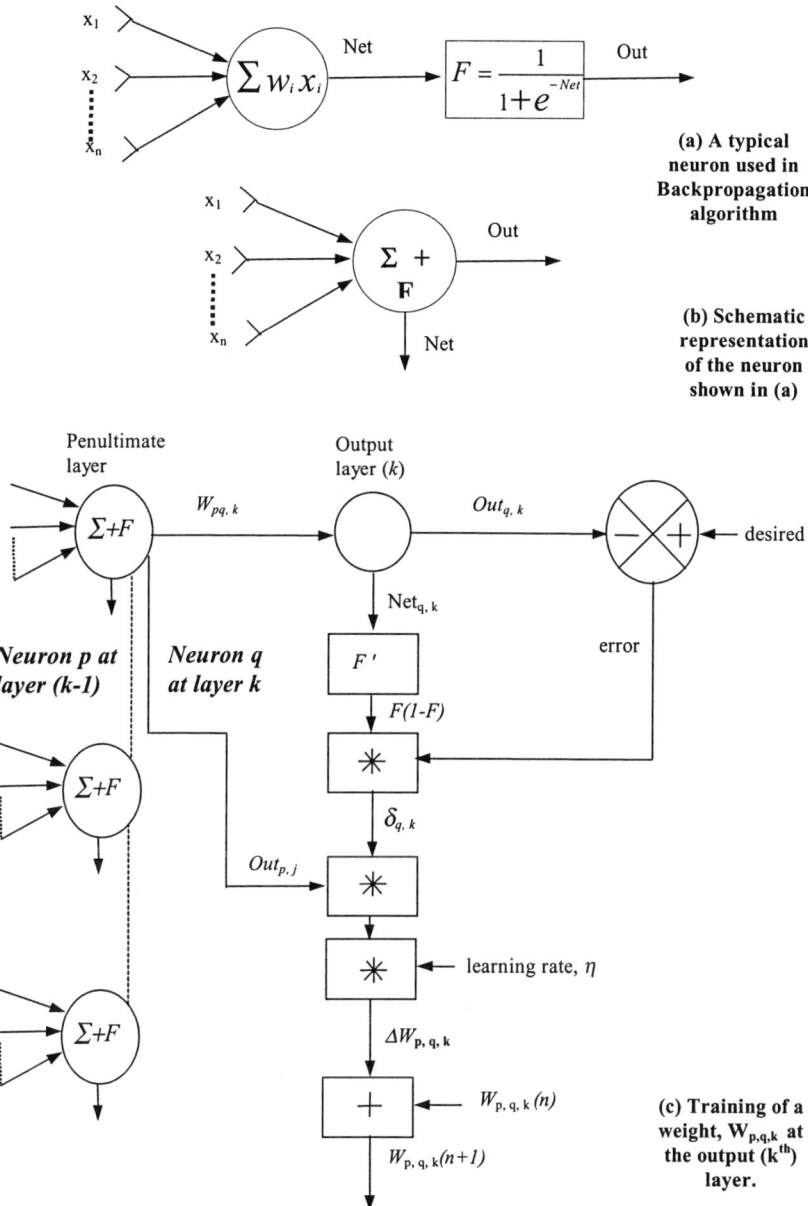

(a) A typical neuron used in Backpropagation algorithm

(b) Schematic representation of the neuron shown in (a)

(c) Training of a weight, $W_{p,q,k}$ at the output (k^{th}) layer.

Figure 3. A schematic diagram of the back-propagation learning algorithm.

Consequently,

$$W_{pq} \leftarrow W_{pq} + \eta(t_q - Out_q)\, Out_p \quad \text{[by (17)]} \tag{20}$$

Denoting $(t_q - Out_q)$ by δq we have

$$W_{pq} \leftarrow W_{pq} + \eta\, \delta q\, Out_p \tag{21}$$

Now, we consider the case, when the neurons in all the layers including the output layer contain sigmoid type non-linearity. The network structure with two cascaded weights is given in Figure 4.

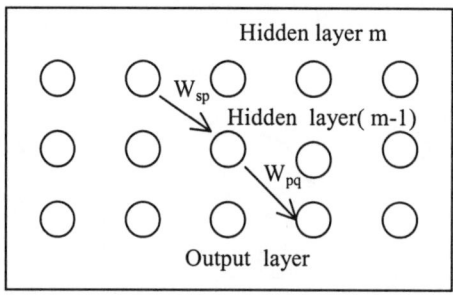

Figure 4. Defining the weights in a feed-forward topology of neurons.

Here,

$$Out_p = 1/(1+e^{-Netp})$$

$$Net_p = \Sigma_r\, W_{rp}.Out_r$$

where the index r corresponds to neurons in the hidden layer m.

$$Out_r = 1/(1+e^{-Net\,r})$$

$$Net_r = \Sigma_i\, Wir.Out_i$$

where w_{ir} are the weights connected to neuron r from its preceding layer.

Now, for the output layer with sigmoid type non-linearity we have

$$
\begin{aligned}
\partial E/\partial W_{pq} &= (\partial E/\partial Out_q)\,(\partial Out_q/\partial Net_q)\,(\partial Net_q/\partial W_{pq}) \\
&= - (t_q - Out_q)\, Out_q\, (1 - Out_q)\, Out_p \\
&= - \{(t_q - Out_q)\, Out_q\, (1 - Out_q)\}\, Out_p \\
&= - \delta q\, Out_p \text{ (say).}
\end{aligned}
\tag{22}
$$

The readers may now compare this result with that given in Figure 3, where this is written as $\delta_{q,k} \, Out_{p,j}$.

Now, we compute the updating rule for W_{sp}:

$$\partial E/\partial W_{sp} = \Sigma_r \, (\partial E/\partial Out_r) \, (\partial Out_r/\partial Net_r) \, (\partial Net_r/\partial Out_p)$$

$$(\partial Out_p/\partial Net_p) \, (\partial Net_p/\partial W_{sp})$$

$$= \Sigma_r - (t_r - Out_r) \, Out_r \, (1 - Out_r) \, W_{pr} \, Out_p \, (1 - Out_p) \, Out_s$$

$$= - \, \Sigma_r \, \delta r \, W_{pr} \, Out_p \, (1 - Out_p) \, Out_s$$

$$= - \, Out_p \, (1 - Out_p) \, Out_s \, \Sigma_r \, \delta r \, W_{pr} \qquad (23)$$

The schematic diagrams explaining the computation of weights and the evaluation of the propagated error [35], [36] at layer j are given in Figures 3 and 5 respectively.

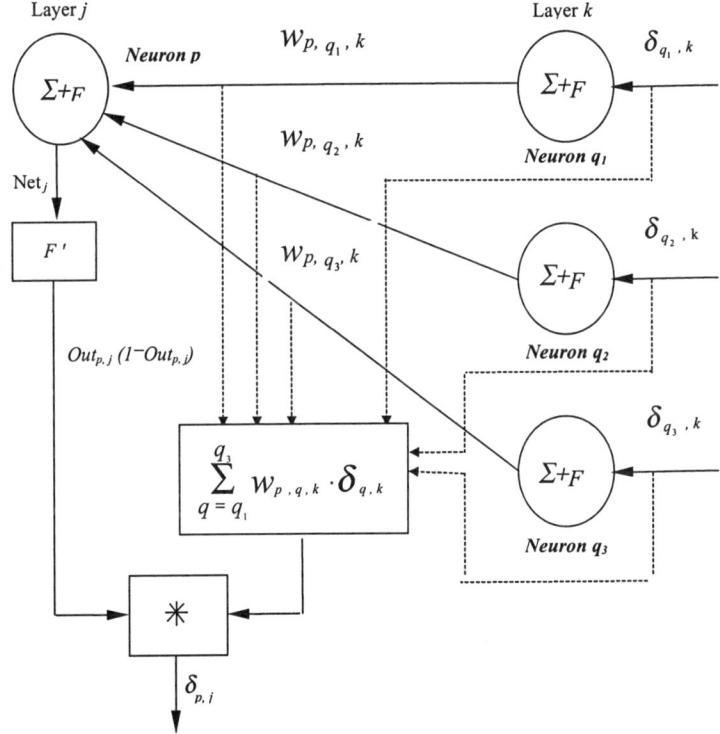

Figure 5. Computation of the back-propagated error at layer j.

3.2 Hopfield Nets

Among the unsupervised learning algorithms, Hopfield neural nets [22], [23] need special mention. In fact, in the nineties, nearly one third of the research papers on neural network include works on Hopfield nets. A Hopfield network employs a recurrent neural topology, where each node in the network receives signal from all other nodes (Figure 6). Hopfield nets can be of two common types, namely,

- **Binary Hopfield net**
- **Continuous Hopfield net**

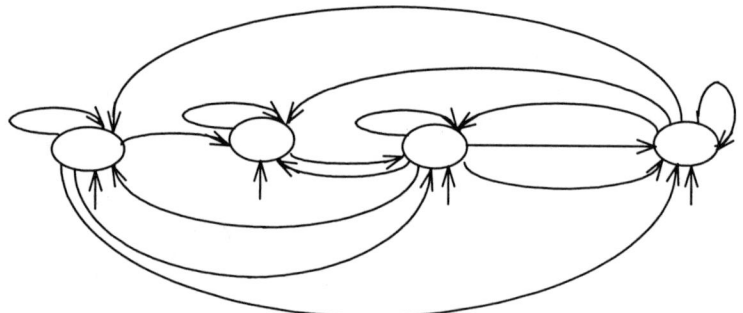

Figure 6. Topology of a Hopfield neural network.

In a binary Hopfield net, the input and output of the neurons are binary, whereas in a continuous Hopfield net, they could assume any continuous values between 0 and 1. Further, a continuous Hopfield net has instantaneous inputs (denoted by vertical arrows in Figure 6), but they are absent from a binary Hopfield network. A Hopfield neural net finds application in many problems, especially in realizing ADCs as proposed by Hopfield in his early papers. The principle of weight adjustment in Hopfield net is done by optimizing a function of weights and signal values at nodes, popularly known as Liapunov function. Liapunov functions are energy functions, used to identify system states, where the function yields the minimum value of energy.

3.2.1 Binary Hopfield Net

In a binary Hopfield model, each neuron can have two output states: $'0'$ and $'1'$ (sometimes denoted as n_i^0 and n_i^1). Let us consider a neuron $'i'$, then the total input to neuron $'i'$, denoted by H_i is given below:

$$H_i = \sum_{j \neq i} w_{ij} n_j + l_i \tag{24}$$

where

l_i = external input to neuron 'i'
n_j = inputs from neuron 'j'
w_{ij} = synaptic interconnection strength from neuron 'j' to neuron 'i'
n_i^0 = '0' output (non-firing)
n_i^1 = '1' output (firing)

Each neuron has a fixed threshold, say, th_i for neuron 'i'. Each neuron readjusts its states randomly by using the following equations.

For neuron i:

$$\text{Output } n_i = 1 \qquad if \sum_{j \neq i} w_{ij} n_j > th_i \;;$$

$$= 0 \qquad \text{otherwise.} \tag{25}$$

The information storage for such system is normally described by

$$w_{ij} = \sum_s (2n_i^s - 1)(2n_j^s - 1) \tag{26}$$

where, n_i^s represents set of states for s being an integer 1,2,....., n.

To analyze the stability of such neural network, Hopfield proposed a special kind of Liapunov energy function given in expression (27)

$$E = -(1/2) \, \Sigma \, \Sigma \, w_{ij} \, n_i \, n_j - \Sigma \, l_i \, n_i + \Sigma \, th_i \, n_i \tag{27}$$

The change in energy ΔE due to change in the state of neuron 'i' by an amount Δn_i is given by

$$\Delta E = -\Sigma \, (w_{ij} \, n_j + l_i - th_i) \, \Delta n_i \tag{28}$$

As Δn_i is positive only when the bracketed term is positive, thus any change ΔE in E under expression (28) is negative. Since E is bounded, so the iteration of the Hopfield neural algorithm, given by expression (25) must lead to stable states.

3.2.2 Continuous Hopfield Net

The continuous Hopfield net can be best described by the following differential equation,

$$C_i \frac{du_i}{dt} = \sum_j w_{ij} n_j - u_i / R_i + l_i \qquad (29)$$

where w_{ij} = synaptic interconnection strength (weight) from neuron 'j' to neuron 'i',

n_j = output variable for neuron 'j',
u_i = instantaneous input to neuron 'i',
l_i = external input to neuron 'i',
g_i = n_i / u_i : output-input sigmoidal relationship,
C_i and R_i represent dimensional constants.

Such symbols are used in the above nomenclature to keep these comparable with parameters of electric networks. The continuous Hopfield model uses an energy like function [23], the time derivative of which is given by:

$$\frac{dE}{dt} = \sum_{i=1}^{n} \frac{dn_i}{dt} \left(\sum_j w_{ij} n_j - u_i / R_i + l_i \right)$$

$$= \sum_{j=1}^{n} C_i \frac{dn_i}{dt} \frac{du_j}{dt} \qquad (30)$$

$$= -\sum_{j=1}^{n} C_i g_i^{-1}(n_i) \left(\frac{dn_i}{dt} \right)^2$$

where $g_i^{-1}(n_i)$ is a monotonically increasing term and C_i is a positive constant. It can be shown easily that if $w_{ij} = w_{ji}$, the Hopfield net can evolve towards an equilibrium state.

3.3 Self-Organizing Feature Map

Kohonen [24] proposed a new technique for mapping a given input pattern onto a 2-dimensional spatial organization of neurons. In fact, he considered a set of weights, connected between the positional elements of the input pattern and a given neuron, located at position (i, j) in a two dimensional plane. Let $\underline{w}_{ij}$ be the weight vector for neuron N_{ij}. Thus, for a set of $(n \times n)$ points on the 2-D plane, we would have n^2 such weight vectors, denoted by $\underline{w}_{ij}$, $1 \leq i,j \leq n$. In Kohonen's model, the neuron with minimum distance between its weight vector $\underline{w}_{ij}$ and the input vector $\underline{X}$ is first identified by using the following criterion [25]:

Find the $(k, l)^{th}$ neuron for which

$$\| \underline{X} - \underline{w}_{kl} \| = \min_{1 \leq i < n} (\min_{1 \leq j < n} \| \underline{X} - \underline{w}_{ij} \|) \qquad (31)$$

After the $(k, l)^{th}$ neuron in the 2-D plane is located, the weights of its neighboring neurons are adjusted by using

$$\underline{w}_{ij}(t + 1) = \underline{w}_{ij}(t) + \alpha \| \underline{X} - \underline{w}_{ij} \| \qquad (32)$$

until the weight vector reaches equilibrium, i.e.,

$$\underline{w}_{ij}(t + 1) = \underline{w}_{ij}(t). \qquad (33)$$

A question, which now may be raised is, how to select the neighborhood neuron $N_{i,j}$. In fact, this is done by randomly selecting a square or circular zone around the neuron $N_{i,j}$, where the furthest neuronal distance with respect to $N_{i,j}$ is arbitrarily chosen.

Once the training of the neighborhood neuron around $N_{i,j}$ is over, the process of selection of the next neuron by criterion (31) is performed. After repetition of the selection process of neuron, followed by weight adaptation of the neighborhood neurons for a long time, the system reaches an equilibrium with all weight vectors $\underline{w}_{ij}$ for $1 \leq i,j \leq n$, being constant for a given dimension of the neighborhood of neurons. The neighborhood of the neurons is now decreased and the process of selection of neuron and weight adaptation of its neighboring neurons are continued until the equilibrium condition is reached. By gradually reducing the neighborhood of the neuron, selected for weight adaptation, ultimately, a steady-state neighborhood of neurons in the 2-D plane is obtained. The plane of neurons thus obtained represents a spatial mapping of the neurons, corresponding to the input pattern.

The algorithm for self-organizing neural adaptation for a input vector $\underline{X}$ and the corresponding weight vector $\underline{w}_{ij}$ is presented below.

Procedure Self-Organization ($\underset{\sim}{X}$, $\underset{\sim}{w}_{ij}$)

Begin

  Repeat

  For i := 1 to n **do**

  Begin

    For j := 1 to n **do**

    Begin

      If $\|\underset{\sim}{X} - \underset{\sim}{w}_{kl}\| = \underset{1 \leq i < n}{\min} (\underset{1 \leq j < n}{\min} \|\underset{\sim}{X} - \underset{\sim}{w}_{ij}\|)$

      Then Adjust weights of neighboring neurons of $N_{k,l}$
          by the following rule:

      For i' := (k − δ) to (k + δ) **do**

      Begin

        For j' := (l − δ) to (l + δ) **do**

        Begin

          $\underset{\sim}{w}_{i'j'}(t + 1) = \underset{\sim}{w}_{i'j'}(t) + \alpha \|\underset{\sim}{X} - \underset{\sim}{w}_{i'j'}\|$

        End For;

      End For;

      δ := δ − ε; // **Space-organization** //

    End For;

  End For;

  Until δ ≤ pre-assigned quantity;

End.

The above algorithm should be repeated for all input patterns. Consequently, all the input patterns will be mapped on the 2D plane as an n-dimensional point, each having a $\underset{\sim}{w}_{ij}$.

3.4   Reinforcement Learning

The reinforcement learning presumes that the agent receives a response from the environment but can determine its status (rewarding/punishable) only at the end of its activity, called the terminal state. We also assume that initially the agent is at a state S_0 and after performing an action on the environment, it moves to a new state S_1. If the action is denoted by a_0, we say

$$S_0 \xrightarrow{\ a_0\ } S_1, \tag{34}$$

i.e., because of action a_0, the agent changes its state from S_0 to S_1. Further, the reward of an agent can be represented by a **utility function**. For example, the points of a ping-pong agent could be its utility.

The agent in reinforcement learning could be either passive or active. A **passive learner** attempts to learn the utility through its presence in different states. An **active learner**, on the other hand, can infer the utility at unknown states from its knowledge, gained through learning.

How can we compute the utility value of being in a state? Suppose, if we reach the goal state, the utility value should be high, say 1. But what would be the utility value in other states? One simple way to compute static utility values in a system with the known starting and the goal state is given here. Suppose the agent reaches the goal S_7 from S_1 (Figure 7) through a state say S_2. Now we repeat the experiment and find how many times S_2 has been visited. If we assume that out of 100 experiments, S_2 is visited 5 times, then we assign the utility of state S_2 as $5/100 = 0.05$. Further we may assume that the agent can move from one state to its neighboring state (diagonal movements not allowed) with an unbiased probability. For example, the agent can move from S_1 to S_2 or S_6 (but not to S_5) with a probability of 0.5. If it is in S_5, it could move to S_2, S_4, S_8 or S_6 with a probability of 0.25.

S_3	S_4	**Goal** S_7
S_2	S_5	S_8
Start S_1	S_6	S_9

S_i denotes the i-th state.

Figure 7. A simple stochastic environment.

We here make an important assumption on utility. *"The utility of sequence is the sum of the rewards accumulated in the states of the sequence"* [68]. The static utility values are difficult to extract as it requires large number of experiments. The key to reinforcement learning is to update the utility values, given the training sequences [32].

In adaptive dynamic programming, we compute utility $U(i)$ of state i by using the following expression:

$$U(i) = R(i) + \sum_{\forall j} M_{ij} U(j) \qquad (35)$$

where $R(i)$ is the reward of being in state i, M_{ij} is the probability of transition from state i to state j.

In adaptive dynamic programming, we presume the agent to be passive. So, we do not want to maximize the $\Sigma M_{ij} U(j)$ term.

For a small stochastic system, we can evaluate the $U(i)$, $\forall i$ by solving the set of all utility equations like (35) for all states. But when the state space is large, it becomes somewhat intractable.

3.4.1 Temporal Difference Learning

To avoid solving the constraint equations like (35), we make an alternative formulation to compute U(i) by the following expression.

$$U(i) \leftarrow U(i) + \alpha \left[R(i) + U(j) - U(i) \right] \qquad (36)$$

where α is the learning rate, that lies in the interval $0 < \alpha < 1$.

In the last expression, we updated $U(i)$ by considering the fact that we should allow transition to state j from state i, when $U(j) >> U(i)$. Since we consider temporal difference of utilities, we call this kind of learning temporal difference (TD) learning.

It seems that when a rare transition occurs from state j to state i, $U(j) - U(i)$ will be too large, causing $U(i)$ large by (36). However, it should be kept in mind that the average value of $U(i)$ will not change much, though its instantaneous value seems to be large occasionally.

3.4.2   Active Learning

For passive learner, we considered M to be a constant matrix. But for an active learner, it must be a variable matrix. So, we redefine the utility equation of (35) as follows.

$$U(i) = R(i) + \max_a \sum_{\forall j} M_{ij}^a U(j) \tag{37}$$

where M_{ij}^a denotes the probability of reaching state j through an action 'a' performed at state i. The agent will now choose the action a for which M_{ij}^a is maximum. Consequently, U(i) will be maximum.

3.4.3   Q-Learning

In Q-learning, instead of utility values, we use q-values. We employ Q(a, i) to denote the Q-value of doing an action a at state i. The utility values and Q-values are related by the following expression:

$$U(i) = \max_a Q(a, i). \tag{38}$$

Like utilities, we can construct a constraint equation that holds at equilibrium, when the Q-values are correct [68]:

$$Q(a, i) = R(i) + \sum M_{ij}^a . \max_{a'} Q(a', i). \tag{39}$$

The corresponding temporal-difference updating equation is given by

$$Q(a, i) \leftarrow Q(a, i) + \alpha[R(i) + \max(a', j) - Q(a, i)] \tag{40}$$

which is to be evaluated after every transition from state i to state j.

The Q-learning continues following expression (40) until the Q-values at each state i in the space reach a steady value.

4   Genetic Algorithms

A Genetic algorithm (GA) is a stochastic algorithm [51] that models the evolutionary process of biological species through natural selection. Proposed by Holland [57] in early 1960s, this algorithm is gaining its importance for its wide acceptance in solving three classical problems, such as learning, search and optimization. A number of researchers throughout the world have developed their own ways to prove the convergence of the algorithm. Among these the work by Goldberg [55], De Jong [52], Davis [51], Muehlenbein [61], Chakraborti [48]-[50], Fogel [53], and Vose [64]-[65] need special mention. In this section we will briefly outline this algorithm with an example and prove the convergence of the algorithm by Markov chain analysis.

A GA operates through a simple cycle of stages [54]:

  (i)   Creation of a "population" of strings,
  (ii)   Evaluation of each string,
  (iii)   Selection of best strings and
  (iv)   Genetic manipulation to create new population of strings.

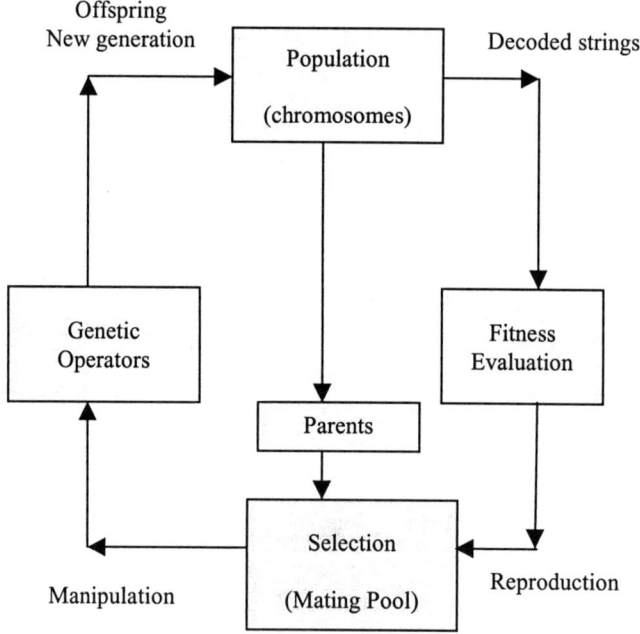

Figure 8. The cycle of genetic algorithms.

The cycle of a GA is presented in Figure 8. Each cycle in GA produces a new generation of possible solutions for a given problem. In the first phase, an initial population, describing representatives of the potential solution, is created to initiate the search process. The elements of the population are encoded into bit-strings, called chromosomes. The performance of the strings, often called fitness, is then evaluated with the help of some functions, representing the constraints of the problem. Depending on the fitness of the chromosomes, they are selected for a subsequent genetic manipulation process. It should be noted that the **selection** process is mainly responsible for assuring survival of the best-fit individuals. After selection of the population strings is over, the genetic manipulation process consisting of two steps is carried out. In the first step, the **crossover** operation that recombines the bits (genes) of each two selected strings (chromosomes) is executed. Various types of crossover operators are reported in the literature. The single point and two points crossover operations are illustrated in Figures 9 and 10 respectively. The **crossover points** of any two chromosomes are selected randomly. The second step in the genetic manipulation process is termed **mutation**, where the bits at one or more randomly selected positions of the chromosomes are altered (Figure 11). The mutation process helps to overcome trapping at local maxima. The offsprings produced by the genetic manipulation process are the next population to be evaluated.

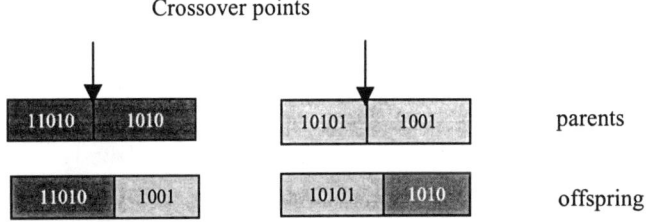

Figure 9. A single point crossover after the 3-rd bit position from the L.S.B.

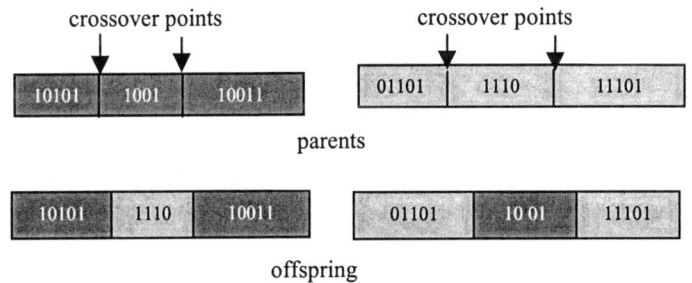

Figure 10. Two point crossover: one after the 4th and the other after the 8th bit positions from the L.S.B.

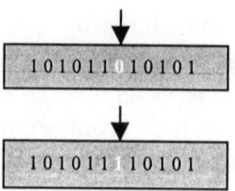

Figure 11. Mutation of a chromosome at the 5th bit position.

Example 3: The GA cycle is illustrated in this example for maximizing a function $f(x) = x^2$ in the interval $0 \leq x \leq 31$. In this example the fitness function is f (x) itself. The larger is the functional value, the better is the fitness of the string. In this example, we start with 4 initial strings. The fitness value of the strings and the percentage fitness of the total are estimated in Table 2. Since fitness of the second string is large, we select 2 copies of the second string and one each for the first and fourth string in the mating pool. The selection of the partners in the mating pool is also done randomly. Here in Table 3, we selected partner of string 1 to be the 2-nd string and partner of 4-th string to be the 2nd string. The crossover points for the first-second and second-fourth strings have been selected after o-th and 2-nd bit positions respectively in Table 3. The second generation of the population without mutation in the first generation is presented in Table 4.

Table 2. Initial population and their fitness values.

string no.	initial population	x	f(x) (fitness)	strength (% of total)
1	01101	13	169	14.4
2	11000	24	576	49.2
3	01000	08	64	5.5
4	10011	19	361	30.9
sum-fitness			1170	100.00

Table 3. Mating pool strings and crossover.

string no.	mating pool	mates string	swapping	new population
1	01101	2	0110[1]	01100
2	11000	1	1100[0]	11001
2	11000	4	11[000]	11011
4	10011	2	10[011]	10000

Table 4. Fitness value in second generation.

Initial population	x	f(x) (fitness)	strength (% of total)
01100	12	144	8.2
11001	25	625	35.6
11011	27	729	41.5
10000	16	256	14.7
sum-fitness		1754	100.00

A **Schema** (or schemata in plural form) / **hyperplane** or **similarity template** [50] is a genetic pattern with fixed values of 1 or 0 at some designated bit positions. For example, S = 01?1??1 is a 7-bit schema with fixed values at 4-bits and don't care values, represented by ?, at the remaining 3 positions. Since 4 positions matter for this schema, we say that the schema contains 4 **genes**.

A basic observation made by Holland is that "*a schema with an above average fitness tends to increase at an exponential rate until it becomes a significant portion of the population.*"

4.1 Deterministic Explanation of Holland's Observation

To explain Holland's observation in a deterministic manner let us presume the following assumptions [47]:

(i) There are no recombination or alternations to genes.
(ii) Initially, a fraction f of the population possesses the schema S and those individuals reproduce at a fixed rate r.
(iii) All other individuals lacking schema S reproduce at a rate s < r.

Thus with an initial population size of N, after t generations, we find $N f r^t$ individuals possessing schema S and the population of the rest of the individuals is $N(1 - f) s^t$. Therefore, the fraction of the individuals with schema S is given by

$$(N f r^t) / [N (1 - f) s^t + N f r^t] = f (r / s)^t / [1 + f \{(r / s)^t - 1\}]. \quad (41)$$

For small t and f, the above fraction reduces to $f (r / s)^t$, which means the population having the schema S increases exponentially at a rate (r / s). A stochastic proof of the above property will be presented shortly, vide a well-known theorem, called the *fundamental theorem of Genetic algorithm*.

4.2 Stochastic Explanation of GA

For presentation of the fundamental theorem of GA, the following terminologies are defined in order.

The **order of a schema H**, denoted by **O(H)**, is the number of fixed positions in the schema. For example, the order of schema H = ?001?1? is 4, since it contains 4 fixed positions.

The **defining length of a schema**, denoted by d(H), is the difference between the leftmost and rightmost specific (i.e., non-don't care) string positions. For example, the schema ?1?001 has a defining length d(H) = 4 – 0 = 4, while the d(H) of ???1?? is zero.

The schemas defined over L-bit strings may be geometrically interpreted as **hyperplanes in an L-dimensional hyperspace** (a binary vector space) with each L-bit string representing one corner point in an n-dimensional cube.

4.3 The Fundamental Theorem of Genetic Algorithms (Schema Theorem)

Let the population size be N, which contains $m_H(t)$ samples of schema H at generation t. Among the selection strategies, the most common is the *proportional selection*. In proportional selection, the number of copies of chromosomes selected for mating is proportional to their respective fitness values. Thus, following the principles of proportional selection [62], a string i is selected with probability

$$f_i / \sum_{i=1}^{N} f_i \qquad (42)$$

where f_i is the fitness of string i. Now, the probability that in a single selection, a sample of schema H is chosen is described by

$$\sum_{i=1}^{m_H(t)} f_i / \sum_{i=1}^{N} f_i$$

$$= m_H(t) f_H / \sum_{i=1}^{N} f_i \tag{43}$$

where f_H is the average fitness of the $m_H(t)$ samples of H.

Thus, in a total of N selections with replacement, the expected number of samples of schema H in the next generation is given by

$$m_H(t+1) = N m_H(t) f_H / \sum_{i=1}^{N} f_i \tag{44}$$

$$= m_H(t) f_H / f_{av}$$

where
$$f_{av} = \sum_{i=1}^{N} f_i / N \tag{45}$$

is the population average fitness in generation t. The last expression describes that the genetic algorithm allocates over an increasing number of trials to an above average schema. The effect of crossover and mutation can be incorporated into the analysis by computing the schema survival probabilities of the above average schema. In crossover operation, a schema H survives if the cross-site falls outside the defining length d(H). If p_c is the probability of crossover and L is the word-length of the chromosomes, then the disruption probability of schema H due to crossover is

$$p_c \, d(H) / (L - 1). \tag{46}$$

A schema H, on the other hand, survives mutation, when none of its fixed positions is mutated. If p_m is the probability of mutation and O(H) is the order of schema H, then the probability that the schema survives is given by

$$(1 - p_m)^{O(H)}$$

$$= 1 - p_m \, O(H). \tag{47}$$

Therefore, under selection, crossover and mutation, the sample size of schema H in generation (t + 1) is given by

$$m_H(t + 1) \geq (m_H(t) \, f_H / f_{av}) \, [1 - p_c \, d(H) / (L - 1) - p_m \, O(H)]. \tag{48}$$

The above theorem is called the **fundamental theorem of GA** or the **Schema theorem**. It is evident from the Schema theorem that for a given set of values of $d(H)$, $O(H)$, L, p_c and p_m, the population of schema H at the subsequent generations increases exponentially when $f_H > f_{av}$. This, in fact, directly follows from the difference equation:

$$m_H(t + 1) - m_H(t) \geq (f_H / f_{av} - 1 / K) \, K \, m_H(t) \tag{49}$$

where

$$K = 1 - p_c \, d(H) / (L - 1) - p_m \, O(H). \tag{50}$$

$$\Rightarrow \Delta m_H(t) \geq K \, (f_H / f_{av} - 1 / K) \, m_H(t). \tag{51}$$

Replacing Δ by $(E - 1)$, where E is the extended difference operator, we find

$$(E - 1 - K_1) \, m_H(t) \geq 0 \tag{52}$$

where

$$K_1 = K \, (f_H / f_{av} - 1 / K). \tag{53}$$

Since $m_H(t)$ in equation (51) is positive, $E \geq (1 + K_1)$. Thus, the solution of (52) is given by

$$m_H(t) \geq A \, (1 + K_1)^t \tag{54}$$

where A is a constant. Setting the boundary condition at $t = 0$, and substituting the value of K_1 by (54) therein, we finally have:

$$m_H(t) \geq m_H(0) \, (K \, f_H / f_{av})^t \tag{55}$$

Since K is a positive number, and $f_H / f_{av} > 1$, $m_H(t)$ grows exponentially with iterations. The process of exponential increase of $m_H(t)$ continues until some iteration r, when f_H approaches f_{av}. This is all about the proof of the schema theorem.

4.4   The Markov Model for Convergence Analysis

To study the convergence of the GA, let us consider an exhaustive set of population states, where 'state' means possible members (chromosomes) that evolve at any GA cycle. As an illustration, let us consider 2-bit chromosomes and population size = 2, which means at any GA cycle we select only two chromosomes. Under this circumstance, the possible states that can evolve at any iteration are the members of the set S, where

$$S = \{(00, 00), (00, 01), (00, 10), (00, 11), (01, 00), (01, 01),$$
$$(01, 10), (01, 11), (10, 00), (10, 01), (10, 10), (10, 11),$$
$$(11, 00), (11, 01), (11, 10), (11, 11)\}$$

For the sake of understanding, let us now consider the population size = 3 and the chromosomes are 2-bit patterns, as presumed earlier. The set S now takes the following form:

$$S = \{(00, 00, 00), (00, 00, 01), (00, 00, 10), (00, 00, 11),$$
$$(00, 01, 00), (00, 01, 01), (00, 01, 10), (00, 01, 11),$$
$$\ldots\ldots\ldots \quad \ldots\ldots\ldots \quad \ldots\ldots\ldots \quad \ldots\ldots\ldots$$
$$(11, 11, 00), (11, 11, 01), (11, 11, 10), (11, 11, 11)\}$$

It may be noted that the number of elements of the last set S is 64. In general, if the chromosomes have the word length of m bits and the number of chromosomes selected in each GA cycle is n, then the cardinality of the set S is $2^{m\,n}$.

The Markov transition probability matrix P for 2-bit strings of population size 2, thus, will have a dimension of (16×16), where the element p_{ij} of the matrix denotes the probability of transition from i-th to j-th state. A clear idea about the states and their transitions can be formed from Figure 12.

It needs mention that since from a given i-th state, there could be a transition to any 16 j-th states, therefore the row sum of P matrix must be 1. Formally,

$$\sum_{\forall j} P_{ij} = 1, \tag{56}$$

for a given i.

Now, let us assume a row vector π_t, whose k-th element denotes the probability of occurrence of the k-th state at a given genetic iteration (cycle) t; then π_{t+1}, can be evaluated by

$$\pi_{t+1} = \pi_t \cdot P \tag{57}$$

Thus starting with a given initial row vector π_0, one can evaluate the state probability vector after n-th iteration π_n by

$$\pi_n = \pi_0 \cdot P^n \tag{58}$$

where P^n is evaluated by multiplying P matrix with itself $(n-1)$ times.

Identification of a P matrix for a GA that allows selection, crossover and mutation, undoubtedly, is a complex problem. Goldberg [55], Davis [51], Fogel [53] and Chakraborty [48]-[50] have done independent work in this regard. For simplicity of our analysis, let us now consider the GA without mutation.

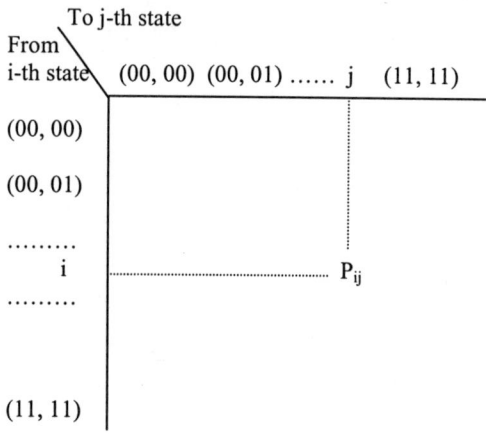

Figure 12. The Markov state-transition matrix P.

The behavior of GA without mutation can be of the following three types:

(i) The GA may converge to one or more absorbing states (i.e., states wherefrom the GA has no transitions to other states).

(ii) The GA may have transition to some states, wherefrom it may terminate to one or more absorbing states.

(iii) The GA never reaches an absorbing state.

Taking all the above into account, we thus construct P as a partitioned matrix of the following form:

$$P = \left(\begin{array}{c|c} I & 0 \\ \hline R & Q \end{array} \right) \tag{59}$$

where I is an identity matrix of dimension (a × a) that corresponds to the absorbing states; R is a (t × a) transition sub-matrix describing transition to an absorbing state; Q is a (t × t) transition sub-matrix describing transition to transient states and not to an absorbing state and 0 is a null matrix of dimension (t × t).

It can be easily shown that P^n for the above matrix P can be found as follows.

$$P^n = \begin{pmatrix} I & 0 \\ N_n R & Q^n \end{pmatrix} \tag{60}$$

where the n-step transition matrix N_n is given by

$$N_n = I + Q + Q^2 + Q^3 + \ldots + Q^{n-1} \tag{61}$$

As n approaches infinity,

$$\lim_{n \to \infty} N_n = (I - Q)^{-1}. \tag{62}$$

Consequently, as n approaches infinity,

$$P = \begin{pmatrix} I & 0 \\ (I-Q)^{-1} R & 0 \end{pmatrix} \tag{63}$$

Goodman [53] has shown that the matrix $(I - Q)^{-1}$ is guaranteed to exist. Thus given an initial probability vector π_0, the chain will have a transition to an absorbing state with probability 1. Further, there exists a non-zero probability that absorbing state will be the globally optimal state.

We now explain: why the chain will finally terminate to an absorbing state. Since the first 'a' columns for the matrix P^n, for $n \to \infty$, are non-zero and the remaining columns are zero, therefore, the chain must have transition to one of the absorbing states. Further, note that the first 'a' columns of the row vector π_n for $n \to \infty$ denote the probability of absorption at different states, and the rest of the columns denote that the probability of transition to non-absorbing states is zero. Thus probability of transition to absorbing states is one. Formally,

$$\sum_{i=1}^{a} \lim_{n\to\infty}(\pi_n)_i = \sum_{i=1}^{a} \lim_{n\to\infty}(\pi_0 P^n)_i$$

$$= \sum_{i=1}^{a} \lim_{n\to\infty}\left(\pi_0 \binom{I}{(I-Q)^{-1}R}\right)_i \qquad (64)$$

$$= 1$$

5   Belief Networks

A Bayesian belief network [2], [3], [7], [8] is represented by a directed acyclic graph or tree, where the nodes denote the events and the arcs denote the cause-effect relationship between the parent and the child nodes. Each node, here, may assume a number of possible values. For instance, a node A may have n number of possible values, denoted by $A_1, A_2, ..., A_n$. For any two nodes, A and B, when there exists a dependence A→B, we assign a conditional probability matrix [P (B/A)] to the directed arc from node A to B. The element at the jth row and i^{th} column of P(B/A), denoted by $P(B_j/A_i)$, represents the conditional probability of B_j assuming the prior occurrence of A_i. This is described in Figure 13.

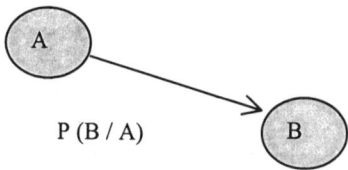

P (B / A)

Figure 13. Assigning a conditional probability matrix in the directed arc connected from A to B.

Given the probability distribution of A, denoted by $[P(A_1) \ P(A_2) \ ... \ P(A_n)]^T$, we can compute the probability distribution of event B by using the following expression:

$$P(B) = [P(B_1) \ P(B_2) \ .... P(B_m)]^T_{m \times 1}$$

$$= [P(B/A)]_{m \times n} \ [P(A_1) \ P(A_2) \ ... \ P(A_n)]^T_{n \times 1}$$

$$= [P(B/A)]_{m \times n} \times [P(A)]_{n \times 1}. \qquad (65)$$

We now illustrate the computation of P(B) with an example.

Example 4: Consider a Bayesian belief tree describing the possible causes of a defective car.

Here, each event in the tree (Figure 14) can have two possible values: true or false. Thus the matrices associated with the arcs will have dimensions (2 x 2). Now, given $P(A) = [P(A = true) \ P(A = false)]^T$, we can easily compute P(B), P(C), P(D), P(E), ..., P(I) provided we know the transition probability matrices connected with the links. As an illustrative example, we compute P(B) with P(B/A) and P(A).

$$\text{Let } P(A) = [P(A = true) \ P(A = false)]^T$$
$$= [\quad 0.7 \qquad 0.3 \quad]^T$$

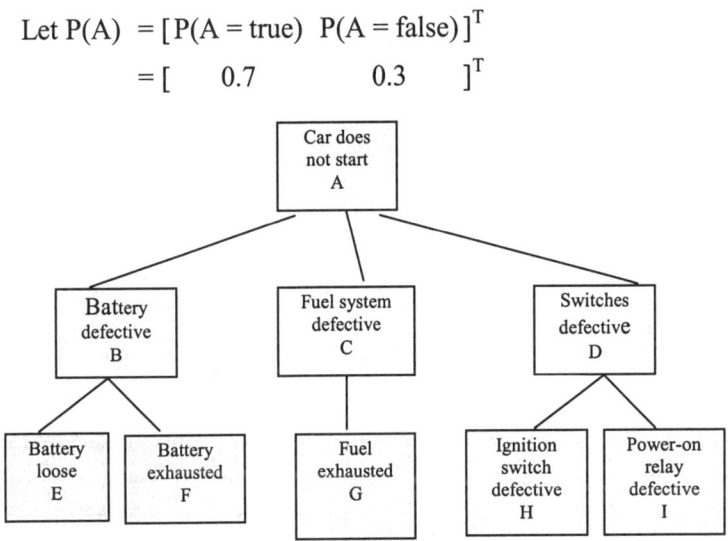

Figure 14. A diagnostic tree for a car.

		$B_j = true$	$B_j = false$
$P(B/A) =$	$A_i = true$	0.8	0.2
	$A_i = false$	0.4	0.6

So, $\qquad P(B) = P(B / A) \cdot P(A) = [0.62 \ \ 0.46]^T$.

One interesting property of Bayesian network is that we can compute the probability of the joint occurrence easily with the help of the topology. For instance, the probability of joint occurrence of A, B, C, D, E, F, G, H, I (see Figure 14) is given by

P(A, B, C, D, E, F, G, H, I)

$$= P(A/B) \cdot P(A/C) \cdot P(A/D) \cdot P(B/E,F) \cdot P(C/G) \cdot P(D/H,I) \qquad (66)$$

Further, if E and F are independent, and H and I are independent, the above result reduces to

P(A, B, C, D, E, F, G, H, I)

$$= P(A/B) \cdot P(A/C) \cdot P(A/D) \cdot P(B/E) \cdot P(B/F) \cdot P(C/G) \cdot P(D/H) \cdot P(D/I)$$

Thus, given A,B,C,...,H all true except I, we would substitute the conditional probabilities for P(B = true / A = true), P(A = true / C = true), ..., and finally P(D = true / I = false) in the last expression to compute P(A = true, B = true, ..., H = true, I = false).

Judea Pearl [2] proposed a scheme for propagating beliefs of evidence in a Bayesian network. We shall first demonstrate his scheme with a Bayesian tree like that in Figure 14. It may, however, be noted that like the tree of Figure 14, each variable, say A,B,..., need not have only two possible values. For example, if a node in a tree denotes German Measles (GM), it could have three possible values like severe-GM, little-GM, moderate-GM.

In Pearl's scheme for evidential reasoning, he considered both the causal effect and the diagnostic effect to compute the **belief function** at a given node in the Bayesian belief tree. For computing belief at a node, say V, he partitioned the tree into two parts: (i) the subtree rooted at V and (ii) the rest of the tree. Let us denote the subset of the evidence, residing at the subtree of V by e_v^- and the subset of the evidence from the rest of the tree by e_v^+. We denote the belief function of the node V by Bel(V), where it is defined as

$$\text{Bel(V)} = P(V/e_v^+, e_v^-)$$

$$= P(e_v^-/V) \cdot P(V/e_v^+) / \alpha$$

$$= \lambda(V) \, \Pi(V) / \alpha \qquad (67)$$

where $\lambda(V) = P(e_v^-/V),$

$\Pi(V) = P(V/e_v^+),$ $\qquad\qquad\qquad (68)$

and α is a normalizing constant, determined by

$$\alpha = \sum_{v \in (true, false)} P(e_v^- / V) \cdot P(V/e_v^+) \qquad (69)$$

It seems from the last expression that v could assume only two values: true and false. It is just an illustrative notation. In fact, v can have a number of possible values.

Let node V have n offsprings, see Figure 15. For computing $\lambda(V)$, we divide e_v^- into n disjoint subsets e_{Zi}, $1 \le i \le n$, where Z_i is a child of V.

So, $\lambda(V) = P(e_v^- / V)$.

$$= P(e_{Z1}^- / V, e_{Z2}^- / V, \ldots, e_{Zn}^- / V)$$

$$= P(e_{Z1}^- / V) \cdot P(e_{Z2}^- / V) \cdot \ldots \cdot P(e_{Zn}^- / V)$$

$$= \prod_{i=1}^{n} \lambda_{Zi}(V). \qquad (70)$$

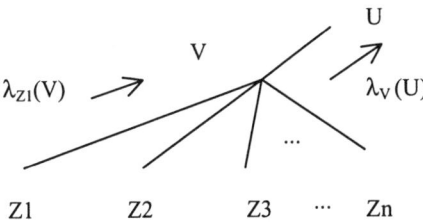

Figure 15. Propagation of λs from the children to the parent in an illustrative tree.

We now compute $\Pi(V)$ using the message $\Pi_V(U) = P(U|e_v^+)$ from the parent U of V.

$$\Pi(V) = P(U|e_v^+)$$

$$= \sum_{u \in (true, false)} P(V |e_v^+, U = u) \cdot P(U = u |e_v^+)$$

$$= \sum_{u \in (true, false)} P(V |U = u) \cdot P(U = u |e_v^+)$$

$$= \sum_{u \in (true, false)} P(V |U = u) \cdot \Pi_V(U = u)$$

$$= [P(V|U)]^T_{2 \times 2} \times [\Pi_V(0) \; \Pi_V(1)]^T_{2 \times 1} \qquad (71)$$

We now compute the messages that node V sends to its parents U and each of its children Z_1, Z_2, ..., Z_n to update their values. Each of these two messages is a conditional probability, given that the condition holds and the probability given that it does not.

Now, the message from V to parent U, denoted by $\lambda_v(U)$, is computed as

$$\lambda_v(U) = \sum_{v \in (\text{true,false})} P(e_v^- | U, V = v)\, P(V = v | U)$$
$$= \sum_{v \in (\text{true,false})} P(e_v^- | V = v)\, P(V = v | U)$$
$$= \sum_{v \in (\text{true,false})} P(V = v | U)\, \lambda(V = v)$$
$$= [P(V | U)]_{2 \times 2} \times [\lambda(0)\ \lambda(1)]^T_{2 \times 1} \tag{72}$$

Finally, the message from V to its child Z_j is given by

$$\Pi_{Zj}(V) = P(V | e_{zj}^+)$$
$$= P(V | e_v^+, e_{Z1}^-, e_{Z2}^-, ..., e_{Zi-1}^-, e_{Zi+1}^-, ..., e_{Zn}^-)$$
$$= \beta\, \Pi_{j \neq i}\, P(e_{zj}^- | V, e_v^+)\, P(V | e_v^+)$$
$$= \beta\, \Pi_{j \neq i} P(e_{zj}^- | V)\, P(V | e_v^+)$$
$$= \beta(\Pi_{j \neq i}\lambda_{Zj}(V))\, \Pi(V)$$
$$= \beta(\lambda(V) / \lambda_{Zj}(V))\, \Pi(V)$$
$$= \beta\, Bel(V) / \lambda_{Zj}(V) \tag{73}$$

where β is a normalizing constant computed similarly as α.

The belief updating process at a given node B (in Figure 16) has been illustrated based on the above expressions for computing the λ and Π messages. We here assumed that at each node and link of the tree (Figure 15) we have one processor [7]. We call these node and link processor respectively. The functions of the node and the link processors are described in Figure 16.

The main steps [2]-[3] of the belief–propagation algorithm of Pearl are outlined below.

1. During initialization, we set all λ and Π messages to 1 and set $\Pi_B(A)$ messages from root to the prior probability $[P(A_1)\ P(A_2), ..., P(A_m)]^T$ and define the conditional probability matrices. Then

estimate the prior probabilities at all nodes, starting from the children of the root by taking the product of transpose of the conditional probability matrix at the link and the prior probability vector of the parent. Repeat this for all nodes up to the leaves.

2. Generally, the variables at the leaves of the tree are instantiated. Suppose, the variable $E = E_2$ is instantiated. In that case, we set [75]

$$\lambda_E(B) = [0 \ 1 \ 0 \ 0 \ 0 \ 0 \dots 0],$$

where the second element corresponds to instantiation of $E = E_2$.

3. When a node variable is not instantiated, we calculate its λ values following the formula, outlined in Figure 16.

4. The λ and $\prod$ messages are sent to the parents and the children of the instantiated node. For the leaf node there is no need to send the $\prod$ message. Similarly, the root node need not send the λ message.

5. The propagation continues from the leaf to its parent, then from the parent to the grandparent, until the root is reached. Then down stream propagation starts from the root to its children, then from the children to grandchildren of the root and so on until the leaves are reached. This is called an equilibrium condition, when the λ and $\prod$ messages do not change, unless instantiated further. The belief value at the nodes now reflects the belief of the respective nodes for 'the car does not start' (in our example tree).

6. When we want to fuse the beliefs of more than one evidence, we can submit the corresponding λ messages at the respective leaves one after another, and repeat from step 3, otherwise stop.

The resulting beliefs at each node now appear to be the fusion of the joint effect of two or more observed evidences.

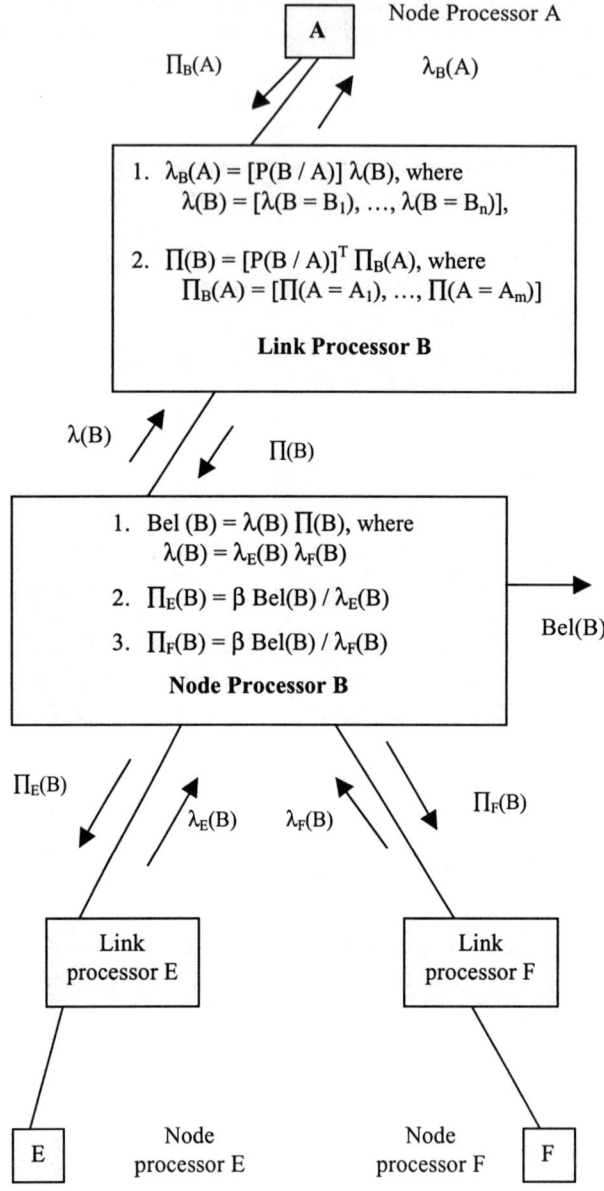

Figure 16. The computation and propagation of λ and Π messages from and to node B of Figure 15.

We presented Pearl's scheme for evidential reasoning for a tree structure only. However, the belief propagation scheme of Pearl can also be extended to **polytrees**, i.e., graphs where the nodes can have more than one parent, but there must be a single arc between each parent to a child and the graph should not have any cycles. The details of belief revision on a polytree are available in [75].

6   Computational Learning Theory

The main question on machine learning is how does one know that his/her learning algorithm has generated a concept, appropriate for predicting the future correctly? For instance, in inductive learning how can we assert that our hypothesis h is sufficiently close to the target function f, when we do not know f? These questions can be answered with the help of computational learning theory.

The principle of computational learning theory states that any hypothesis, which is sufficiently incorrect, will be detected with a high probability after experimenting with a small number of training instances. Consequently, a hypothesis that is supported by a large number of problem instances will be unlikely to be wrong and hence should be **Probably Approximately Correct (PAC).**

The PAC learning was defined in the last paragraph w.r.t. training instances. But what about the validity of PAC learning on the test set (not the training set) of data. The assumption made here is that the training and the test data are selected randomly from the population space with the same probability distribution. This in PAC learning theory is referred to as **stationary assumption.**

In order to formalize the PAC learning theory, we need a few notations [68].

Let   X = exhaustive set of examples,
      D = distribution by which the sample examples are drawn,
      m = cardinality of examples in the training set,
      H = the set of possible hypothesis, and
      f = function that is approximated by the hypothesis h.

We now define an error of hypothesis h by

$$\text{Error}(h) = P(h(x) \neq f(x) \mid x \in D) \tag{74}$$

A hypothesis h is said to be **approximately correct** when

$$\text{error}(h) \leq \varepsilon, \tag{75}$$

where ε is a small positive quantity.

When an approximate hypothesis h is true, it must lie within the ε-ball around f. When h lies outside the ε-ball, we call it a bad hypothesis [68].

Now, suppose a hypothesis $h_b \in H_{bad}$ is supported by first m examples. The probability that a bad hypothesis is consistent with an example $\leq (1 - \varepsilon)$. If we consider m examples to be independent, the probability that all m samples will be consistent with hypothesis h_b is $\leq (1 - \varepsilon)^m$. Now, if H_{bad} has to contain a consistent hypothesis, at least one of the hypothesis of H_{bad} should be consistent. The probability of this happening is bounded by the sum of individual probabilities.

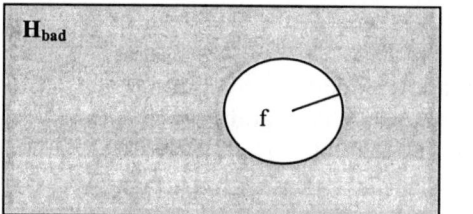

Figure 17. The ball around f.

Thus, $P(\text{a consistent } h_b \in H_{bad}) \leq |H_{bad}| \, (1 - \varepsilon)^m$ (76)

$$\leq |H| \, (1 - \varepsilon)^m \tag{77}$$

where $|H_{bad}|$ and $|H|$ denote their respective cardinalities. If we put a small positive upper bound δ to the above quantity, we find

$$|H| \, (1 - \varepsilon)^m \leq \delta \tag{78}$$

$$\Rightarrow m \geq (1/\varepsilon) \, [\ln(1/\delta) + \ln(H)] \tag{79}$$

Consequently, if a learning algorithm asserts an hypothesis that is supported by m number of examples, then it must be correct with a probability $\geq (1 - \delta)$, when the error is $\leq \varepsilon$. So, we can call it probably approximately correct.

7   Synergism of the Computational Intelligence Paradigms

At the beginning of this chapter it was mentioned that the computational power of the computing algorithms increases to a very high extent when two or more computational tools are used together. In fact the composite use of the tools far exceeds the sum of the computational power of the individual tools. This is called synergism because of its similarity with many environmental and biological systems. For instance, the presence of unsaturated nitrogen oxides (NO_x), carbon oxides (CO_x), and particulate matters in humid weather far exceeds the pollution of the sum of these items. This is well known as synergism in the environmental literature. Unfortunately, this is an example of the harmful effect of the synergism, the synergism of computing tools is useful indeed! But how the synergism of the computational tools exactly take place? This is very difficult to generalize at present, as there is no such formalization on these issues till date. But a few examples of neuro-fuzzy, neuro-GA, Fuzzy-GA and neuro-Belief systems, discussed below, will help us to understand the synergistic behavior of these tools.

7.1   Neuro-Fuzzy Synergism

We already know that an ANN is good enough for autonomous machine learning, while a fuzzy system has a significant potential in reasoning with inexact (approximate) data and knowledge. A neuro-fuzzy system, which integrates the behavioral properties of these two systems, thus is capable of learning from approximate data/ knowledge and use it for futuristic reasoning. We can broadly classify the neuro-fuzzy systems into two types: (i) **weakly coupled systems** and (ii) **tightly coupled systems**. In a weakly coupled system, the ANNs and the fuzzy systems maintain their identity. A tightly coupled system, on the other hand, employs neurons based on the characteristics of fuzzy sets; thus the basic elements in the network have the composite characteristics of both neural nets and fuzzy sets. A brief outline of these two type of neuro-fuzzy systems are presented in order.

7.1.1 Weakly Coupled Neuro-Fuzzy Systems

A weakly coupled neuro-fuzzy system employs both a neural net unit and a fuzzy system unit in cascade. Depending on the type of applications the ordering of these 2 units in a composite system may take different forms. In this section we present 2 most common configurations. The first configuration, given in Figure 18 is employed in recognizing an object from its approximate features. Suppose we have the approximate measurements about the features of an object, which are transformed to membership values by using intuitively constructed membership functions in the fuzzifier module. A fuzzy inference engine then maps these primary fuzzified features to other secondary fuzzy features by using a set of fuzzy production rules. The secondary features are usually less affected by the measurement noise of the primary features. The membership values of the secondary features are then supplied to the input of a pre-trained neural net for classification of objects. The output of the neural net in the present context is a vector with one output = 1 and the rest outputs being zero. The output that comes up with a one corresponds to a particular object assigned to that output.

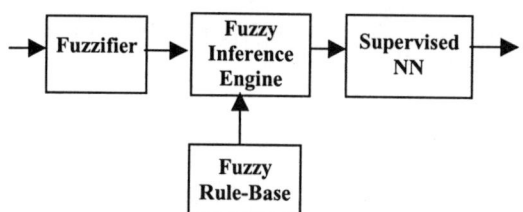

Figure 18. A neural net system followed by a fuzzy logic unit.

In the second configuration a neural net is essentially placed before the fuzzy logic unit in a neuro-fuzzy system (Figure 19). Here, the neural net first transforms a given set of measurements to a target pattern, which is then fuzzified by a set of membership functions. A fuzzy inference engine in consultation with a set of fuzzy production rules is then used to transform the calculated membership values to a desired set of fuzzy inferences. A defuzzification algorithm [15] then may be invoked to transform the fuzzy variables to the desired non-fuzzy variables. Such systems have major applications in control systems, where the measurement signals and controller input parameters are connected through some nonlinear functions, whose closed form representation is not known.

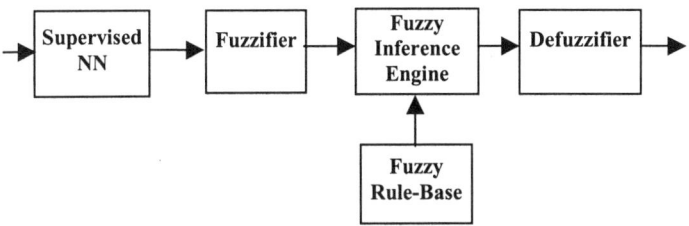

Figure 19. A fuzzy logic system followed by a neural net.

7.1.2 Tightly Coupled Neuro-Fuzzy Systems

In a tightly coupled system the neurons, the basic elements of a neural net, are constructed by amalgamating the composite characteristics of both a neuronal element and fuzzy logic. A large number of tightly coupled neuro-fuzzy systems are referred to in the current literature. Pedrycz's AND-OR neuronal model [34], Konar and Pal's fuzzy Petri net model [27], Paul, Konar and Mandal's Fuzzy ADALINE model [33], Pal and Mitra's fuzzy perceptron model [74] are some of the well-known examples of the tightly coupled neuro-fuzzy system.

Among the classical fuzzy neural nets, OR-AND neuron model of Pedrycz [34] needs special mention. In an OR-AND neuron, we have two OR nodes and one AND node (see Figure 20).

Here, $Z_1 = \wedge\, (W_{ij} \vee x_j)$, $1 \leq j \leq n$,

$Z_2 = \vee\, (V_{ij} \wedge x_j)$, $1 \leq j \leq n$,

and $y = (S_{11} \wedge Z_1) \vee (S_{12} \wedge Z_2)$.   (80)

where '$\wedge$' and '$\vee$' denote fuzzy 't' and 's' norm operators. The above two operations are executed by using expressions (81) and (82).

$$x_1 \wedge x_2 = x_1 \cdot x_2 \tag{81}$$

$$x_1 \vee x_2 = (x_1 + x_2 - x_1 \cdot x_2) \tag{82}$$

where '·' and '+' are typical algebraic multiplication and summation operations.

Pedrycz devised a new algorithm for training an AND-OR neuron, when input membership distributions x_i for $i = 1, ..., n$ and target scalar y are given. The training algorithm attempts to adjust the weights, w_{ij} so that a predefined performance index, is optimized.

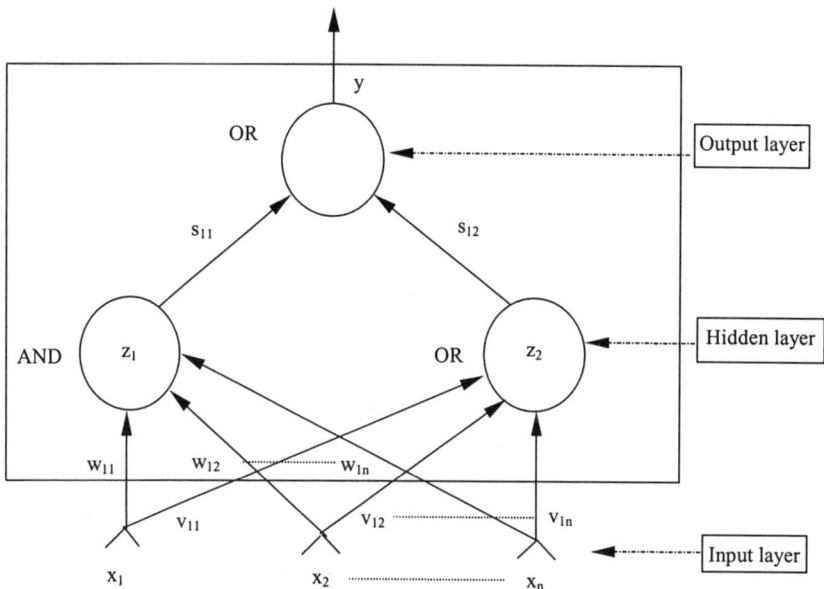

Figure 20. Architecture of an AND-OR neuron.

Pedrycz also designed a pseudo-median filter for using AND-OR neurons, which has many applications in median filtering under image processing. The typical organization of a pseudo-median filter for five data points is given in Figure 21.

In the figure, the pseudo-median is defined as:

$$pseudomedian(x) =$$
$$(1/2) \max[\min(x_1,x_2,x_3), \min(x_2,x_3,x_4), \min(x_3,x_4,x_5)] +$$
$$(1/2) \min[\max(x_1,x_2,x_3), \max(x_2,x_3,x_4), \max(x_3,x_4,x_5)], \qquad (83)$$

which loses little accuracy w.r.t. typical median filter.

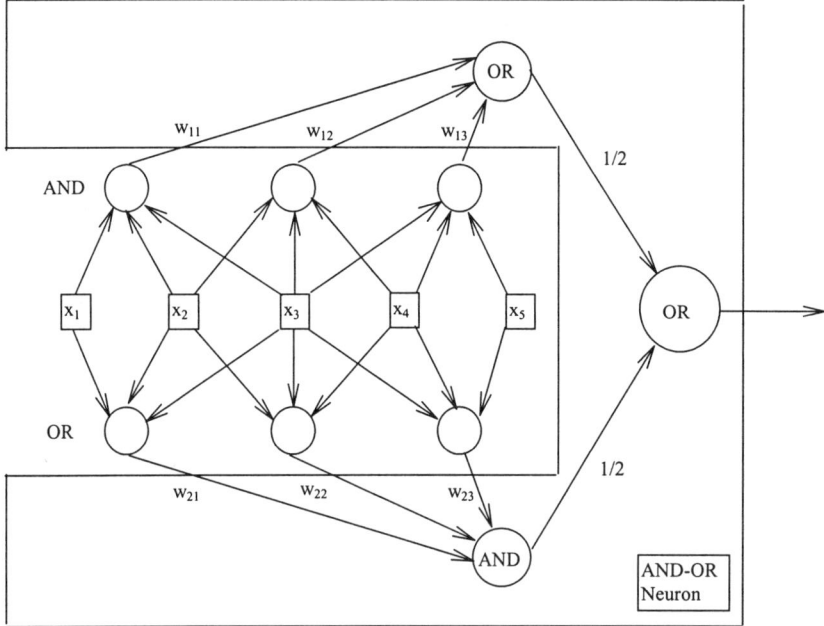

Figure 21. A typical pseudo-median filter.

A set of 10 input-output patterns have been used to train pseudo-median filter. The weights w_{ij}, after training the net are saved and used for subsequent recognition phase. In the recognition phase, the neurons can compute the pseudo-median, when the input data points are given.

7.2   Fuzzy-GA Synergism

Fuzzy logic and genetic algorithm can be used in a composite system for one or more of the following reasons. First, the membership functions, which are usually chosen intuitively in a fuzzy system can be optimized by using GA. Secondly, the GA may be employed to adapt the parameters of a fuzzy relational system. Thirdly, the evolutionary process of GA can be realized with fuzzy operators and logic. We briefly outline the first type of synergism of a Fuzzy-GA system below.

Let x, y and z be three fuzzy variables and $\mu_x(A)$, $\mu_x(B)$, $\mu_x(C)$ are three fuzzy membership distributions with respect to fuzzy sets A, B and C respectively. Similarly, we have $\mu_y(A)$, $\mu_y(B)$, $\mu_y(C)$ and $\mu_z(A)$, $\mu_z(B)$,

$\mu_z(C)$ for variables y and z respectively. Assume that the membership distributions mentioned above, have been constructed intuitively. Now we want to optimize the membership functions by using GA. We can realize it by adjusting the parameters of the membership functions. For brevity of our representation of the chromosomes, let us assume that the membership curves are isosceles triangles with an internal angle between the 2 equal sides = θ. For convenience let us denote such angle for the curve $\mu_x(A)$ by $\theta_x(A)$, say. The same nomenclature is also applicable to θ of all curves. The chromosomes in the present context thus has 9 fields, one each for each membership curves. The crossover and mutation operations in the present context are realized in a conventional way, and thus they have no special characteristics.

A question now may arise: how should we select the fitness function? We remember that here we want to select the membership curves to optimize the system behavior. One approach to implement this is to compare the system responses $F(x, y, z)$ with the desired system responses $F_d(x, y, z)$ for a given i-th input vector $(x, y, z)_i$, for all known input-output instances i. So, we must have a prior knowledge about the desired system response to tune the membership curves. A brief schematic overview of the system is presented in Figure 22.

7.3   Neuro-GA Synergism

The first point that appears in our mind about an ANN is its capability of machine learning. The most important aspect of a GA, on the other hand, is perhaps its application in optimization problems. Fusion of these 2 characteristics of ANN and GA together can give rise to the development of many intelligent machines. This is usually referred to as the neuro-GA synergism. Though GA and neural nets have other form of symbiosis, building machines by utilization of the last 2 characteristics is the most common in the current literature [59], [60]. In this section, we briefly outline the scope of the above form of symbiosis in an optimal learning of a neural net.

7.3.1   Adaptation of a Neural Learning Algorithm Using GA

A supervised neural net trained with input vectors $I = [I_1 \ I_2]$ and target vector $T = [T_1 \ T_2]$ is usually trained following a learning rule that adapts the weight vector W (and threshold vector Th) (Figure 23) of the neural net. Generally, the learning rule includes some intuitively selected parameters like A, B and C. A GA may be employed to optimize the learning algorithm by judiciously adjusting the parameters A, B and C. The chromosomes in the present context thus should have 3 fields: A, B and C. The fitness function in the present context is the sum of the norm of the error vectors for all possible combination of the input and the target vectors I and T respectively. The norm of an error vector E here is evaluated by taking the square root of the sum of its squared components E_1 and E_2. The details of the scheme are available in [75].

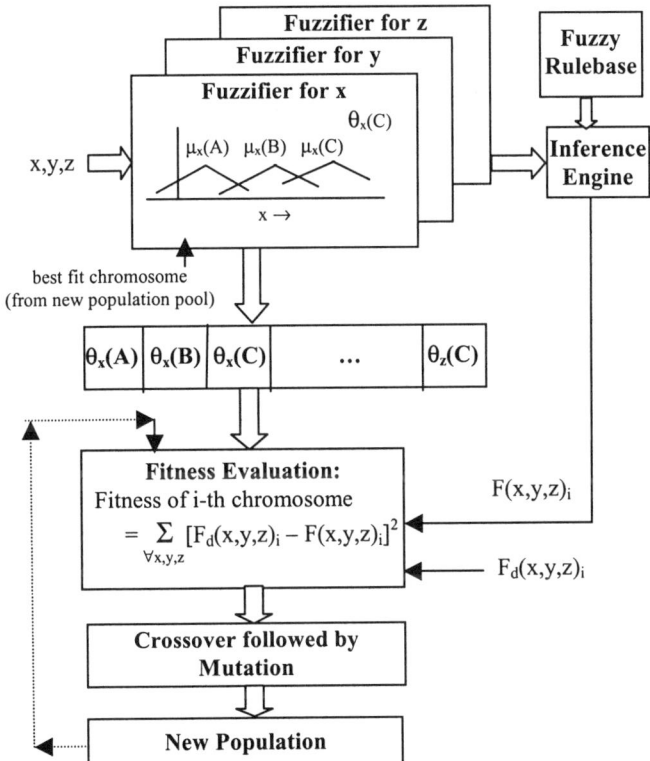

Figure 22. Parameter adjustment of the fuzzy membership functions by a GA.

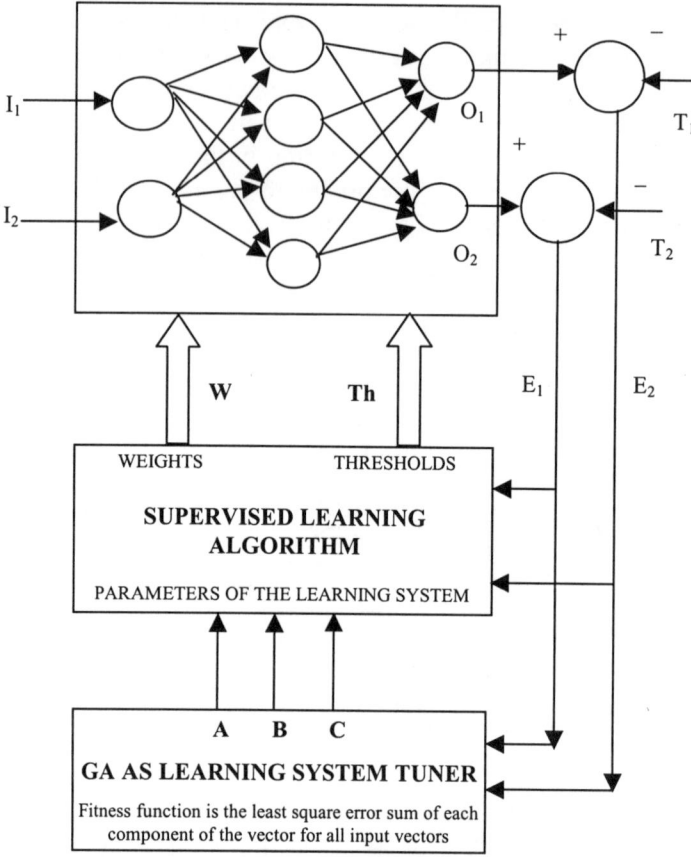

Figure 23. A GA as a learning system tuner.

7.4 GA-Belief Network Synergism

It is clear from our previous discussion on Belief networks that for deriving beliefs of an inference we must have a prior knowledge of the conditional probabilities of the reasoning system. Generally, experts in a domain assign the conditional probabilities of a reasoning system. But on occasions the conditional probabilities need to be adjusted [75]. A genetic algorithm may be employed for an on-line tuning of the conditional probabilities by comparing the response (here beliefs) of the reasoning system with that of an expert. A mean square error norm fitness function constructed from the system's response and that of the expert may be used to adapt the chromosomes comprising of the conditional probabilities of the system. The tuning range of the probabilities should be small (say 0.1) so that the analysis of a particular case history cannot influence the results much.

8 Conclusions and Future Directions

A number of computational intelligence tools and their synergism have been briefly outlined in this chapter. It can now easily be visualized that fuzzy logic is most useful for reasoning with approximate data and knowledge. Neural nets have much scope in machine learning, and GA is used mainly for the purpose of search, optimization and machine learning as well. Belief network model is an alternative approach to approximate reasoning that works following the fundamental principles of Bayesian statistics. Computational learning theory, which has been briefly outlined in the chapter, seeks to answer questions such as "Under what conditions a particular learning algorithm is assured of learning successfully?" It has not yet taken a shape but we believe that in the coming years it will have a great impact on the next generation intelligent systems.

It may be noted that in spite of the characteristic differences of the computational models of intelligence, there are some applications where more than one model is equally applicable. For instance consider the problem of classification. We can formulate and solve the problem by fuzzy logic and various models of neural nets. A common question then naturally arises – which model should we select when? In fact there is no hard and fast rule to sort this out, and thus intuition and experience plays a great role in selection of the right tools for a given problem. How we can select the tools intuitively is outlined below.

For instance for classification by fuzzy c-means clustering algorithm we should know the number of classes and the approximate membership of the data points (features) in the classes. So, when these parameters are known fuzzy c-means clustering algorithm may be used for pattern classification in general. In case the number of data points and classes are too large, obviously the algorithm will be too expensive for classification in real time. When we have no background about the number of classes, we feel from our experience that SOM or the Hopfield net is a good choice. The novice readers may wonder: why?

The answer is obvious as SOM has its inherent characteristics of classification of similar patterns, and the Hopfield net (continuous) may have a number of stable points where a number of instantaneous input vector force the system to one of many stable points. So, SOM or Hopfield net should be preferred when the number of classes is not known.

Let us now consider the problem of optimization. It can be formulated and solved both by GA and neural net models. The question is which one should we select when? It is needless to mention that GA can be employed in most search/ optimization problems, only if we can represent the constraints of the problem by the fitness function. Still there are some problems like the well-known traveling salesperson problem (TSP), which can be nicely solved using Hopfield net (binary) as well. The advantage of the Hopfield net-based formulation is that it is most efficient in real time; unfortunately GA is not so efficient for real-time search/ optimization problems.

There is a great debate in the selection of belief networks and fuzzy logic networks for applications in reasoning with inexact/ incomplete data and knowledge bases. In diagnostic applications belief networks have proved itself successful, where scope of fuzzy logic is yet to be ascertained. Konar in one of his recent books [75], however, presented a new approach to backward reasoning on a specialized fuzzy logic network using their own formulation [78] of fuzzy compositional inverse matrices [79]. Undoubtedly, if the conditional probabilities of a reasoning system can be correctly ascertained, belief networks should yield more accurate results than fuzzy network models. But determining the conditional probabilities accurately itself is a complex and unsolved problem till date. Reasoning using fuzzy network models with intuitively defined membership functions, therefore, is much safe.

Because of the page restrictions we could not discuss much on the possible scope of applications of the computational models presented in this chapter. The subsequent chapters of the book, however, will address these issues in greater detail. Interested readers may also find a number of resourceful texts and monographs [70]-[80], covering a wide range of applications of the soft computational tools. The scope of computational intelligence in biometry [70], medicine [40], control systems [71], character and face recognition [70], [73], fingerprint identification [74], image matching [76] and mobile robotics [80] are just a few to mention. These applications will find a significant role in the service of the mankind. We believe that the next generation computational models of machine intelligence will be equipped with more sophisticated tools, thus improving the quality in every spell of our life.

References

[1] Patterson, D.W. (1990), *Introduction to Artificial Intelligence and Expert Systems*, Prentice-Hall, Englewood Cliffs, NJ, pp. 107-119.

[2] Pearl, J. (1986), "Fusion, propagation and structuring in belief networks," *Artificial Intelligence*, vol. 29, pp. 241-288.

[3] Pearl, J. (1987), "Distributed revision of composite beliefs," *Artificial Intelligence*, vol. 33, pp. 173-213.

[4] Peng, Y. and Reggia, J.A. (1987), "A probabilistic causal model for diagnostic problem solving," *IEEE Trans. on Systems, Man and Cybernetics*, SMC- 17, no. 3, pp. 395-408, May-June.

[5] Shafer, G. (1976), *A Mathematical Theory of Evidence*, Princeton University Press, Princeton, NJ.

[6] Shafer, G. and Logan, R. (1987), "Implementing Dempster's rule for hierarchical evidence," *Artificial Intelligence*, vol. 33, pp. 271-298.

[7] Shenoy, P.P. and Shafer, G. (1986), "Propagating belief functions with local computations," *IEEE Expert*, pp. 43-52, Fall.

[8] Shoham, Y. (1994), *Artificial Intelligence Techniques in PROLOG*, Morgan Kaufmann, San Mateo, CA, pp. 183-185.

[9] Shortliffe, E.H. (1976), *Computer Based Medical Consultations: MYCIN*, American Elsevier, New York.

[10] Shortliffe, E.H. and Buchanan, B.G. (1975), "A model of inexact reasoning," *Mathematical Biosciences*, vol. 23, pp. 351-379.

[11] Bezdek, J.C. (1973), *Fuzzy Mathematics in Pattern Classification*, Ph.D. thesis, Applied Mathematics Center, Cornell University, Ithaca.

[12] Ross, T.J. (1995), *Fuzzy Logic with Engineering Applications*, McGraw-Hill.

[13] Zadeh, L.A. (1965), "Fuzzy sets," *Information and Control*, vol. 8, pp. 338-353.

[14] Zadeh, L.A. (1973), "Outline of a new approach to the analysis of complex systems and decision processes," *IEEE Trans. Systems, Man and Cybernetics*, vol. 3, pp. 28-45.

[15] Zimmerman, H.J. (1996), *Fuzzy Set Theory and Its Applications*, Kluwer Academic, Dordrecht, The Netherlands, pp. 131-162.

[16] Klir, G.J. and Yuan, B. (1995), *Fuzzy Sets and Fuzzy Logic: Theory and Applications*, Prentice-Hall, NJ.

[17] Anderson, J.A. (1972), "A simple neural network generating an associative memory," *Mathematical Biosciences*, vol. 14, pp. 197-220.

[18] Carpenter, G.A. and Grossberg, S. (1987), "A massively parallel architecture for a self-organizing neural pattern recognition machine," *Computer Vision, Graphics and Image Processing*, vol. 37, pp. 54-115.

[19] Carpenter, G.A. and Grossberg, S. (1987), "ART2: Self-organization of stable category recognitioncodes for analog input patterns," *Applied Optics*, vol. 23, pp. 4919-4930, December.

[20] Fu, L.M. (1994), *Neural Networks in Computer Intelligence*, McGraw-Hill, NewYork.

[21] Hertz, J., Krogn, A., and Palmer, G.R. (1990), *Introduction to the Theory of Neural Computation*, Addison-Wesley, Reading, MA.

[22] Hopfield, J. (1982), "Neural nets and physical systems with emergent collective computational abilities," *Proc. of the National Academy of Sciences*, vol. 79, pp. 2554-2558.

[23] Hopfield, J.J. (1984), "Neural networks with graded response have collective computational properties like those of two state neurons," *Proc. of the National Academy of Sciences*, vol. 81, pp. 3088-3092, May.

[24] Kohonen, T. (1989), *Self-organization and Associative Memory*, Springer-Verlag, Berlin.

[25] Kohonen, T., Barna, G., and Chrisley, R., "Statistical pattern recognition using neural networks: Benchmarking studies," *IEEE Conf. on Neural Networks*, San Diego, vol. 1, pp. 61-68.

[26] Konar, A. (1994), *Uncertainty Management in Expert Systems Using Fuzzy Petri Nets*, Ph.D. thesis, Jadavpur University.

[27] Konar, A. and Pal, S. (1999), "Modeling cognition with fuzzy neural nets," in Leondes, C.T. (Ed.), *Neural Network Systems: Techniques and Applications*, Academic Press, New York.

[28] Kosko, B. (1987), "Adaptive bi-directional associative memories," *Applied Optics*, vol. 26, pp. 4947-4960.

[29] Kosko, B. (1988), "Bi-directional associative memories," *IEEE Trans. on Systems, Man and Cybernetics*, vol. SMC-18, pp. 49-60, January.

[30] Kosko, B. (1991), *Neural Networks and Fuzzy Systems: a Dynamical Systems Approach to Machine Intelligence*, Prentice-Hall, Englewood Cliffs, NJ.

[31] Luo, F.L and Unbehauen, R. (1997), *Applied Neural Networks for Signal Processing*, Cambridge University Press, London, pp. 1-31.

[32] Mitchell, M.M. (1997), *Machine Learning*, McGraw-Hill, New York, pp. 81-127.

[33] Paul, B., Konar, A., and Mandal, A.K. (1999), "Fuzzy ADALINEs for gray image recognition," *Neurocomputing*, vol. 24, pp. 207-223.

[34] Pedrycz, W. (1996), *Fuzzy Sets Engineering*, CRC Press, Boca Raton, FL, pp. 73-106.

[35] Rumelhart, D.E. and McClelland, J.L. (1986), *Parallel Distributed Processing: Exploring in the Microstructure of Cognition*, MIT Press, Cambridge, MA.

[36] Rumelhart, D.E., Hinton, G.E. and Williams, R.J. (1986), "Learning representations by back-propagation errors," *Nature*, vol. 323, pp. 533-536.

[37] Haykin, S. (1999), *Neural Networks: a Comprehensive Foundation*, Prentice-Hall, NJ.

[38] Schalkoff, R.J. (1997), *Artificial Neural Networks*, McGraw-Hill, New York, pp. 146-188.

[39] Tank, D.W. and Hopfield, J.J. (1986), "Simple neural optimization networks: An A/D converter, signal decision circuit and a linear programming circuit," *IEEE Trans. on Circuits and Systems*, vol. 33, pp. 533-541.

[40] Teodorescu, H.N., Kandel, A. and Jain, L.C., Eds. (1999), *Fuzzy and Neuro-Fuzzy Systems in Medicine*, CRC Press, London.

[41] Wasserrman, P.D. (1989), *Neural Computing: Theory and Practice*, Van Nostrand Reinhold, New York, pp. 49-85.

[42] Widrow, B. (1962), "Generalization and information storage in networks of ADALINE neurons," in Yovits, M.C., Jacobi, G.T., and Goldstein, G.D. (Eds.), *Self-Organizing Systems*, pp. 435-461.

[43] Widrow, B. and Hoff, M,E. (1960), "Adaptive switching circuits," *1960 IRE WESCON Convention Record*, Part 4, pp. 96-104, NY.

[44] Williams, R.J. (1988), "On the use of back-propagation in associative reinforcement learning," *IEEE Int. Conf. on Neural Networks*, NY, vol. 1, pp. 263-270.

[45] Williams, R.J. and Peng, J. (1989), "Reinforcement learning algorithm as function optimization," *Proc. of Int. Joint Conf. on Neural Networks*, NY, vol. II, pp. 89-95.

[46] Altshuler, E.E. and Linden, D.S. (1997), "Wire–antenna designs using genetic algorithms," *IEEE Antennas and Propagation Magazines*, vol. 39, no. 2, April.

[47] Bender, E.A. (1996), *Mathematical Methods in Artificial Intelligence*, IEEE Computer Society Press, Los Alamitos, pp. 589-593.

[48] Chakraborty, U.K., Deb, K., and Chakraborty, M. (1996), "Analysis of selection algorithms: a Markov chain approach," *Evolutionary Computation*, vol. 4, no. 2, pp. 133-167.

[49] Chakraborty, U.K. and Muehlenbein, H. (1997), "Linkage equilibrium and genetic algorithms," *Proc. 4th IEEE Int. Conf. On Evolutionary Computation*, Indianapolis, pp. 25-29.

[50] Chakraborty, U.K. and Dastidar, D.G. (1993), "Using reliability analysis to estimate the number of generations to convergence in genetic algorithm," *Information Processing Letters*, vol. 46, pp. 199-209.

[51] Davis, T.E. and Principa, J.C. (1993), "A Markov chain framework for the simple genetic algorithm," *Evolutionary Computation*, vol. 1, no. 3, pp. 269-288.

[52] De Jong, K.A. (1975), *An Analysis of Behavior of a Class of Genetic Adaptive Systems*, Doctoral dissertation, University of Michigan.

[53] Fogel, D.B. (1995), *Evolutionary Computation*, IEEE Press, Piscataway, NJ.

[54] Filho, J.L.R. and Treleven, P.C. (1994), *Genetic Algorithm Programming Environment*, IEEE Computer Society Press, pp. 28-43, June.

[55] Goldberg, D.E. (1989), *Genetic Algorithms in Search, Optimization and Machine Learning*, Addison-Wesley, Reading, MA.

[56] Gupta, B. (1999), "Bandwidth enhancement of microstrip antenna through optimal feed using GA," *Seminar on Seekers and Aerospace Sensors*, Hyderabad, India.

[57] Holland, J.H. (1975), *Adaptation in Natural and Artificial Systems*, University of Michigan Press, Ann Arbor.

[58] Koza, J.R. (1992), *Genetic Programming: on the Programming of Computers by Means of Natural Selection*, MIT Press.

[59] McDonell, J.R. (1998), "Control," in Back, T., Fogel, D.B., and Michalewicz, Z. (Eds.), *Handbook of Evolutionary Computation*, IOP and Oxford University Press, New York.

[60] Mitchell, M. (1996), *An Introduction to Genetic Algorithms*, MIT Press, Cambridge, MA.

[61] Muehlenbein, H. and Chakraborty, U.K. (1997), "Gene pool recombination genetic algorithm and the onemax function," *Journal of Computing and Information Technology*, vol. 5, no. 3, pp. 167-182.

[62] Michalewicz, Z. (1992), *Genetic Algorithms + Data Structures = Evolution Programs*, Springer-Verlag, Berlin.

[63] Srinivas, M. and Patnaik, L.M. (1996), "Genetic search: analysis using fitness moments," *IEEE Trans. on Knowledge and Data Engg.*, vol. 8, no. 1, pp. 120-133.

[64] Vose, M.D. and Liepins, G.E. (1991), "Punctuated equilibrium in genetic search," *Complex Systems*, vol. 5, pp. 31-44.

[65] Vose, M.D. (1999), *Genetic Algorithms*, MIT Press.

[66] Antoniou, G. (1997), *Nonmonotonic Reasoning*, MIT Press.

[67] Quinlan, J.R., "Induction of decision trees," *Machine Learning*, vol. 1, no. 1, pp. 81-106.

[68] Russel, S. and Norvig, P. (1995), *Artificial Intelligence: a Modern Approach*, Prentice-Hall, Englewood Cliffs, NJ, pp. 598-644.

[69] Winston, P. (1970), *Learning Structural Descriptions from Examples*, Ph.D. Dissertation, MIT Technical Report AI-TR-231.

[70] Jain, L.C. (Ed.) (1999), *Intelligent Biometric Techniques in Fingerprint and Face Recognition*, CRC Press, Boca Raton.

[71] Jain, L.C. and De Silva, C.W. (Eds.) (1998), *Intelligent Adaptive Control: Industrial Applications*, CRC Press, Boca Raton.

[72] Jain, L.C. and Martin, N.M. (Eds.) (1998), *Fusion of Neural Networks, Fuzzy Sets and Genetic Algorithms: Industry Applications*, CRC Press.

[73] Jain, L.C. and Lazzerini, B. (Eds.) (1999), *Knowledge-Based Intelligent Techniques in Character Recognition*, CRC Press, Boca Raton.

[74] Pal, S.K. and Mitra, S. (1999), *Neuro-Fuzzy Pattern Recognition: Methods in Soft Computing*, John Wiley & Sons, Inc.

[75] Konar, A. (1999), *Artificial Intelligence and Soft Computing: Behavioral and Cognitive Modeling of the Human Brain*, CRC Press, Boca Raton.

[76] Biswas, B., Konar, A., and Mukherjee, A.K., "Image matching with fuzzy moment descriptors," *Engineering Applications of Artificial Intelligence*. (To appear).

[77] Sil, J. and Konar, A., "Reasoning with probabilistic predicate/ transition nets," *IASTED J. of Modeling and Simulations*. (To appear).

[78] Konar, A. and Mandal, A.K. (1996), "Uncertainty management in expert systems using fuzzy petri nets," *IEEE Trans. on Knowledge and Data Engineering*, vol. 8, no. 1.

[79] Saha, P. and Konar, A., "A heuristic approach to computing inverse fuzzy relation," *J. of Approximate Reasoning*. (To appear).

[80] Jamshidi, M., Titli, A., Zadeh, L., and Boverie, S., Eds. (1997), *Applications of Fuzzy Logic: Towards High Machine Intelligence Quotient Systems*, Prentice-Hall, Englewood Cliffs, NJ.

Chapter 2

Networked Virtual Park

N. Magnenat-Thalmann, C. Joslin, and U. Berner

In this chapter we introduce the Networked Virtual Park. This park consists of two parts, the content creating Attraction Builder and the actual Network Virtual Environment System. We describe not only the basics of both systems, but also the techniques used to bring reality to such a Virtual Park, including Virtual Humans and the available interactions.

1   Introduction

Network Virtual Environments (NVEs) hold the key to interactivity in virtual worlds. They provide a framework with which one can immerse oneself in a completely virtual situation and communicate with other users. The virtual world is completely synthetic and the computer generates all visual/auditory aspects. The complexity of the world is limited only to the computing power of the machine on which it is running. Therefore the simplest of worlds can be made up of simple shapes and sounds, whereas more complex scenes can include physical models to recreate gravity, reactions and collisions, to give but a few examples.

Only one user can use simple virtual environments, but to recreate an experience more like reality (where the world is shared with other users) it is necessary to connect these users together. Sharing these virtual environments via a network provides each user with a copy of the same world on their computer, thus the network providing the link between them. The users can be situated all in the same room, or in many different continents (as long as their interconnecting networks are able to transfer the required information).

Therefore these virtual environments enable users to sit in remote locations, connected via simple networks (such as the Internet) and share any virtual experience that can be perceived (providing the computer hardware used can support the complexity of the situation). In the following chapter we discuss a specific example of an NVE, the Virtual Park.

This Virtual Park is slightly more than just a park with trees and flowers as it contains interactive Attractions. The Attractions themselves are also generic, although in the following example we use a Virtual Dance to demonstrate the current possibilities. An Attraction is made up of several autonomous actions that are either triggered by a time event or by responses by a user. These autonomous actions can be anything from switch on a light, to virtual humans that automatically communicates with a user. An Attraction is normally a combination of these actions, assembled to create an ensemble that is interesting to watch and interact with.

In the following paragraphs we describe the two parts of an NVE; the first part is the construction of the Scene and more importantly, the Attraction. The second is the actual NVE, the software that allows a user to interact and watch the Attraction along with other users.

2   The Attraction Builder

2.1   Introduction

The main aim of the Attraction Builder was to provide a complete tool for the creation of attractions in virtual environments where users can interact with the scene. In general, an attraction is the combination of avatars (a graphical representation of the user as a virtual human, which are called actors when they are pre-programmed and not the actual users) and objects that can be animated, plus additional attributes which interact with a user. The Attraction Builder's purpose is to give the user the tools to create an attraction, with all its contents (such as animated avatars (called actors), objects and sounds), and output the resulting "Interactive Scene" into a file format understandable by the actual NVE System (as described later).

The Attraction itself is, as previously described, made up of a set of temporal descriptions that allow a user interaction with them. The users themselves are given equivalent tools to allow an interaction with the Attraction, but this will be explained in Section 3. In this work we have focused on the following items:

- Model and animation data acquisition
- Real-time performance in rendering and animation
- Control on virtual human actions
- Effective and easy-to-use way of attraction creation
- Support for standards (VRML, MPEG-4) for scalability and flexibility

It should be noted that the Attraction Builders purpose is not to create the Avatars and Objects represented in the scene, but to create the interactive links between these items and the user of the NVE System. This is done using high-level actions in order to avoid low-level descriptions for each interaction or movement (for example you have a body action called "bow-down" and not a description of the actual torso movement).

2.2   Virtual Avatars

Avatar is a generic term used to mean an embodiment. A Virtual Avatar is a computer representation of a human. This term is extended to Virtual Actor to mean an Avatar that is solely controlled by the computer and therefore, is autonomous. The visual and behavioural realism of virtual avatars is one of the key features of our system. Realistic human representations in NVE's have a role in the perception within the virtual environment and therefore highly realistic representations are used in order to provide as much realism in this virtual world as possible. Figure 1 shows a snapshot of two Virtual Avatars in a Scene. Greater attention is paid to the avatars in comparison with the surrounding environment, as these avatars are the main interactive points of the environment. In addition to a realistic representation of a virtual avatar, a large amount of consideration is given to face, body and speech animation, as they play an essential role in our everyday communication.

The Attraction Builder provides tools to integrate individualised face, body and speech animation into a scene, for each avatar (These animations are described in more detail in Sections 2.2.2 to 2.2.4). They are directed by simple selection from the menus of provide a set of predefined actions. These predefined actions can be produced by other programs that can create these standards for animation.

Figure 1. In Virtual Park system, representative avatars and virtual actors co-exist.

2.2.1   Avatar Realism

Each Avatar is dealt with as two parts, the body and the face. Each part is considered/developed separately and the amount of detail required for each part is approximately the same. The avatars themselves are textured polygonal meshes that provide more realism to a scene.

The face models are produced either by using a modelling tool [1] or from an automatic method of generating heads from two orthogonal photos [2]. The latter method allows participants to represent themselves in an efficient and more aesthetically pleasing way. In any case, face models are modified from a generic model that can be animated with facial expressions (including speech).

The body section is also a textured polygonal mesh, but uses VRML97 format as a basis for their basic representation description. The body, unlike the head, is composed of individual sections. This method supports the H-ANIM format [3] models being free available from the Internet. The users can reuse models made available from many sources. The generation of H-ANIM individualised body models from two photos is an on-going research area.

2.2.2 Face Animation

High-level descriptions of facial expressions are provided through a graphical user interface (or GUI). In fact, the animation module is multi-layered: A high-level action such as 'smile' or 'surprise' can be decomposed into a sequence of Facial Animation Parameters (or FAPs from the MPEG4 Specification [4]). Each expression is composed of up to 68 different FAP values describing the movements of specific "feature" points the face. These "feature" points are approximate end points of facial muscles that allow the animation of the face to create expressions. By manipulating these feature points, virtually any facial expression can be achieved. These expressions are obtained either by our interactive software (which allows the specific movement of the aforementioned end points), or from processing of video frame sequence from a camera [5]. Figure 2 shows some of the predefined expressions applied to one of a face model.

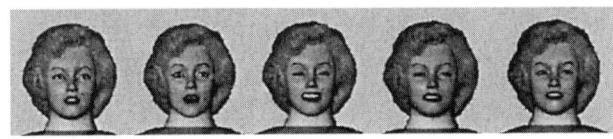

Figure 2. Predefined facial expressions applied to a virtual actor (Angry, surprised, hilarious, disgusting and happy).

2.2.3 Body Animation

The animation of the body of a virtual avatar is achieved by moving each segment of the body according to some rotational value. Each segment can then be moved over a period of time to create a gesture or specific body movement. These animations, like the body description itself, are done using the H-ANIM specification (both body description and animation formats must match for the animation to be correct). The H-ANIM format is also directly compliant with the newly released MPEG4 specification. The animation of the body is done using Body Animation Parameters (or BAPs); this BAP format consists of multiple rotational values for each body segment provided on a frame-by-frame basis. As each frame is played, one or more segments are moved. There is also a translation value that positions the entire avatar in space. The GUI, however, is designed on a high-level basis that provides the user with a list of gestures (body animations) that can be initiated at any time. Access to low-level animations (e.g. arm rotation) is extremely difficult due to large amount of data required to create a more complex animation. Figure 3 shows examples of the postures resulting from certain gestures.

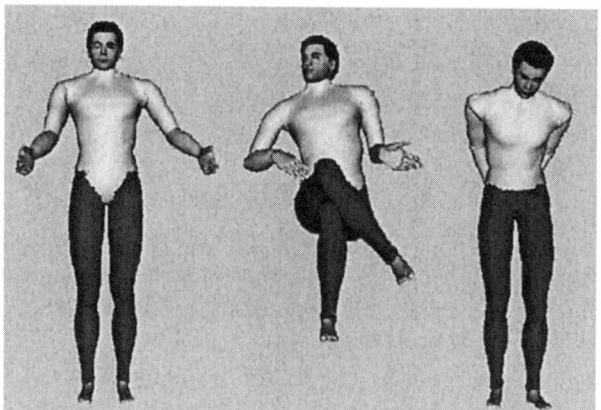

Figure 3. Predefined body actions applied to a virtual actor (Welcome, sit and talk, curious).

2.2.4 Speech Animation

Speech animation consists of two parts: the visual and the audible. As with a real situation, the lips of the virtual avatar must synchronise while producing sound. As text is input into the module via the User

Interface, it is first converted into phonemes (sections of speech) using a text-to-phoneme converter [6], and the visual equivalent of the phonemes (often referred to as visemes) is calculated. The two streams are played together to produce the ensemble that is Speech Animation.

2.3   The Scene

The actual scene or the visual representations of the objects/avatars themselves are imported from other software (some in-house, some external). The Attraction Builders purpose is to combine these imported objects and avatars to create an Attraction. These imported objects can also be other non-3D objects, such as audio and html files (displayed in a separate browser). Figure 4 shows the imported objects and avatars (and their components), and how they are grouped together. Also the output file for the NVE System is the Animation File (*.anm).

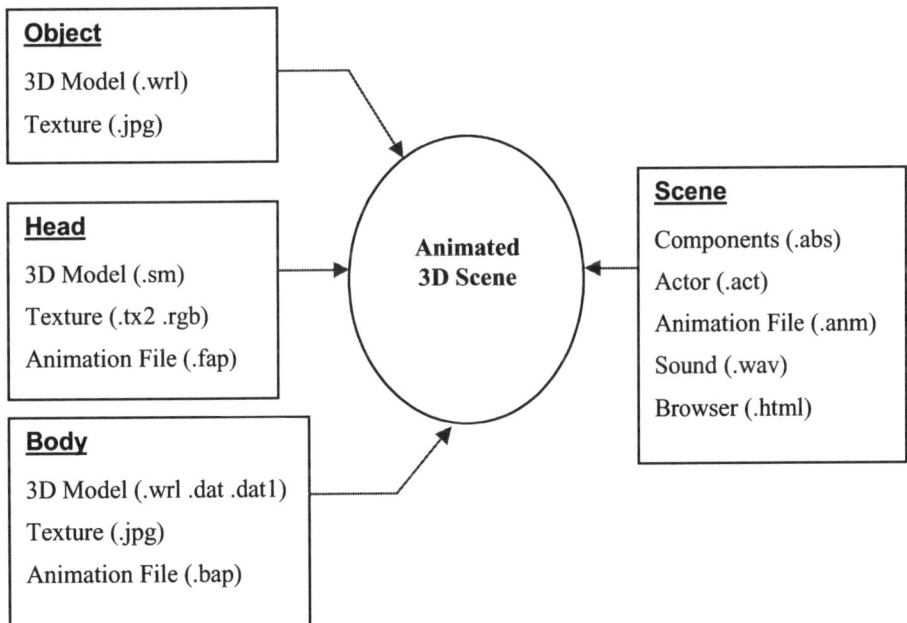

Figure 4. Composition of Scene according to Type.

2.4   Adding Interactivity

There were two interfaces that could have been used: a script-language based interface or just a direct GUI. The scripting language option, although powerful, was dropped since it requires specialised computer programming expertise from the user (which is not common). The GUI option was chosen, in combination with high-level action specification, as it provided the better approach for a user to get quickly acquainted with the system and producing simple Attractions straight away. In this approach, the user can direct an actor by simply choosing an action from the menu. The selected action is then simulated by one of the animations engines provided.

In this context, an action may be an emotion, a gesture, or a sentence. Each high level action is considered to be a basic unit of animation. Collections of this basic units are then assembled into an animation sequence, which is to compose an attraction. Users are aided with tools that enable them to adjust the time duration of the animation, move in the timeline, edit animation units, play back the animation to see current status of the animation at any time, and load predefined animations as well as save current ones. Figure 5 shows a snapshot of the Attraction Builder interface. (Note – The red numbers are used for the below references). It is not the intention to explain all the details of the interface, but to give an overview of the main functionalities:

1. New Scene, Load Scene and Save Scene
2. Add an object or an actor to the scene
3. Select an object to apply an action
4. Translate, scale or rotate an object
5. Record all actor actions and let them walk
6. Different pre-recorded actions for body and face
7. Speech animation for actors
8. Different display options for the rendering window
9. Options for manipulating the content of the rendering window
10. Record different camera positions with time and name
11. Manipulate the position of the camera
12. Invoke and edit triggers
13. Set the current time, play/record an attraction (inc. Key-Frames)
14. Edit indices
15. Edit breakpoints
16. Rendering window

Figure 6 shows an animation sequence composed of six action units based on a timeline. Note that several actions are applicable to an actor at the same time. The key to linking the whole animation sequence and animation engines lies in a timer that maintains the sequence of actions to feed them into an animation engine at a desired time (see Figure 7). The whole animation for all actors and objects inside an attraction is driven by a single and global time.

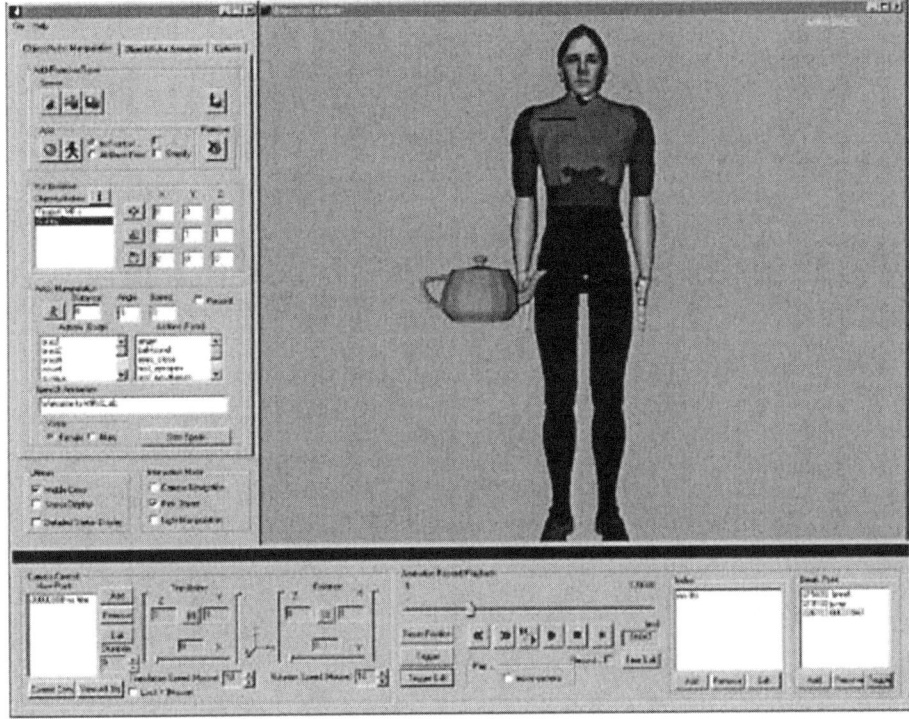

Figure 5. Snapshot of Attraction Builder GUI.

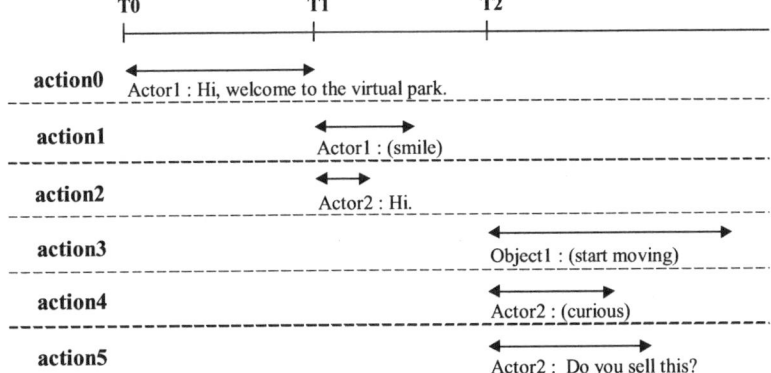

Figure 6. Timeline based scenario representation.

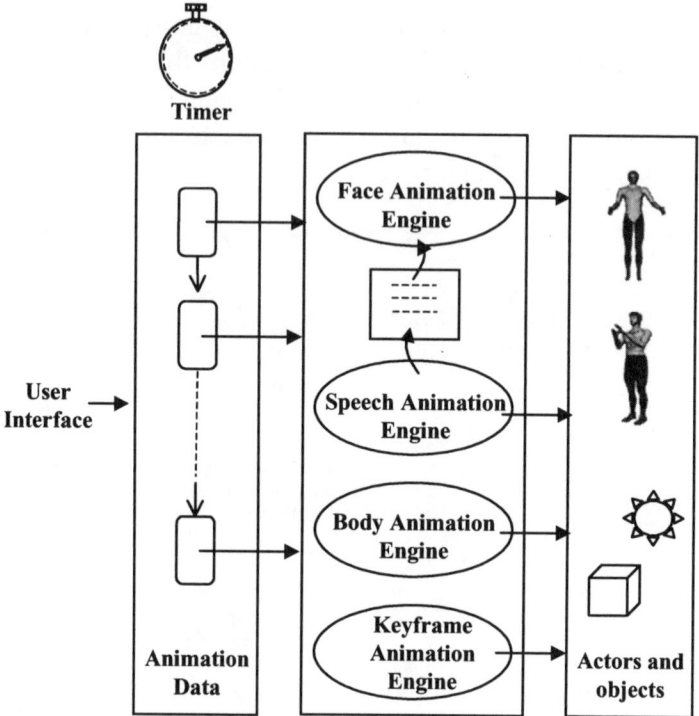

Figure 7. Animations through time management.

A special feature is to set breakpoints in the timeline on which the animation stops. The user can then continue the animation at the same time or choose another time to continue. This makes it possible to select between different animation parts. Another possibility is to trigger the animation time by selecting an object inside a scene or to go inside a proximity range of an object to interact with the animation.

The overall principle to animating a scene and adding interactivity is as follows:

- First record all the actions of the actors by choosing a time on the time line and activating an action (face, body, or speech animation). Or record different key-frames to animate an object.
- The result is a sequence of actions recorded like a video recorder, with the usual play and stop buttons.
- Breakpoints can be added. If the recorder reaches a breakpoint, it can stop or go to another point on the time line. Now the order of actions is non-linear as there are different parts on the timeline like tracks on a CD.

- Also different types of sensors can be added. Once a User triggers a sensor, the sequence is moved to another point on the time-line.
- The different options for triggering the timeline are:

1. **Proximity Sensor** - The timeline is triggered when an avatar (real users) enters a certain proximity range of an object or a virtual actor.

2. **Touch Sensor** – In this instance the timeline is triggered by an avatar touching something.

3. **Index** – A selection box appears (automatically) and then the user can choose one of the given options. The time line is then adjusted according to the selection.

All the triggering possibilities can be personalised for each avatar and will therefore only work for one specific avatar. Another feature of triggers is the ability to enable and disable the triggers depending on the current time. All actions and interactivity features are co-ordinated by a single timeline. The reason for this is the origin of the animation: the video recorder metaphor.

2.5 Possible Attractions

What Interactive Environments, including attractions, can this system be used to create? In principle, every interactive 3D-world can be created using this software. In reality the basic restriction is the global time that controls the attraction. That means that for every attraction its timeline and the environment must be synchronised to react in unity. The user triggers the animations inside the attraction and influences all the actors and the objects. Thus it is suitable to make a rock concert, a virtual theatre or a dance session within which the interactivity is minimal. The user controls parts of the entire scene, and therefore it becomes very event driven (i.e. highly interactive).

But it is currently impossible for one avatar to interact with one independent avatar (actor) without influencing all the other components of the animation. However, this is required for highly interactive environments or games such as a first-person role-playing. A possible solution is that every logged in user has their own position on the timeline and if there is any conflict for an avatar (for example they are required to smile and weep at the same time), then there is a mechanism to solve this collision of animations. Although this is not in the current scope of the project, it is an option for the further development.

3   Networked Virtual Environment System

3.1   Introduction

Once an Attraction has been created, it requires a separate piece of software to enable users to have viewing and interaction capabilities. This software is basically the Networked Virtual Environment System (or NVES).

The NVES software that contains a virtual copy of a world/scene and any changes made in the scene are communicated via a network (such as the Internet). These updates create interactivity in the scene with other users, as in moving an object or communicating with another user, for example.

The NVES that is presented in the following sections is a real software package that works hand in hand with the Attraction Builder mentioned in the first section. It works on the basis of Server and connecting Clients. Each client is a connecting user that joins the NVE from a remote location, via some network connection. The Server, in its fundamental form, is much like a Web Server; the connecting Client requests information and the data is sent from Server to Client. This information, passed from Server to Client, is in the form of 3D World description (object/avatars) and the data produced by the Attraction Builder. The analogy of the Web Server differs in that the Server and Client continue to exchange information, and the Server allows the connecting Client to have knowledge of the other connecting Clients. The information exchange is in the form of updates to the scene; if an object is moved, or an avatar sends some speech or arm motions. The Server processes this information and distributes the data to the other Clients, if necessary.

The system itself is hereon referred to as **W-VLNET**, which is short for **W**indows **V**irtual **L**ife **Net**work. The prefix of Windows is used to distinguish between this software and its predecessor [7], which was based on the UNIX Operating System. The following paragraphs briefly explain the main aspects of the W-VLNET software.

3.2   Overview of System Architecture

The W-VLNET system is based on the Client-Server architecture. Each Client, as it joins or leaves a Virtual World, opens or closes a connection to a Server. The server, acting essentially as a giant intelligent network switch, ensures that all the data in or around the world that a client requires, is sent to all other Clients. As Clients connect and disconnect, their data is automatically sent to or deleted from other clients.

In the following sections we shall describe the following system components:

- The Client (The connecting users)
- The Server
- The Overall Communication Protocol

3.3   The Client

3.3.1   Introduction

The Client itself contains many components that make it into the required software package; the following is list of the important components (roughly ordered in importance):

- Scene Manager – Controls what is seen, including Animation
- Network Management – Handles the data connection between itself and the Server.
- Navigation – Enables the user to move around the world
- Audio Manager – Handles all audible aspects of the scene.
- Speech System – Generates Speech for a given text input.
- Devices – External Devices for handling other user input.
- System Communication – The surrounding architecture.

3.3.2   System Communication

Although the System communication is one of the least important components in terms of actual visibility in the system, it is the most important component with regards to holding the entire system together.

The System Communication Layer provides a method for all the other modules to communicate with each other. It also controls the multi-processing (concurrent tasks) aspect of the system and provides a plugin mechanism.

- **Multi-processing** – One of the many requirements of an NVE system is that it perform many tasks at once. Hence the system architecture must be able to support multi-processing. The System Communication layer does just this, and therefore concurrent tasks (for instance, animating one human and whilst another is speaking) are handled in such a way as to make the user unaware of the fact that they are separate tasks running at the same time. The system can make use of multiple processors in order to distribute the load and it also has a large limit for the number of concurrent tasks that can be performed at once. Figure 8 shows a simple example of a multitasking situation.
- **Communication** – Since the System was designed to be multitasking; a specific communication architecture was designed to enable each task to communicate with each other or one another. The communication layer allows a balanced buffering mechanism that allows the tasks to communicate with each other without blocking or upsetting the multiprocessing mechanism. Figure 9 shows an overview of the communication between modules.
- **Plugins** – One of the most important aspects of the NVE system has been the use of Plugins. As the NVE Systems goal has been to create virtual instances of life, it is continually expanding or being improved, both for the effects of realism and for the incorporation of additional aspects (such as different or specialized situations). Each of the following modules are created as plugins, and can be exchanged, upgraded or removed without the need to change anything in the main system. Also new plugins can be written, both by the developers and users alike, to add new functionality to the system.

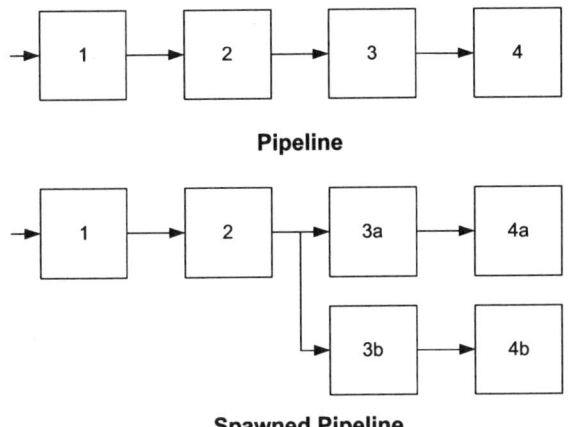

Pipeline

Spawned Pipeline

Figure 8. Multitasking Pipeline and Spawned Pipeline.

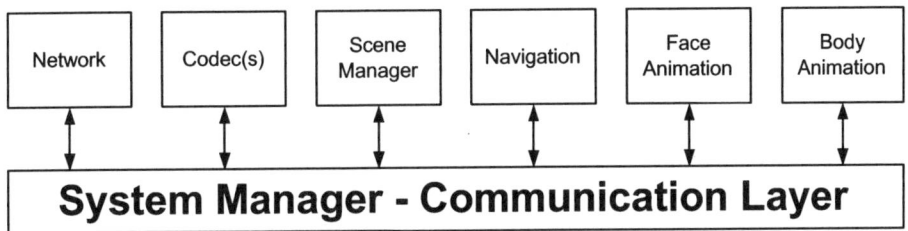

Figure 9. Module Communication.

3.3.3   Scene Management

The Scene Manager (SM) manages the Scene and controls all the data associated with the Graphical Scene, plus the supplementary data used by the external modules. The main tasks of the Scene Manager are as follows:

- Loading of main Scene (described in VRML97)
- Adding/Deleting of VRML97 Objects
- Adding/Deleting of MPEG4 Compatible Avatars (Face and Body)
- Navigation and Camera viewpoint around Scene
- Control and Manipulation of Objects in Scene
- Identity of Avatars/Objects in Scene
- Caching and updating of information in Scene
- Control of information specifying the update of positions of Avatars and Objects

The SM is based on the OpenGL Optimizer Scene Graph from Silicon Graphics [8]; this library contains not only the rendering functions provided by OpenGL, but also the Scene Graph functions provided by the Optimizer library. The SM uses the current standards for Scene and Object representation (VRML97) and of Body/Face Representation (MPEG4 Body Description Parameters and Face Description Parameters). This allows the user to have access to large databases of information created by not only our research group, but also other institutions.

The SM also controls the manipulation of scene data, as external modules can adjust the scene (such as Navigation devices, or devices used to move objects). Therefore it is up to the SM, not only to manipulate the Scene accordingly, but also to control the updates to the Network in order to apply certain constraints (such as collision detection and gravity). Figure 10 shows the information flow between modules and the Scene Manager.

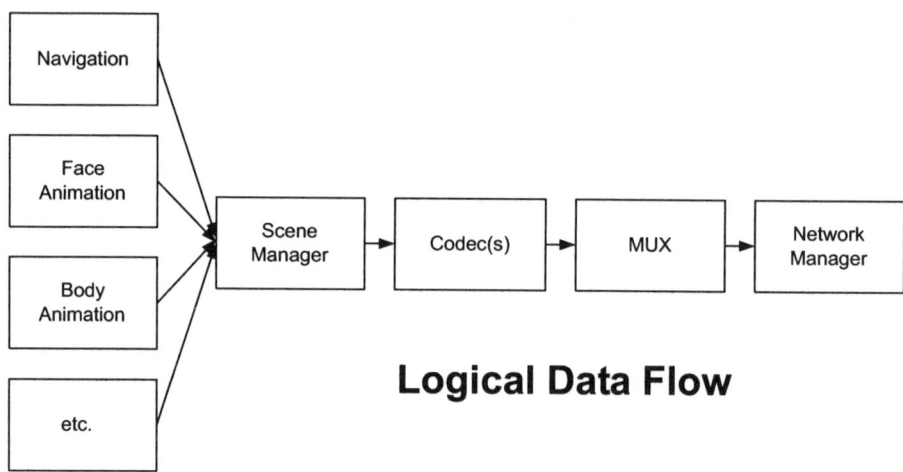

Figure 10. Information Flow to Scene Manager.

3.3.4 Avatar Representation and Animation

The Avatars used in the NVE System are exactly the same as the ones described in Section 2.2. They also use the same animation methods; this is so that there is continuity between the two pieces of software described here and also the rest of the world.

3.3.5 Navigation

Navigation around the virtual world is most important to enable the users the ability to move around. Navigation can come in many forms (from the simplest to the most advanced), but the basic principle is the much the same, only the complexity varies.

One of the simplest navigation methods that can be used is by using the mouse and keyboard. The two together enable a user to navigate around a virtual world with ease. This navigation module enables more complex connections, such as space-ball for more intuitive 3D movements, or items such as a Magnetic Tracking system. The Magnetic Tracking System (which uses a steady magnetic field to track the position and rotation of its specific units. Its is often referred to as a Flock of Birds, due to its resemblance to a lot of small birds) enables not only the tracking of the real humans position in the virtual space, but also the position of each limb. This means that the entire virtual avatar can be animated in real time. Figure 11 shows an example of this.

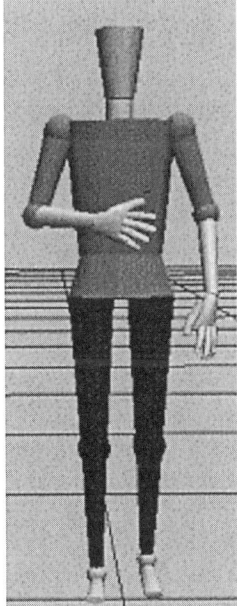

Figure 11. Magnetic Tracking System and the resulting Virtual Avatar.

3.3.6   Audio Communication

The audio within the system is generated via two sources (microphone/file) and outputted via one (speaker), this is done to make it possible for users to speak with each other, but also for audio to be attributed to objects (such as a personal stereo or such like). All three are attached to objects (usually Avatars) to enable the perception of real audio positioning. The Audio is sent/received in a standard raw format of 128Kbits/s 8KHz. This is compressed with a standard ITU codec (either G.711 [9] or G.728 [10] providing either 64Kbits/s and 16Kbits/s bandwidth usage respectively), depending on the network being used and the quality required.

The Audio module provides the 3D Audio effects by requiring the Source and Destination positions. This enables a more in-depth audio experience and brings a greater sense of reality to the environment.

3.3.7   Speech

Section 2.2.4 describes the speech in its basic form; the only additional item to mention would be the transmission of text as opposed to audio. As can be seen from the previous Section, the sending of real audio is still bandwidth consuming even in its most compressed form. However, for even a large quantity of text, resulting in the animation of the avatar and the audio, consumes very little bandwidth when it is sent. A text transmission, regardless of gender, would consume a bandwidth of approximately 256 bytes/s on average, which is less than 1% of a 33.6K Baud modem.

3.3.8   Devices

To complete the description of the NVE System, a brief mention of the connecting devices is given. Like the Magnetic Tracking system, any device can be connected to NVE System to enable linkage with the any of the above modules. New optical tracking systems (using optical recognition to track points on a real human) could in fact be attached to the system to be directly linked to the animation module. As the animation module already exists, it uses a specific standard protocol (BAP from MPEG4) to animate the body; hence as more devices become available they can be incorporated into the system for their own specific purpose. The actual modules can take input from any device using their format and are therefore not tied directly to any specific device.

3.3.9   Network Manager

The network is based on the Client-Server architecture (see Figure 12).

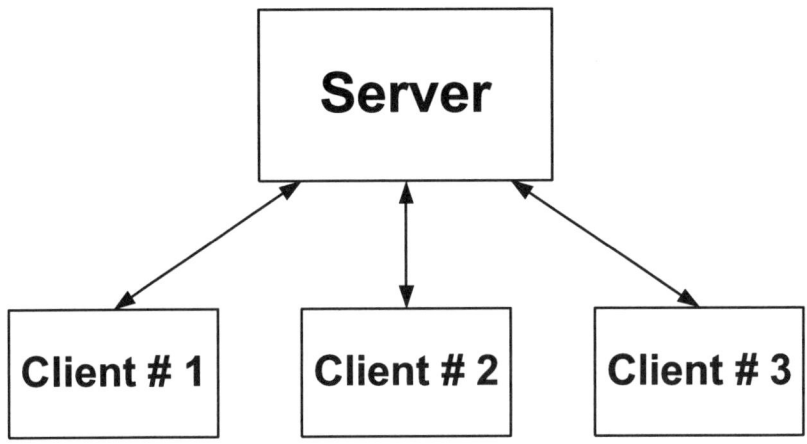

Figure 12. Client-Server Architecture.

The Network Manager (NM) controls the data flow to and from the Server. The tasks that are assigned to this module are as follows:

- Setting up of connection with Server
- Creation of data Channels (Stream, Update, File and Control)
- Flow control of data to/from Server
- Error correction of data to/from Server
- Control of disconnection with Server

The Network Manager connects to the Server from which it then downloads the information about the Server, extra sockets, etc. The Server and Client negotiate to obtain 4 channels of communication (Stream, Update, File and Control) using two basic Internet Protocols (Transmission Control Protocol, TCP and User Datagram Protocol, UDP [11]):

- Stream Channel is basically a fast UDP channel that is used for streaming the audio or visual data at a normally constant rate. If a packet of data is lost, no error correction is provided (i.e. feedback to sender) as it is assumed too late to correct.

- The Update Channel, using UDP, has control over the data that is passed through it; the data is corrected using an error control mechanism (check for lost/corrupt packets). This channel is mainly used for Scene Updates.
- The File Channel, using the TCP Protocol, is used only for the transfer of large files; it is basically able to download data over 1024 bytes. This enables the error-free transfer of files between Client and Server (in both directions).
- The Control Channel, again using TCP, is basically used as a channel of communication between the Client and Server (e.g. warning of shutdowns, disconnection, Is-Alive requests etc).

4   The Server

4.1   Server Overview

The Server is a multithreaded application that from the outwardly acts likes a giant network switch. However, the Server is more intelligent than this and provides error checking, a database and a data-control mechanism, which is extremely important in NVE systems.

Once the Server has been started, the Server negotiates with the Client as to the channels that the Client wishes to use and the bandwidth of the connection. Also transmitted is the extra information to/from the Client that allows both Client and Server Users to know what is happening in the virtual environment. Each server contains information about both the World and the Attractions. The world is the basic environment (grass, trees and lakes etc). It is the basic world in which the attraction is based, and this environment can be used without the Attraction for generic purposes. The Attraction basically is the Scene that is movable and animated in the NVE. It contains the objects, digital avatars, and audio files which are played in the attraction, it also contains descriptions of the animations and timings. The Attraction is a self-contained (i.e. contains all necessary information for the Attraction) package of data outputted by the Attraction Builder (See Section 2).

Both World and Attraction are compressed packages sent to the Client by the Server when the Client connects (to be generic the transmission of the Attraction can be switched on or off). Both packages contain

Objects and Avatars that are owned by the Server (i.e. their identifiers are referenced to the Server), however all animations are sent to the animation modules by the Client. The reason for this is that less data is then transmitted for each piece of animation; this can only be done for Server animations because all Clients download this information when they connect. The only data that is sent to the Clients by the Server with respect to these animations are timing signals used to synchronize the animation events.

4.2   Server Database

Once the Client has connected and the World and the Attraction packages have been transmitted, the Clients can then send/receive information to and from each other. The Server stores some of this information in a Database, this is done for two reasons: Firstly, to enable the Clients connecting after any data has been sent to the already connected Clients, to have an up-to-date version of the scene (i.e. if an Object or Avatar has moved) and Secondly to enable error control. For Error control, the Server acts as a Master for all Clients. Therefore, if a packet gets lost between Server and Client, it can reference the database to obtain the correct data (which was lost).

The Database is updated with the information on the Avatars/Objects loaded and their positions (and obviously to which Client they belong). Therefore when a Client connects, it downloads information about other Clients that are present in that scene. Once the scene has been constructed correctly according to the latest database on the Server, data is sent to/from the Server to update the scene-graph on each Client. Data sent to and from the Server might include the following:

- Avatar transformations
- Body or Facial animations
- Object transformations
- Control data
- Video and Audio Data

The Server forwards data to other clients and also performs the filtering functions.

4.3   Client-Server Communication Protocol

Real-time multimedia applications require certain levels of QoS (Quality of Service). We have developed a communication protocol that runs on the top of UDP. The protocol was inspired with real-time protocols. The following functionalities are provided:

- Content identification: The transmitted data is transparent to the networking module. Inside, each packet there is a field that provides the identification of kind of data that is transmitted: BAP, FAP, files, objects related data.
- Source identification: There are two levels of source identification. First is the identification of the user inside the system: Server ID, Client ID, and Object/Avatar ID's. Second is the identification of the plugin module from which the data is sent, so the receiver is able to determine to which module the packet should be passed (as each module, such as animation, has its own identifying code).
- Timestamps recognition: This is an absolute value that is accurate to within 10ms. Once the connecting computers are synchronized with each other the timestamps allow the Server and Clients to determine the time delay of a data packet.
- Packet loss detection.
- Feedback information to server.

The Communication protocol does not guarantee the QoS, but allow a possibility to monitor the data delivering.

5   Conclusion

These two systems (the NVES and the Attraction Builder) provide a complete system from design to usage of a Networked Virtual Environment. Originally, this system was to be used to design and run a Virtual Park, made up of various attractions. However the system was designed in such a way that many scenarios can be created outside the context of a virtual park. These scenarios could be corporate meetings where Autonomous Avatars could ask or present questions, or teaching situations where the class could provide feedback. In the context of entertainment, the virtual park is the most obvious choice, but this could be extended to longer, more elaborate, scenarios with deeper role-playing capabilities.

The question of compatibility is always in question in the world of graphics and it is always found most convenient for users when the tools they are using for creation are compatible with other tools. In this system one of the main goals was to provide this type of compatibility to do just this. Using the most popular standards in graphics has enabled the system design/building not to be slowed by having to recreate tools that already exist and permit concentration on new tools (i.e. the ability to build attractions and to immerse in them).

Acknowledgements

We would like to thank the following people for their contribution to the work described above: Hyewon Seo, Sumedha Kshirsagar and Tom Molet. We would also like to thank Gabby Rieder for reading this chapter.

References

[1] Kalra, P., Magnenat-Thalmann, N., Moccozet, L., Sannier, G., Aubel, A., and Thalmann, D. (1998), "Real-time animation of realistic virtual humans," *IEEE Computer Graphics and Animation*.

[2] Lee, W. and Magnenat-Thalmann, N. (1998), "From real faces to virtual faces: problems and solutions," *Proc. of 3IA'98*, Limoges (FRANCE), pp. 5-19.

[3] H-ANIM Humanoid Animation Working Group, "Specification for a Standard Humanoid Version1.1," `http://ece.uwaterloo .ca/~h-anim/spec1.1/`.

[4] Moving Picture Experts Group, "MPEG 4 Specification," `http: //www.cselt.it/mpeg/`.

[5] Goto, T. *et al.* (1999), "MPEG-4 based animation with face feature tracking," *CAS '99 (Eurographics Workshop on Animation and Simulation)*, pp. 89-98.

[6] Microsoft Speech Software Development Kit, Microsoft Corporation, http://www.microsoft.com/IIT/projects/ sapisdk.htm.

[7] Pandzic, I.S., Capin, T.K., Magnenat-Thalmann, N., and Thalmann, D. (1995), "VLNET: a networked multimedia 3D environment with virtual humans," *Proc. Multi-Media Modeling MMM`95*, World Scientific Press, Singapore.

[8] Open GL Optimizer Specification, Silicon Graphics Limited, http://www.sgi.com/software/optimizer/.

[9] International Telecommunication Union (1988), "Recommendation G.711 (11/88) – Pulse code modulation (PCM) of voice frequencies," http://www.itu.int/itudoc/itu-t/rec/g/ g700-799/g711.html.

[10] International Telecommunication Union (1992), "Recommendation G.728 (09/92) – Coding of speech at 16 kbit/s using low-delay code excited linear prediction," http://www.itu.int/itudoc /itu-t/rec/g/g700-799/g728.html.

[11] Stevens, W.R. (1994), *TCP/IP Illustrated Volume 1*, Addison and Wesley.

Chapter 3

Commercial Coin Recognisers Using Neural and Fuzzy Techniques

J.M. Moreno, J. Madrenas, and J. Cabestany

In this chapter we address the applicability of artificial neural network and fuzzy logic models to real tasks in industrial environments. For this purpose, we shall present a general methodology which will be outlined by means of a case study which consists in the implementation of a classification/decision engine included in an automatic coin recogniser. This coin recogniser is a part of currently available commercial vending machines. The methodology presented can be considered as divided in three main tasks: database compilation, selection of the proper neural or fuzzy model and implementation. A wide range of models, including classical as well as evolutionary algorithms, has been considered. The experimental results demonstrate that the use of artificial neural and fuzzy models overcomes some of the limitations inherent in the traditional techniques considered when solving this task.

1 Introduction

Artificial neural network paradigms have proven to be good candidates for handling signal processing tasks such as pattern recognition, time series prediction, data compression, and function approximation, especially when non-linear or highly complex (in terms of input space dimension) mappings have to be considered. On the other hand, fuzzy models have demonstrated their usefulness for handling tasks characterised by uncertainty in the input data, or when the knowledge about the system functionality is expressed by means of approximate rules. Therefore, these paradigms can be considered as complementary to adaptive signal processing tasks.

In this chapter we shall address the applicability of artificial neural and fuzzy models to real world tasks following a general methodology which considers the specific features associated with industrial and commercial developments. The application we shall consider consists of implementing the decision/classification engine of an automatic coin recogniser included in commercial vending machines.

The methodology we shall use to solve this application can be divided in three main steps. First of all, it is necessary to compile a representative database that includes the main features of the problem to be handled. Then an appropriate neural or fuzzy model has to be selected in order to fulfil the constraints specified for the application. Finally, an efficient implementation is provided for the final structure resulting from the two previous steps. In our analysis we have considered a wide range of models, encompassing classical (i.e., models with fixed structure) as well as evolutive (i.e., those able to find the proper network structure for a given task) algorithms. Furthermore, fuzzy logic models have also been tested as candidates for solving the proposed task.

The sections are structured as described hereafter. Following this brief introduction, we shall consider the specific characteristics of the problem to be handled, as well as the restrictions to be met in order to provide a competitive commercial product. Bearing in mind these features, we shall then review the procedure chosen to construct a representative database to be used as a benchmark suite for the different candidate neural and fuzzy models. The experimental results derived from these benchmarks will allow for the selection of the proper models able to handle this application. Once the models are selected, we shall present the particular implementation alternative chosen in order to meet the product specification. Finally, the conclusions will be outlined.

1.1   Problem Statement

Vending machines based on automatic coin recognisers constitute a high-volume market niche due to the large number of potential applications they can cover. Among these applications we could consider for instance product dispensers (tobacco, drinks, tickets, etc), or automatic toll payment systems.

The coin recognisers which constitute the main core of these machines can be considered as divided in two main sections: the feature extraction block, which obtains several measures related with the physical properties of the coins inserted in the vending machine, and the classification/decision engine, which, based on the features extracted previously, decides whether the coin is rejected or accepted and, in this last case, to which coin category it belongs to.

There exists a series of specific constraints to be met by the coin classifier system:

- The high-volume and competitive market where the final product has to be introduced imposes very restrictive commercial margins. Thus the cost of the system is one of the most limiting factors to be considered. As a consequence, coin classifiers are usually based on low-cost sensors, whose tolerance is affected by the conditions of the environment, and by factors related to the fabrication process.

- Due to the different physical places where vending machines have to operate, they are exposed to a wide range of variations in the environment (temperature, humidity, etc.), some of them affecting directly the behaviour of the components included in the classifier. Therefore, proper operation of the system should be guaranteed even for severe variations in its parameters.

- The objects to be recognised (coins) may show a high degree of variability, even for coins of the same type, due to external factors such as material degradation due to ageing or manipulation. Therefore, a high degree of robustness is to be expected from the classifier intended to recognise these objects.

The above mentioned factors make the development of commercial automatic electronic coin classifiers a very hard task, where several parameters of the system have to be tuned individually before the system is inserted in real operating conditions. However, the intrinsic characteristics of these limitations suggest that artificial neural or fuzzy models represent good candidates for handling the classification tasks associated with automatic coin recognisers. This fact will be proven through the results provided in this work.

Bearing in mind the considerations stated previously, the specifications for the automatic coin classifier could be summarised as follows:

- The classifier/decision engine has to be included in a commercial coin recogniser which is the core of a vending machine.

- Six different types of coins, corresponding to the spanish currency, have to be identified by the classifier.

- The classification/decision task is to be performed based on nine different features extracted from each coin by the sensors included in the coin classifier. These features provide information about the size, weight and composition (metal alloy) of the coin to be classified.

- To provide robustness against variability in the characteristics of the coins, ten different prototypes will be used for each type of coin.

- The system should operate correctly in a temperature margin ranging from 0°C to +40°C.

- The classifier engine has to consider the tolerances affecting the components of the system due to the fabrication process.

Taking into account these system specifications, in the next section we shall explain the procedure used to construct a consistent and representative database in order to train a classifier and to check its performance for the proposed task.

2 Problem Analysis and Database Compilation

Coin recognition requires not only a classification but also a verification to decide that a given coin belongs to the legal currency. The system has to distinguish not only among different coins of a given country, but also from other countries and non-legal coins.

Information collected from a coin dropped into a coin recogniser comprises several parameters concerning its physical characteristics. Given financial constraints of this application, the performed measures should use only low-cost sensors. For instance, typical measurements that can be made are: weight, dimensions and material (metal alloy), this last feature being estimated from dielectric, ferromagnetic and/or piezoelectric properties. Other alternatives, as for instance image analysis, are impractical because they seem not to be very effective. Furthermore, image acquisition and analysis is expensive and complicated to perform in a reduced space.

2.1   Problem Analysis

In Figure 1, a coin recogniser containing commonly used sensors is represented. The coin is dropped over the weight sensor **W**, normally a gauge. The coin then rolls down following a ramp, screened by a set of optocouplers **L1**, **L2** and **L3** during the journey. It is also common to use a set of only two optocouplers. The coin magnetic properties are sensed when it crosses the core gap of the coil sensor **M**. Additionally, capacitive, piezoelectric or even acoustic measurements could be performed.

When the coin leaves the ramp, the system has about 30 ms to perform calculations from measurements before the coin reaches gate **G**. If the coin is recognised as a valid one, it is accepted by opening gate **G**. Otherwise, the coin is rejected.

Given a set of valid coin classes, and a measurement vector obtained from the particular coin, the classical approach to recognise it as belonging to one of the valid classes is as follows:

- Each coin recogniser is individually calibrated by dropping through it a set of master coins several times. A calibration table containing the average of the measurements obtained for each coin class is stored into the device memory. This task is performed in the factory.

During operation, when a coin is dropped, the obtained measurement vector is compared on a component-by-component basis to the vector table. For a coin to be accepted as belonging to a valid coin class, *all* its components have to fit with a very small degree of variation with that valid coin vector that is stored in the calibration table.

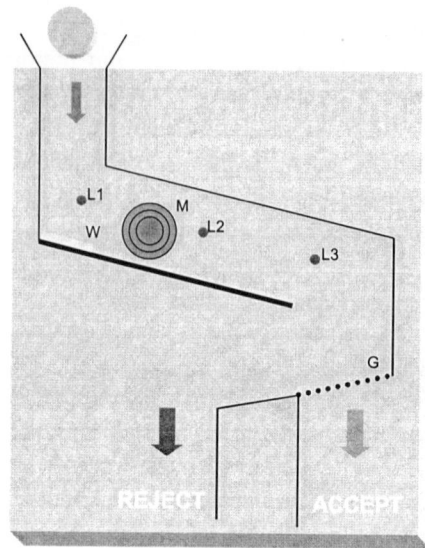

Figure 1. Basic diagram of a coin recogniser.

This approach can be seen as defining a set of hypercubes having as center the vectors of the calibration table. The size of the hypercubes depends on the programmed maximum variation degree allowed for each measurement component. The measurement vector has to be inside one hypercube to accept a given coin.

Although effective, the table-based recognition method requires an expensive and cumbersome process of calibration for each manufactured device. Furthermore, this calibration only guarantees the compensation of the static tolerances of the transducers, but not variations due to ageing or temperature changes, for instance.

In order to overcome the limitations of the classical table-based approach, adaptive decision models able to cope with variations of different coin recognisers were analysed.

The first task was to generate a database statistically significant in terms of coin recognisers, coins and temperatures. A part of the database was used for learning, and the rest to evaluate performance during execution, as will be explained in next sections. This database was generated for the six widely used types of spanish coins.

2.2   Database Compilation

There are two main goals to be met when constructing a database to be used for training and testing purposes corresponding to a real-world signal processing task. First of all, this database should be representative, that is, it has to include relevant information about the problem or task to be handled. This means that in our case the database should take into account the intrinsic variability of the classification/ decision task to be performed by the coin recogniser, as it was explained in Section 1. Furthermore, to avoid the well-known curse of dimensionality phenomenon [1], a careful trade-off has to be obtained between the number of vectors constituting the database and the number of features describing each vector.

As it was stated previously, the sensors included in the coin recogniser provide for each coin drop a feature vector composed of 9 measurement components and 6 components used to identify the coin category. Concerning measurement components, 3 of them are related to material measurements, one is related to weight and the remaining 5 are for optical measurements.

The variability in the characteristics of the coins was represented by selecting 10 prototypes for each category. Each coin was dropped 10 times to deal with the different ways the coin is dropped. The resulting 600 drops were repeated for 10 randomly selected different recognisers and under 3 different temperatures: $0°$ C, $20°$ C and $40°$ C. Therefore, the total database size is 18,000 15-component feature vectors.

For the second step once this database was constructed it is necessary to reduce the database input dimension for two reasons: First to guarantee database representativeness, and second to reduce the recognition engine complexity to obtain a reliable answer rapidly. This led us to divide the recognition process into two steps: classification and validation.

The classification step is intended to predict the category of the dropped coin. From statistical analysis of the 9 components, it was observed that only three features (two related to the material of the coin and the remaining one related to its weight) are enough to classify the coin.

After the coin has been classified, the validation stage checks that the coin features are consistent with its assigned class. Since the rejection mechanism has to be very accurate, this stage will consider only those features that show the highest confidence for the proposed task. As a consequence, only the weight of the coin and the readouts derived from the information extracted of the optical sensors will be considered. A careful analysis has demonstrated that the information provided from the 5 optical readouts can be compressed into 2 relevant features, thus alleviating the computational requirements for the decision task. In the next sub-section we shall explain in detail this pre-processing stage.

2.3   Optical Measurements Preprocessing

The coin diameter can be analytically calculated from the optical features by means of a non-linear function. Calculations involve products and divisions with 20-bit precision. These calculations require almost all the available processing time of the existing processor, 30 ms as indicated before.

From an exhaustive data analysis, a much simpler diameter estimation method was developed. Figure 2 shows the database projection onto the two most significant optical parameters (labelled as *Var 5* and *Var 6* in this figure) for the three larger size coins, where label **A** represents the largest coin. A similar figure is obtained for the remaining three smaller coins. Parameters *Var 5* and *Var 6* correspond to delays obtained from the crossing times of L2 and L3 optocouplers. For each coin, a high degree of linearity is observed between both parameters. The reason is that the initial speed and acceleration ranges are inside a quasi-linear segment of the original non-linear function.

Bearing in mind the previous considerations, the calculation is reduced to a slope calculation:

$$a = \frac{v_6}{v_5} \tag{1}$$

The computationally costly division can be avoided by calculating slope a for each coin. Therefore, during run-time, only a product and addition is necessary to obtain parameter p_1:

$$p_1 = v_6 - a \cdot v_5 \tag{2}$$

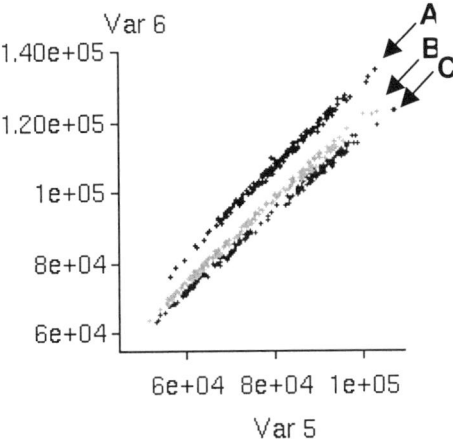

Figure 2. Database projection onto the two most significant optical parameters for the three larger size coins. In both axes, units represent time steps.

Furthermore, the product has been simplified to a maximum of four shift-add operations by selecting an appropriate estimate for the slopes.

A comparison between the proposed calculation simplifications and the exact diameter calculation lead to very similar data distributions. After performing offset correction and normalisation in Figure 3, data regions for the three larger coins are shown projected onto the p_1 (*Var 16*) and weight (*Var 4*) components. The regions are very easily separable.

For verification purposes, a second parameter p_2 was calculated in a similar way as p_1 from a different projection of the optical features.

Summarising, the database has been reduced to 5 input features since the 5 optical measurements were reduced to two (p_1 and p_2). This reduced number of components enables an effective processing using adaptive neural and fuzzy models.

Now that the database has been compiled, we shall explain our design methodology in the following sections. These steps will determine first the proper model to handle the proposed task. Thereafter an efficient implementation of this model will be provided, taking into account the constraints such as speed and cost posed for the proposed application.

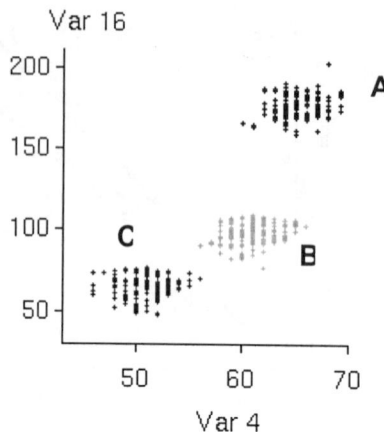

Figure 3. Data regions for the three larger coins projected onto the p_1 (Var 16) and weight (Var 4) components.

3   Approach Using Artificial Neural Networks Models

After the compilation of the database which includes the relevant features of the problem, the next step in the design flow can be started. This consists of selecting the appropriate classification/decision architecture. In the first stage, artificial neural network models have been selected as potential candidates for handling this task. Their intrinsic robustness and ability to be adapted to the inherent variability of the problem to be tackled makes them attractive for this purpose. A Number of constraints such as cost and processing speed have to be met in the final implementation and a careful model selection process has to be performed.

In the next sections we shall describe the neural model selection process used in order to define the appropriate parameters of the required neural paradigm. After defining the final network structure, we shall provide the details corresponding to its physical implementation. Alternatives have been considered for the physical realisation of the classification/decision engine. The first approach consists of emulating the functionality of the selected neural network model by means of a program running on a microcontroller included in the coin recogniser.

Because of security and privacy considerations, a further evaluation has been made for an eventual hardware implementation by means of a dedicated Application Specific Integrated Circuit (ASIC).

3.1   Neural Model Selection

In our search for an efficient neural network structure, we have considered both classical as well as evolutive [2], [3] neural models. By classical neural models we mean those neural paradigms whose structure (number of layers, number of neurons in each layer and connectivity pattern between neurons) is fixed. The learning process is achieved by modifying only the strength of the connections between neurons. Evolutive neural models have the additional capability of redefining the structure and in some times the connectivity pattern during the learning stage. This alternative has the ability to automatically construct a network structure suited to the target problem.

Among the classical neural network models we have considered are the Multilayer Perceptron (MLP) [4] trained by means of the Backpropagation learning algorithm [5], the Learning Vector Quantization (LVQ) models [6], and the Inertia Rated Vector Quantization [7]. The evolutive models used in our analysis are the Restricted Coulomb Energy (RCE) [8] and the Neural Trees algorithm [9].

The training set used during the learning phase of the different models considered in our analysis is composed of the feature vectors corresponding to four different coin recognisers for three different reference temperatures. This training set is therefore composed of 7200 feature vectors. The test set used to validate the performance provided by the models after training is the whole database integrated by the 18000 feature vectors obtained for the ten coin recognisers.

Two main factors have been considered so as to compare the neural models used in this analysis:

- The network size required by the model.
- The correct classification a model provides after training.

The first factor will investigate the possibility of finding a cheap implementation of the classifier for the given timing constraints. The second factor will indicate the robustness of the decision engine against eventual deviations such as the operating conditions, the tolerance of the components which constitute the system and the intrinsic variability in the features extracted from the objects to be classified.

The experiments were commenced with the MLP model, since the parameters which control the learning process can be tuned interactively in an intuitive way. After several initial trials, it was determined that an appropriate network structure for the proposed problem is one hidden layer with ten units and one output layer of six units, one for each coin category. The sigmoid activation function was used for all the units in the network. The weight update during the learning process was performed in batch mode, and 1000 iterations were required to provide an absolute mean error less than 0.01. The learning rate and the momentum are held constant during the network training phase, having values of 1.18 and 0.3 respectively.

Table 1 shows the test results provided by this model for four different training processes, each conducted with a different training set of 7200 vectors. In this table we have indicated the correct classification rate in percentage for the complete 18000-vector database, as well as the minimum and maximum values for the data corresponding to a given coin recogniser.

Table 1. Results provided by the MLP model.

Training process	Correct classification (%)	Min. correct classification (%)	Max. correct classification (%)
1	99.3	96.0	100.0
2	98.7	87.0	100.0
3	98.8	91.0	100.0
4	98.2	87.0	100.0

From this table it can be seen that the MLP model is able to handle the proposed classification task very efficiently, since it always provides classification rates of over 98.0 %. Furthermore, the worst case results are commonly provided for a few classifiers, and always at +40° C, where the tolerance of the components which constitute the system causes a substantial change in its behaviour.

The second neural network considered for implementing the classification/decision engine is the family of Learning Vector Quantization algorithms [10]. A preliminary analysis has demonstrated that LVQ2 provides the best performance for the given data. The values of the optimal parameters which determine the behaviour of this model are the following:

- Initial adaptation gain, $\alpha = 0.3$
- Window width $= 0.3$
- Number of iterations $= 1000$

Table 2 shows the results provided by this model for different codebook sizes (i.e., number of prototype vectors used to perform the clustering of the input space). The results indicate the mean correct classification rate in percentage for the complete database after four learning processes on different training sets, which were the same used for the MLP model. That is, each line contains the mean value calculated from the results obtained from the four training processes.

Table 2. Results provided by the LVQ2 algorithm.

Codebook size	Mean correct classification (%)
20	97.8
40	96.2
80	96.8

It can be seen that the performance of this classifier even when using a larger network structure, is slightly lower than that provided by the MLP model, with a mean correct classification rate of 98.8 % for the same data sets.

Another kind of neural model used in our analysis is the probabilistic neural networks [11]. This model attempts to define an appropriate discriminant function by estimating from the training samples their *state-conditional* probability density function (i.e., the probability density of the samples given that they belong to a certain class) as well as the *prior* probabilities of each class (i.e., the probability of occurrence of a given class). From these models we have chosen for our experiments the Inertial Rated Vector Quantization algorithm [7], which overcomes the main limitations of these models (i.e., memory size) by performing a clustering process in the input space. In table 3 we present the results provided by this model for different codebook sizes, and considering the same training-test methodology as was presented for the MLP and LVQ2 models.

Table 3. Results provided by the IRVQ algorithm.

Codebook size	Mean correct classification (%)
20	97.7
40	98.7
80	98.5

This algorithm presents performance similar to that provided by the MLP model, its main drawback lies in the large network size (40 units against the 16 units used in the MLP network) required to handle the proposed task, thus hindering the possibility of finding a cheap implementation with the given timing constraints.

Evolutive neural models have also been used for implementing the decision tasks. Among these we considered in the first stage the Restricted Coulomb Energy (RCE) algorithm [8]. This neural model attempts to solve a discrimination task by estimating from the training samples the regions of the input space where each defined class is dominant. Also included in the analysis is the Neural Trees model [9]. In this case, the learning phase is in charge of providing in an incremental way a piecewise linear approximation for the desired discriminant function.

Tables 4 and 5 show, respectively, the results indicated as number of units generated by the algorithm and its associated correct classification rate provided by the RCE and Neural Trees algorithms for this classification task. The same methodology used is that presented for the fixed-structure models.

It is important to achieve a solution with the highest possible performance as defined by the correct classification rate. This should be achieved with the lowest possible complexity as defined by the use of the least costly components. A quality factor defined as the ratio between the mean correct classification rate and the number of units constituting the network structure is used to evaluate the models used in our experiments. The model selected for the final implementation should be that able to maximise the quality factor. In Figure 4 we have represented this quality factor for the different neural models considered in our analysis.

Table 4. Results provided by the RCE algorithm.

Number of units	Correct classification (%)
29	93.0
37	95.2
94	92.9
101	93.1

Table 5. Results provided by the Neural Trees algorithm.

Number of units	Correct classification (%)
50	98.0
120	98.5
154	96.1
191	97.2

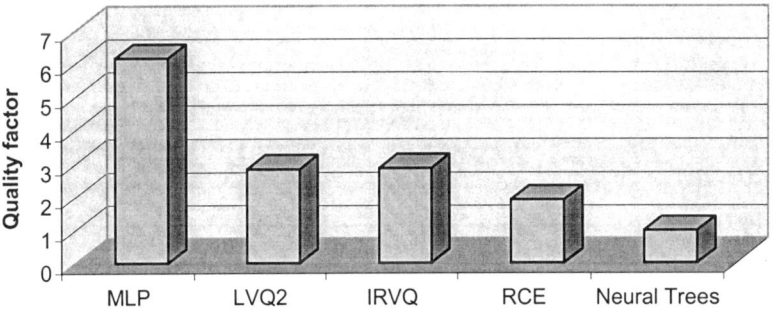

Figure 4. Quality factor for the models considered in the analysis.

From Figure 4, the MLP model demonstrates the best behaviour when the parameters constraining the application to be faced are considered. The LVQ2 and IRVQ algorithms, while showing similar performances in terms of correct classification rates than MLP are seriously limited by the resulting network complexity. Finally, the evolutive models (RCE and Neural Trees) offer rather poor results. Taking into account these results, MLP model has been our choice for the final implementation of the decision engine to be included in the automatic coin recognisers.

It is important to note that up to this point only the "nominal" behaviour of the coin recognisers has been considered. That is, in the databases used in training and test experiments only legal coins have been considered. The ability of neural models to cope with the intrinsic variability of the features extracted from these objects has been studied. However, as has been already pointed out in [12], it is important to consider also the ability to reject "outlier" objects. These are objects that have been intentionally manipulated in order to resemble legal coins. In the next section the design of an additional validation stage for an efficient rejection of non-legal coins is described.

3.2   Validation Stage for the Rejection of Outliers

As explained in the previous section, the MLP model shows the best performance for the classification task in this application. However, one of its major shortcomings is that the classification regions determined after learning are usually open-ended. This means that in some cases the class boundary determined by the discriminant function is not closed. As a consequence, the region where input vectors are assigned to a given class extends from this class boundary to infinity. Malfunction may occur when non-legal coins have to be detected. This particular problem is considered in Figure 5.

Figure 5(a) depicts a classification task where the system has to decide between two categories, both characterised by two features, x_1 and x_2. In this figure the patterns belonging to class 1 are represented with a cross, x, while those vectors belonging to class 2 are shown by a circle, o. Once the training process for the database is complete, the MLP network can provide a discriminant function, similar to that depicted in Figure 5(a). This discriminant function partitions the input space into two open subspaces where vectors lying above the curve are classified as belonging to class 2, while the vectors lying below this curve are classified as class 1.

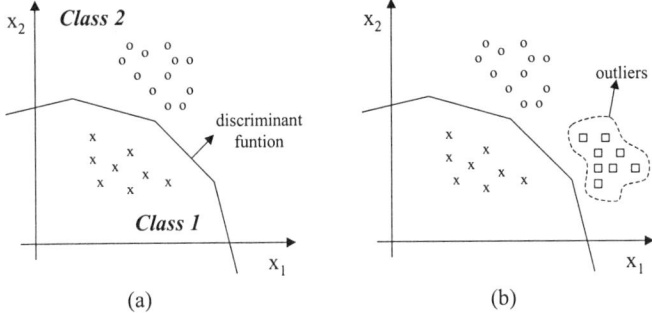

Figure 5. (a) Discriminant function determined by an MLP network. (b) Outlier objects incorrectly classified as belonging to class 2 due to open-ended decision regions

These open decision regions may cause problems if, as in our case, some objects can be intentionally modified so as to provide features similar to those corresponding to a correct category. These items were given the name "outliers" earlier in this chapter. This is shown in Figure 5(b), where outliers are represented with a square. These vectors are obtained on items resembling coins and they will be classified as belonging to class 2. This problem is especially difficult, since the features corresponding to outliers are usually not available when the system is being designed. Consequently it is difficult to include rejection mechanisms during the training phase of the neural model.

Therefore, it is mandatory:

- to classify correctly the patterns included in the training set, and
- to reject those patterns whose features do not resemble those associated with legal objects

These are the main aims of the validation stage included in the decision engine. The validation stage is composed of six different neural networks, one for each legitimate coin category. Each of the six neural networks is trained to activate its output for the training patterns corresponding to one of the legal coins, and to deactivate it otherwise. Patterns corresponding to outliers have been artificially generated to construct a second training database. These have been placed uniformly surrounding the legal vectors, so as to provide completely closed decision regions. All the networks are composed of a hidden layer integrated by 5 neurons, and an output layer with one neuron, whose output decides whether the coin has to be accepted or rejected.

So as to provide an accurate rejection mechanism which aims to reject as few legal coins as possible, features are chosen which show less variability with respect to the tolerance of the components and the operating conditions. As a consequence, three features have been considered for performing the rejection mechanism. These are:

- coin weight, and
- two features related to the coin size, that are extracted from the optical sensors included in the coin recogniser as was explained in Section 2.

As a consequence of this choice, the resulting structure of the decision engine is that represented in Figure 6.

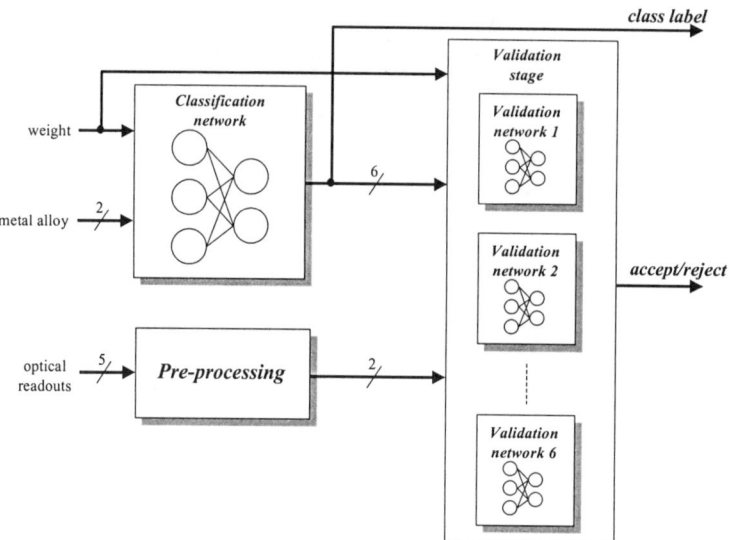

Figure 6. Final structure of the decision engine.

Once the different networks which constitute the system in Figure 6 have been trained, a test phase is performed using data extracted from 12 non-legal coins (outliers) found in commercial vending machines. This database has been constructed by dropping each coin 20 times in each of the 20 coin recognisers available. That is, the 10 recognisers used to construct the training/test database plus 10 additional recognisers left for validation purposes. The mean rejection rate observed has been 89.92%, which outperforms that obtained (76.55%) with the classical rejection mechanism included previously. No performance degradation has been observed in the classification of legal coins. These results thus validate the effectiveness of connectionist models for handling the proposed real-world decision task.

The structure and parameters of the decision engine for the proposed task have now been defined. In the next section the details related to the physical implementation of the system are discussed.

3.3   Implementation

As indicated in previous sections, there are two main constraints which need to be considered for the physical implementation of the decision engine included in the coin recognisers.

The first limiting factor comes from the mechanical structure of the system. It imposes that, once the coin has been introduced and its features extracted by the sensors included in the system, the decision to reject or accept the coin has to be performed in less than 30 ms.

In addition, the very stringent commercial/financial margins associated with the final product requires the cost to be minimised. Therefore, when considering the physical implementation of the neural architecture of the previous section, as many hardware resources as possible from the previous version of the automatic coin recogniser are retained. The classification task in the earlier version was performed by means of classical techniques.

An 8031 microcontroller is the core of this recogniser. This acquires data coming from the different sensors included in the system, and also determines from this data the specific category to which a given coin belongs. The main goal of this application is to emulate in software the neural structures which define the functionality of the decision engine depicted in Figure 3. It means that, for each coin drop, it is necessary to emulate the recall phase of a 3-10-6 MLP network (classification stage) and a 3-5-1 MLP network (validation stage for a specific coin category).

The data provided by the sensors which give the weight and metal alloy composition of the coin are represented with 8-bit precision, and the 5 optical readouts are encoded as 20-bit numbers. The pre-processing stage depicted in Figure 6 compresses the five 20-bit numbers onto two 8-bit data numbers. These, as explained in Section 2, offer an estimation of the coin size, more specifically of the diameter of the coin. The connection weights of the neurons in both the classification and the validation stages have been coded using 16-bit precision. The sigmoid activation function of these neurons has been implemented as an internal 256 x 8-bit look-up table. The emulation of the MLP structures with these precision constraints has rendered exactly the same classification performance as the simulation of the original network where the parameters were handled in 64-bit floating point precision.

The program emulating the neural network has been written in the C programming language and then included in an EPROM which is part of the hardware constituting the coin recogniser. Special care was taken only in the implementation of the multiplication algorithm, so as to meet the timing constraints imposed by the application. The complete system has been verified by means of field tests. These tests consisted of inserting 270 objects (150 legal coins and 120 outliers) into 20 vending machines under different environmental conditions. The results obtained have demonstrated the effectiveness of the approach taken in developing the application, since the final physical system is able to reproduce exactly the expected behaviour.

As the performance of this software implementation is satisfactory, from a commercial point of view it is very important to protect the final product against non-authorised copies or duplications. For this reason, we have also performed an estimation [13] for a dedicated hardware implementation of the decision engine by means of an Application Specific Integrated Circuit (ASIC). The precision (8-bit for the inputs, 16-bit for the weights) requirements and the structure of the neural models to be implemented have now been defined. It is now necessary to determine the hardware architecture required for the efficient implementation of the resulting data flow. Following the results obtained in the analysis performed in [14], we have selected the structure corresponding to the Broadcast Bus Architecture to implement the networks included in the neural decision engine. This choice has been dictated by the fact that this parallel architecture renders the most rapid emulation for the recall phase of the MLP neural model. Figure 7 depicts the basic organisation of this architecture.

It is seen in Figure 7 that the architecture is composed of several processing elements indicated as blocks P1 to PN which share a global input and a global output bus. Both buses are managed by a control unit. The processing elements will emulate the functionality of the units which constitute the networks used in the decision engine. The processing elements will receive their inputs from the global input bus, and, once processed, will deliver their outputs on the global output bus.

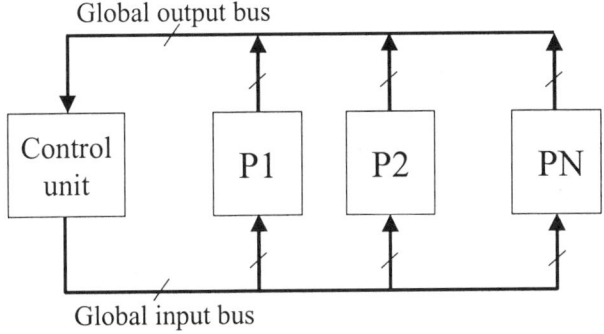

Figure 7. Basic organisation of the Broadcast Bus Architecture.

In order to define the number of processing elements constituting the physical Broadcast Bus Architecture, we have to consider the network structures which are to be emulated. The classification network is composed of a hidden layer of 10 units and an output layer of 6 units. The validation networks are composed of a hidden layer consisting of 5 units and an output layer composed of one unit.

The emulation of the whole neural decision engine follows an inherent sequential data flow. The validation network is to be emulated once the classification network has indicated the category to which the coin potentially belongs. Furthermore, inside each network the emulation of the output layer has to wait until the outputs corresponding to the units in the hidden layer have been obtained. This means that each processing element can emulate the functionality of different units in different time steps. This reduces the physical size of the array to be implemented, which will be imposed by the size of the largest layer in the networks (in our case, 10 units). Figure 8 shows the actual mapping of the units to be emulated on the physical processing elements included in the array.

In this figure, the blocks identified as P1...P10 constitute the processing elements of the physical array, while the blocks labelled N1...N22 correspond to the units which integrate the classification and validation networks. Only the units corresponding to one of the six validation networks is represented in this figure for the sake of clarity.

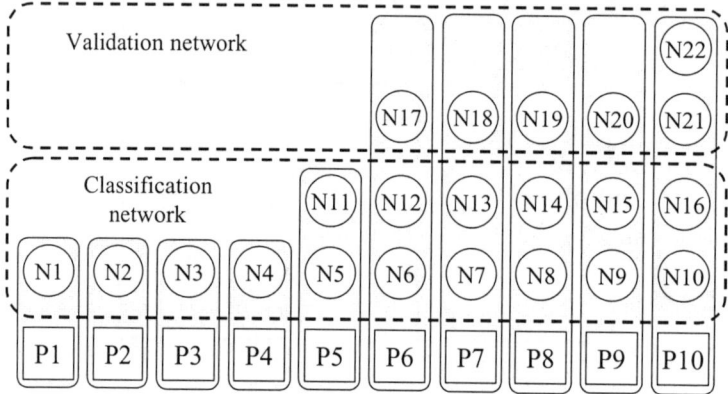

Figure 8. Mapping of the classification and validation networks on the physical array of processing elements.

Once the basic organisation of the hardware architecture to be implemented is determined, the next step is to define the features to be included in the processing units. These are the basic building blocks of this architecture. Bearing in mind the functionality of the neurons which constitute a MLP network in recall phase, it can be seen that the basic operations performed by these units are multiplication, addition and non-linear (sigmoid) function. In order to save area in the final realisation, only one non-linear function block has been implemented in the control unit of the array, being the two remaining operations (multiplication and addition) integrated in each processing element of the array.

Among the different alternatives available to implement the multiplication block [15], we have chosen a systolic scheme, due to the simplicity of the required control and its low area requirements. This multiplication scheme is able to provide the result of an 8-bit input by a 16-bit weight product in 16 clock cycles. This result is generated in a serial fashion, from the least significant to the most significant bit.

By using this multiplier scheme it is possible to further simplify the functionality of the processing elements in the array. Since a bit of the partial products is obtained every clock cycle, it is possible to implement the internal accumulator by means of a shift register and a 1-bit full adder, whose carry output is saved in a register. In order to avoid overflow, a 28-bit shift register is used as accumulator. Consequently, the final organisation of the ten processing units included in the decision engine corresponds to the structure depicted in figure 9. In this figure, the term x_j represents the j-th input value for the neuron being emulated at a given time step, while the term ω_j is the synaptic weight associated with this input.

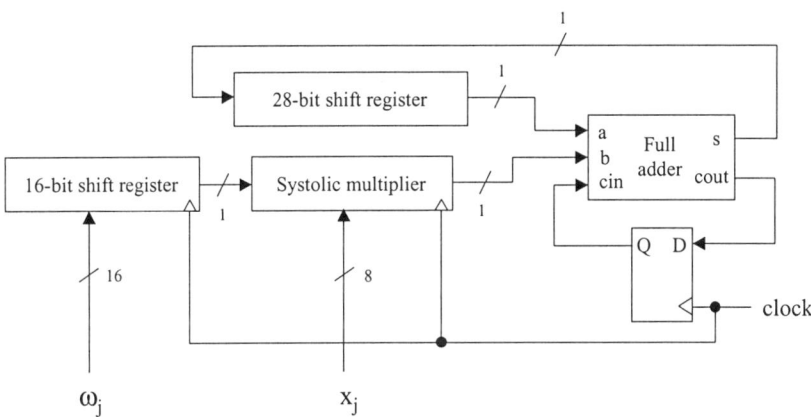

Figure 9. Organisation of the processing units included in the neural decision engine.

A modification has been included in the organisation of the Broadcast Bus Architecture, so as to avoid the tri-state outputs in the processing elements of the array. In this way, instead of using the global common bus to transfer to the control unit the results obtained from the accumulation of the partial products in each processing element, a sequential transfer scheme is implemented. Thus, each processing element transfers this result to its direct neighbour. This process is repeated until all the results are available for the control unit, where they are used as inputs for the non-linear sigmoid function generator, thus obtaining the final outputs corresponding to the neurons of the network structure currently being emulated.

This non-linear function is implemented by means of a 355×8-bit look-up table. Due to the 8-bit precision limit used in the implementation of this function, it is enough to consider 14 bits from the 28-bit result obtained in each processing unit. Another function to be performed by the control unit consists of generating the correct address for the RAM bank included in each processing element, where the connection weights corresponding to the different units emulated by each processing element are stored, according to the mapping scheme depicted in Figure 8.

Finally, this control unit has to provide a fast and efficient interface between the neural decision engine and the microcontroller which is in charge of managing the different components (sensors, accounting system, etc.) which integrate the coin recogniser. This is accomplished by an 8-bit data bus which can be directly connected to the data bus of the microcontroller, together with several dedicated lines which can be used to control and monitor the different stages of the classification and validation processes.

The behaviour of the basic building blocks which constitute the neural decision engine has been described using the VHDL language. After an exhaustive verification of their functionality, a synthesis process for a 0.7 µm CMOS ASIC process has been performed. The synthesis stage has been carried out independently for the main building blocks which constitute the system, so as to detect and improve those parts that may be critical in the physical realisation. Table 4 briefly summarises the synthesis results (in terms of occupied area and critical path delay) for some of the main components of the neural decision engine.

From these results, taking into account that the Broadcast Bus Architecture included in the system consists of processing elements, an area estimate of approximately 3.4 mm^2 (without RAM banks) can be obtained for the system. Including the RAM banks, interconnection area and pads, a quite conservative estimate of less than 10 mm^2 can be obtained for the system.

Table 6. Synthesis results for the blocks included in the neural decision engine.

Unit	Critical path delay	Area
Processing unit	5.29 ns	0.28 mm^2
Control unit	16.11 ns	0.45 mm^2
Look-up table	7.55 ns	0.13 mm^2

Regarding the timing properties of the system, as it was expected the most critical part is the control unit of the system. In order to verify that the system still works correctly after introducing the delays associated with the components used to implement the different building blocks, an extensive functional simulation has been performed with the structural netlist obtained from the automatic synthesis process. In these simulations we have used a quite conservative clock frequency of 25 MHz. Even using this quite slow system clock, the time required to classify and validate the features extracted from a coin is 33080 ns, almost three orders of magnitude faster than the 30 ms limit dictated by the specifications of the coin recogniser.

These results demonstrate the possibility of providing cost-effective solutions either in software or in dedicated hardware form for real-world industrial applications based on principles derived from artificial neural network processing paradigms.

4   Approach Using Fuzzy Logic Models

Since its inception by Zadeh in the 1960s [16], fuzzy set theory has gained acceptance in a wide range of applications, encompassing control, pattern recognition, and signal processing. The application addressed here is characterised by inherent uncertainties, which are caused by internal factors such as tolerances of the components constituting the system and external environment conditions such as variability in the extracted features due to wear and manipulation factors. As a consequence, the application of fuzzy logic techniques has also been evaluated for possible inclusion in commercial coin recognisers.

The global organisation of a system based on fuzzy logic techniques is depicted in Figure 10.

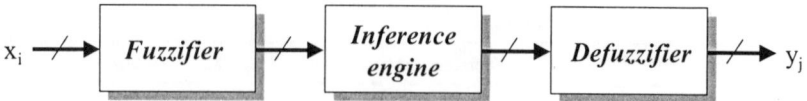

Figure 10. Organisation of a system based on fuzzy logic techniques.

The system processes crisp inputs, x_i, and delivers crisp outputs, y_j. The *fuzzifier* block of this figure converts the crisp inputs to membership degrees of some linguistic variables defined over the input space. Based on these membership degrees, and taking into account a given set of rules (fuzzy relations), the *inference engine* is able to generate fuzzy results by composing the evaluations provided by these rules. Finally, the *defuzzifier* block translates these values onto crisp outputs to be used in the target application.

The details and parameters of the different building blocks integrated in the system will depend on the fuzzy model used which will be selected by taking into account the constraints associated with the application. In the next section it is explained how these parameters have been fixed for the decision task of this application.

4.1 Fuzzy Model Selection and Experimental Results

The constraints to be fulfilled by the final implementation of the coin recogniser system are as follows:

- the system should be able to perform the desired function using inexpensive components, and
- the response time should guarantee that every classification is to be performed in less than 30 ms.

Different fuzzy models have already been proposed [17] and of these the Mamdani fuzzy model [18] has been selected because of its low computational requirements.

In the Mamdani model, the *inference engine* evaluates the fuzzy rules by a max-min composition of the fuzzy values provided by the *fuzzifier*. For the sake of simplicity, the fuzzy membership functions which are responsible for translating the crisp inputs onto fuzzy membership values have been chosen as trapezoidal. Out of the various methods available for implementing the *defuzzifier* block, the Mean of Maximum (MoM) method has been chosen.

Taking into account the considerations explained in Section 3.2, it was decided to implement the decision engine by means of two fuzzy systems. The first system or classification stage will provide a rough estimate of the category to which the coin belongs, while the second or validation stage will specialise in rejecting non-legal coins.

A preliminary analysis has demonstrated that, in order to perform a correct classification process, it is necessary to use five features:

- the three features extracted from the metal alloy of the coin,
- its weight and
- a rough estimation of its size.

This differs from the three features previously required by the neural models. As in the neural approach, the validation stage will handle those features with a higher confidence margin. Therefore, the same three features, namely the weight of the coin and an accurate estimate of its diameter given by the two parameters explained in Section 2 previously used for the neural validation stage will again be provided as inputs for the fuzzy validation engine. The classification and validation stages will provide a total of 7 outputs, indicating if the coin belongs to one of the 6 possible legal categories, or if it is to be rejected. A coin inserted in the recogniser will be accepted if and only if there is an exact agreement between the classification results provided from both systems.

The final parameters of the fuzzy decision engine still to be considered are:

- the number of membership functions per input and
- the number of rules to be evaluated in the *inference engine* of both the classification and the validation stages.

These parameters have been defined after a careful analysis of the features extracted by the sensors for each coin, and paying special attention to the function to be performed by each stage whether it is in the rough classification or accurate rejection mechanisms.

In the classification stage 4 membership functions have been used for the first two features related to the metal alloy of the coin and 5 membership functions for characterising the third feature. The input obtained from the weight of the coin has been described by 6 membership functions, while only 2 functions have been used for the rough estimation of the size of the coin. The *inference engine* used in this classification stage has to evaluate 6 fuzzy rules.

Because of the tighter constraints proposed for the validation stage, its implementation requires a much more detailed evaluation of the information included in the inputs provided for the system. Consequently 10 and 9 membership functions have been used respectively to characterise the weight and the accurate estimation of the diameter which are extracted from a coin.

Once all the parameters of the classification and validation stages have been determined, the performance of the decision engine is evaluated. The fuzzy system is simulated with a database composed of features extracted from the 20 coin recognisers available. The database is composed of 3850 feature vectors, of which 1050 correspond to drops of legal coins and the remaining 2800 vectors were extracted from non-legal objects or outliers. The results which are indicated as mean correct classification in percentage for this decision engine are summarised in Table 7. In this table, the correct classification for non-legal coins means that an object resembling a legal coin has been identified and properly rejected.

Table 7. Results provided by the decision engine implemented by means of fuzzy techniques.

	Correct classification (%)
Legal coins	98.76
Non-legal coins	81.10

It can be seen from this table that the fuzzy system results are similar to those obtained by means of artificial neural networks model. The rejection rate for non-legal objects is slightly worse. Field test results obtained when this fuzzy system has been physically implemented in the coin recognisers are given in the next section.

4.2   Implementation

A prototype physical system for the decision engine has been created. For this purpose, the functionality of the fuzzy systems which define the behaviour of both the classification and decision stages has been written in the C programming language. The resulting code has been compiled for the 8031 microcontroller included in the coin recognisers. The resulting binary program has been included in the EPROM, and then an exhaustive field test procedure has been carried out as described in Section 3.3. Table 8 summarises the results (indicated as correct classification rate, in percentage) obtained after this field strategy has been completed.

The results obtained in this case are similar to those obtained for the decision engine based on artificial neural networks models, and are much better than those obtained previously in the coin recognisers using traditional techniques. The only difference between the two approaches lies in the fact that more features have to be considered in the case of the solution based on fuzzy models. This is due to the inherent approximate reasoning which is the basis for these models, rather than the adaptive non-linear induction process which characterises artificial neural models. These results validate the use of neural and fuzzy techniques for real-world industrial applications, even when the application is seriously limited by hard constraints.

Table 8.   Results obtained after the field test of the fuzzy decision engine performed on the coin recognisers.

	Correct classification (%)
Legal coins	99.10
Non-legal coins	89.50

5   Conclusions

In this chapter we have considered the applicability of neural and fuzzy techniques to real-world industrial applications. The target problem to be solved is the implementation of the classification/decision engine included in the coin recognisers which are part of automatic vending machines. It represents a difficult problem, due to the tight constraints posed by the final system, cost and processing speed, and the inherent variability observed in the objects to be classified caused by the environment conditions and the specific manipulation of non-legal objects.

The methodology used to handle the proposed task has been divided in three stages. Firstly, a careful analysis has been performed on the data to be processed, so as to extract the most relevant features for use in the classification task. As a consequence an exhaustive and representative database has been compiled. A thorough analysis has then been done to select the proper neural model to be used in the implementation of the decision engine. In this study classical fixed-structure, as well as evolutive neural models have been considered. The quality factor defined to choose the proper neural model has revealed the Multilayer Perceptron (MLP) as the most suitable neural model for the problem.

The problems posed by an open-ended decision region have motivated the introduction of a second validation stage also implemented by an MLP network. This provides an accurate rejection mechanism for non-legal objects. Once the parameters of the neural model have been obtained, a physical implementation of the system has been produced and tested. This consists of a software emulation on the microcontroller included in the coin recognisers. The exhaustive field tests performed on the final system show that this approach gives better results than the traditional method used in commercial coin recognisers. In order to improve the privacy of the physical system, a dedicated hardware implementation by means of an ASIC has been considered. A competitive both in terms of cost and processing speed hardware implementation has been developed by the use of careful design strategies.

Since the proposed application has a high degree of uncertainty, the use of fuzzy models has been considered for implementing the classification and validation stages. These would be included as the decision engine to be included in the coin recognisers. In selecting the proper fuzzy model and its associated parameters which include the number of fuzzy membership functions per input, the number of rules, the type of membership functions and defuzzification technique the constraints associated with the application have also been considered. Once the system has been constructed and its functionality validated by means of simulation, a software implementation has been provided. The field tests performed on actual coin recognisers have demonstrated a performance similar to that attainable by means of neural techniques.

The methodology and results provided in this chapter demonstrate the usefulness of neural and fuzzy techniques for solving real-world tasks commonly found in industrial environments. This chapter has shown that if a careful design methodology is used, it is possible to provide commercial products based on these techniques.

References

[1] Bellman, R. (1961), *Adaptive Control Processes: a Guided Tour*, Princeton University Press.

[2] Kwok, T.-Y. and Yeung, D.-Y. (1997), "Constructive algorithms for structure learning in feedforward neural networks for regression problems," *IEEE Trans. Neural Networks*, vol. 8, no. 3, pp. 630-645.

[3] Ash, T. and Cottrell, G. (1995), "Topology modifying neural network algorithms," *Handbook of Brain Theory and Neural Networks*, MIT Press, Cambridge, MA, pp. 990-993.

[4] Rumelhart, D.E. and McClelland J.L. (1986), *Parallel and Distributed Processing: Explorations in the Microstructure of Cognition*, MIT Press, Cambridge, MA.

[5] Werbos, P. (1974), *Beyond Regression: New Tools for Prediction and Analysis in the Behavioral Sciences*, Ph.D. Thesis, Hardvard University.

[6] Kohonen, T. (1989), *Self-Organization and Associative Memory*, Springer-Verlag.

[7] Voz, J.-L., Verleysen, M., and Thissen, P. (1995), "Suboptimal Bayesian classification by vector quantization with small clusters," *Proceedings of the European Symposium on Artificial Neural Networks ESANN'95*, pp. 153-160.

[8]   Reilly, D.L., Cooper, L.N., and Elbaum, C. (1982), "A neural model for category learning," *Biological Cybernetics*, vol. 45, pp. 35-41.

[9]   Sirat, J.A. and Nadal, J.P. (1990), "Neural trees: a new tool for classification," Technical Report, Laboratoires d'Electronique Philips.

[10] Kohonen, T. (1990), "The self-organizing map," *IEEE Proceedings*, vol. 78, no. 9, pp. 1464-1480.

[11] Specht, D.F. (1990), "Probabilistic neural networks," *Neural Networks*, vol. 3, pp. 109-118.

[12] Frosini, A., Gori, M., and Priami, P. (1996), "A neural model for paper currency recognition and verification," *IEEE Trans. on Neural Networks*, vol. 7, no. 6, pp. 1482-1490.

[13] Moreno, J.M., Núñez, J.L., Madrenas, J., Cabestany, J., and Laúna, J.R. (1998), "VLSI implementation of a neural decision engine for commercial coin recognisers," *Proceedings of the 5th International Workshop Mixed Design of Integrated Circuits and Systems MIXDES'98*, pp. 379-384.

[14] Moreno, J.M. (1994), *VLSI Architectures for Evolutive Neural Models*, Ph.D. Thesis, Technical University of Catalunya.

[15] Ma, G.-K. and Taylor, F.J. (1990), "Multiplier policies for digital signal processing," *IEEE ASSP Magazine*, pp. 6-20.

[16] Zadeh, L.A. (1965), "Fuzzy sets," *Information and Control*, vol. 8, pp. 338-353.

[17] Jang, J.-S.R. and Sun, C.-T. (1995), "Neuro-fuzzy modeling and control," *Proceedings of the IEEE*, vol. 83, pp. 378-406.

[18] Mamdani, E.H. and Assilian, S. (1975), "An experiment in linguistic synthesis with a logic controller," *International Journal of Man-Machine Studies*, vol. 7, pp. 1-13.

Chapter 4

Fuzzy Techniques in Intelligent Household Appliances

M. Mraz, N. Zimic, I. Lapanja, J. Virant, and B. Skrt

Several concepts of the use of fuzzy techniques in analysis and design of control systems for household appliances manufactured by Gorenje GA, Slovenia are presented in this chapter. In the past decade this industry has experienced an intensive trend towards the reduction of electrical energy consumption and production costs. American and European markets dictate novel standards which are aimed at a decrease in energy consumption. Adaptation of a rigorous standard usually implies increased production costs which in turn lowers the competitive advantages of products. Our goal was therefore to explore the functional behaviour of several household appliance products and analyse the possibility of using intelligent technologies in their design in order to decrease energy consumption. A 4 to 16 percent energy saving was achieved while the production costs remained unaltered.

1   Introduction

Decreasing the consumption of energy is a trend all over the world in the design of many consumer appliances. Consumers are interested in buying such products not only to satisfy their ecological awareness but also to reduce their living expenses. In the realm of electrical consumer devices terms such as intelligent or smart appliances are used. In general these consume less energy than their equally functional counterparts and prices are normally little higher. The European market has already defined several regulations which classify some specific devices such as refrigerating and freezing appliances into seven consumer classes according to the energy consumption levels. Prices are formed in the open European market with regard to these classes.

Similar tendencies may be observed in the United States market where lower energy consumption levels are being set as pre-condition for the sale of specific appliances. Producers of electrical devices have responded to these market demands by intensive research activities. This includes the search for better materials, parts and possible ways of their use such as plastics, better insulation, improved electromechanical parts and compressors. It may also include new and more optimal control systems which use resources better and thus lead to a decrease in energy consumption.

Over the past few years this group has received financial support from Slovenian Ministry of Science and Technology for the study of the functioning and the possibilities of developing an 'intelligent' control of two household appliances of Gorenje GA, a manufacturer with a strong position in Europe and United States. This work has particularly focused on the electric kitchen oven and refrigerator-freezer. Both products are being offered with various options and approximately 800,000 pieces per a year have been sold. Results presented in this chapter can also be applied to any product. For both appliances a 4 to 16 percents electrical energy saving was achieved by utilising fuzzy logic, intuitive design approaches and approximate knowledge of rules that govern the device system response.

Fuzzy approaches in making 'intelligent' devices are described first. The specific features of the controller design for kitchen oven and refrigerator-freezer are described next. The design of fuzzy controller for a typical one-compressor refrigerator-freezer is then described. Some crude calculations of possible savings from a wider point of view conclude the discussion.

2   Fuzzy Approaches for Intelligent Devices

The object of this chapter is to present the implementation of fuzzy logic to the design of the kitchen oven and refrigerator-freezer. To cover the physical dynamical properties of both devices in their entirety would require an in depth formal treatment. Insteads, the discussion is limited to linguistical processing and building of an approximate knowledge base related to the dynamical aspects. Fuzzy logic proves to be an excellent tool for such an endeavour and [1], [8], [10], and [11] provide a general treatment of its basics. The main idea and purpose of the research was to find some approximate and descriptive rules for the response of observed systems to the control output which were later built-in to the decision procedure as fuzzy rules. The concept of control design can be divided into definition of input and output fuzzy variables, declaration of appropriate membership functions and setting up fuzzy rule database for decision making. The fuzzy control process itself (see Figure 1) can be described by the following steps:

- gathering crisp value of absolute temperature $T(n)$ from the environment,
- fuzzification of crisp input variables,
- making approximate reasoning using the fuzzy rule database,
- defuzzification of fuzzy results.

The meanings of variables in Figure 1 are the absolute temperature $T(n)$ captured by controlled system, the calculated values $x_1(n)$, ..., $x_k(n)$ (for instance $e(n)$ which represents the error on the current step) and finally $\Delta Output$ which represents the change of output power compared to previous step $(n-1)$.

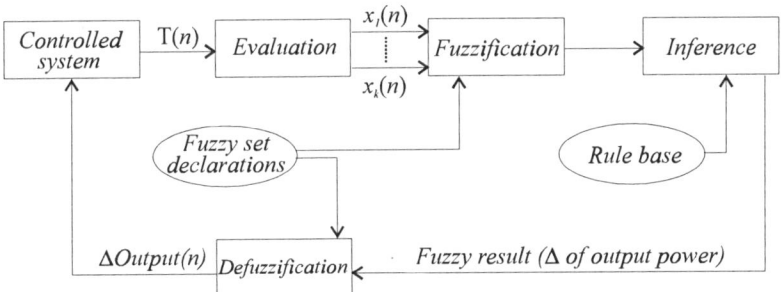

Figure 1. Basic model of the controller.

Hardware equipment gathers crisp input values so the first step of introducing 'intelligence' is the evaluation of all necessary input variables based on gathered input, which is $T(n)$ in our case. This is followed by fuzzification of crisp values with respect to membership functions of terms' values of the observed variable. The process of fuzzification is simply a calculation of the intersection between crisp value and pre-defined membership functions. In Figure 2 we see an example of five membership functions for a given variable $T(n)$ with specified goal temperature of 90°C and an example of calculation for membership function of input crisp value $x = 77.5$. Degree of membership to fuzzy sets *VeryLow* and *Low* is thus $\mu_{VeryLow}(x) = 0.25$ and $\mu_{Low}(x) = 0.75$ respectively.

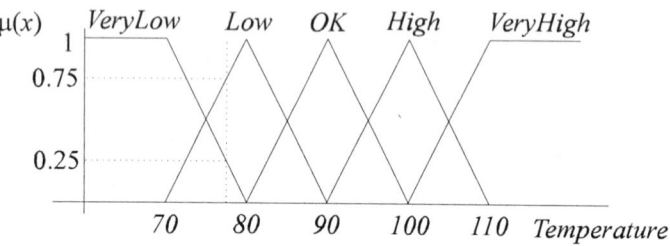

Figure 2. Fuzzy values of variable $T(n)$.

Once all degrees of memberships to individual fuzzy sets $\mu_{Set\_i}(x)$ have been defined it is possible to proceed from *IF* to *THEN* part of the rule in consideration - finally forming the resulting fuzzy set. Assume that there are two rules in the database as in (1) and have output fuzzy sets defined as in Figure 3.

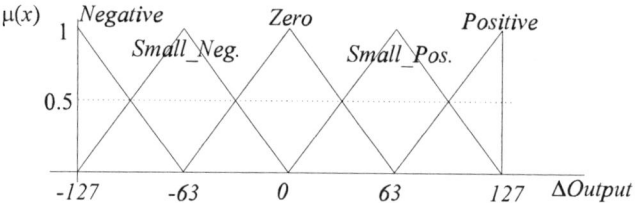

Figure 3. Example of output sets definition.

Minima of trigger factors and output terms are given in Figure 4. They produce a membership shape which is a combination of two cut triangles "*SmallPositive*" and "*Positive*". If certain output term is elected more than once, the one with higher trigger degree from the left is taken.

```
IF  T(n) = "VeryLow" THEN ΔOutput (n) = "Positive"
IF  T(n) = "Low" THEN ΔOutput (n) = "Small_Positive"   (1)
```

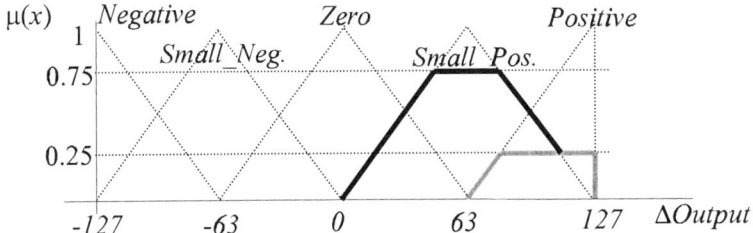

Figure 4. Example of conclusion process based on output sets.

It is now necessary to obtain a crisp output from the resulting set of Figure 4. This is provided by the Centre Of Gravity (COG) defuzzification method. Expression (2) gives a formal definition of COG defuzzification method used to produce crisp values based on output fuzzy terms. Here $\mu(x)$ represents membership of the resulting fuzzy set and x is an element of the output space $\Delta Output$.

$$\Delta Output = \frac{\sum_{i=1}^{n} x * \mu(x)}{\sum_{i=1}^{n} \mu(x)} \qquad (2)$$

3   Introducing Fuzziness to Kitchen Oven

The object of the analysis was the kitchen oven Noblesse of Gorenje GA, Slovenia. In this section a series of steps are presented in which a classical thermostatic oven was converted into a fuzzy controlled, modern oven with improved performances and decreased energy consumption. The project itself was composed of the following phases:

    I.   functional analysis of thermostatically controlled oven,
    II. simulation of fuzzy control on PC, and
    III. prototype design of fuzzy oven controller.

Before the work was started on the oven, an appropriate working environment was simulated on PC-AT. This included software tools for fuzzy processing with rule editing, membership definition and similar functions. Also software and hardware were assembled for temperature read-out and electrical heater regulation. The control consist of interval pulse modulator with a solid state relay.

3.1   Thermostatically Controlled Oven

The oven offers five heating regimes to achieve "*high*" temperatures and one special regime for food defreezing. The heating regimes differ with respect to location and type of heaters. Originally, the temperature is specified by physically adjusting a thermostat. In fuzzy solution, we have two temperature specifications:

- *desired* temperature is the one specified by adjusting a thermostat,
- *chosen* temperature is the one which is actually specified by the thermostat; this allows the temperature nominally below 100°C to be specified with +/- 7°C precision, while temperatures above 100°C are specified with +/- 15°C precision.

As a criteria for the evaluation of control logic two characteristics of temperature as a function of time were observed:

- minimal first temperature overshoot in the centre of the oven, and
- minimal temperature oscillation,

both with respect to the chosen temperature. Temperature in the sense of classifying the quality of control was always measured in the centre of the oven which is supposingly the typical location of the load. Temperature which is input to thermostatical or fuzzy control was measured in upper left corner and close to the back wall of the oven. Each operational regime was tested for four desired temperatures: 60, 80, 100 and 140°C. Considering the chosen regimes the lower heater is involved. Measurement was performed for all four desired temperatures. Figure 5 presents results for thermostatical control.

The results shown are unsatisfactoring from the consumer's point of view. For all four desired temperatures extreme first overshoot with high oscillating amplitude and long time period were observed.

3.2   Design of a Fuzzy Controller

Considering the realisation of a fuzzy controller for the heating regime where only the lower heater is used. This regime exhibited extreme oscillations with a long time period and high deviation from chosen temperature. A fuzzy controller has two inputs based on which output is processed. Inference time period is 0.5 seconds. This means that the control logic must process a new output every half a second. The basic scheme of fuzzy control is illustrated in Figure 1.

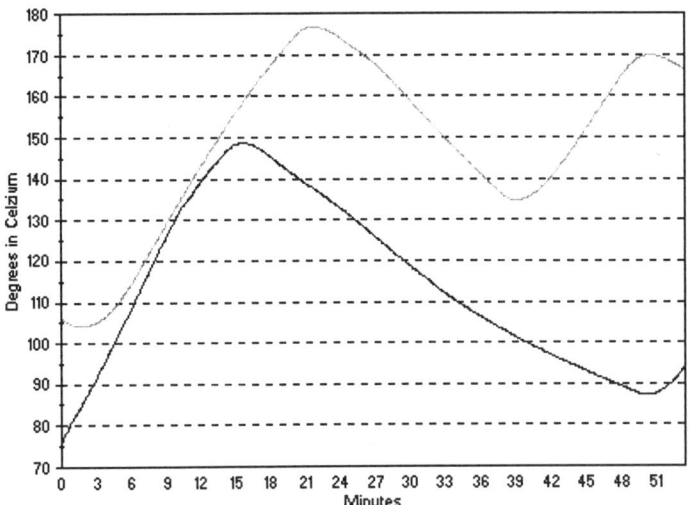

Figure 5. $T_{desired} = 100°C$ (lower curve), $140°C$ (upper curve),
Time = 54 min, thermostatic control.

The input to the controller is temperature, taken from the upper-left back corner of the oven as in original configuration. The temperature value is first fuzzified according to the pre-specified fuzzy sets *A*={*VeryLow, Low, OK, High, VeryHigh*}. This fuzzification step is similar to the one shown in Figure 2. Configuration of membership shapes is created automatically in our prototype with respect to the desired temperature. Three triangles of 20 degrees width are laid around the value chosen (see Figure 2) together with two half trapezoids.

Conventional fuzzy controllers use the difference between the current and previous values of temperature as a secondary input. However, due to the large thermal inertia of the system the temperature $T(n)$ is not a direct reflection of previous inference output $\Delta Output(n-1)$ so that the change of error ($e(n) - e(n-1)$) does not reflect the influence of the previous inference output. For this reason the history of previous system behaviour is used as a secondary input instead of just the temperature difference. Such a technique is described in considerable detail in [2]-[4], [9], and [10] where the emphasis is given to the influence of time in decision processes. The main idea is to observe the history of temperature values at *m* sequential points (the absolute values of temperature read-outs in 'previous' inference points) and fuzzify this history data. In other words, it is necessary to define it with descriptive terms which may be used in current decision process. From the observation of the existing heating system two important conclusions were reached:

- The system has a long time delay. That is the time between the heat transfer phase (action) and evidential temperature rise as recorded by sensor (reaction) is relatively long, and
- The processes of temperature rise and fall can be considered to be approximately linear.

From the above reasons it was decided to use 5 linear regions to classify the past events. Each region is determined by its inclination and width. The history point adds to membership degree in accordance to the region it 'falls into'. The relative region memberships are computed for individual history points. Those are the normalised distances from the centre of the region with respect to region's width and these are then used in simple or weighted sums to give the history membership to all five pre-specified regions. Membership of these regions is described by the following set of descriptive terms B={*Increase, Small_Increase, Stable, Small_Decrease, Decrease*}. This is titled the history function of temperature in oven. The set of coefficients used in this case is (-1.5, -0.5, 0, 0.5, 1.5). The region width is 6 degrees. Depth of history points used is 50 which translates to 25 seconds of past observation time window. Using such descriptive values it is possible to apply inference to eliminate the system's inertia. According to Figure 1 the value of variable *History(n)* may be used in the evaluation and fuzzification phase. The meaning of *y* axis in Figures 6 and 7 is an absolute temperature and the meaning of *x* axis is a time.

The rule matrix is two-dimensional and contains 25 rules. In this case the matrix does not change with different desired temperatures through one regime. An example of the use of matrix is given in Figure 8. In the first column of the matrix are the descriptive values of variable *History*. In the matrix header the descriptive values of the input variable temperature *T* are given. Matrix cells contain descriptive values of the output. Output variable Δ*Output* can also be described with relation to Figure 4. To explain the contents let us make an example of inference.

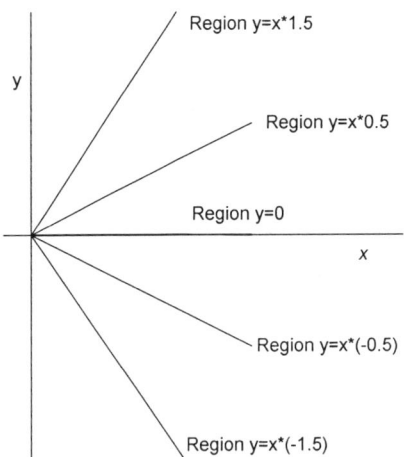

Figure 6. Fuzzy regions of *History(n)*.

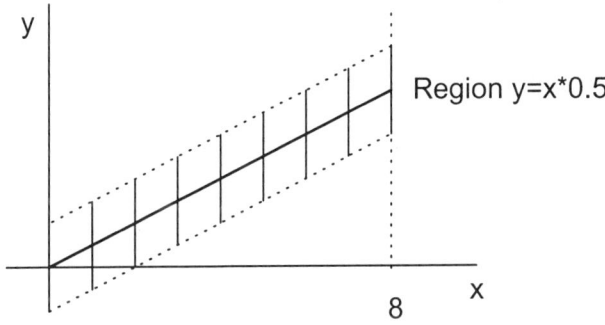

Figure 7. An example of fuzzy region, $m=8$, Width=1.

	VeryLow	Low	OK	High	VeryHigh
Increase	Small_Pos.	Small_Neg.	Small_Neg.	Negative	Negative
Small_Increase	Positive	Small_Pos.	Small_Neg.	Small_Neg.	Negative
Stable	Positive	Positive	Zero	Small_Neg.	Negative
Small_Decrease	Positive	Positive	Small_Pos	Small_Neg.	Small_Neg.
Decrease	Positive	Positive	Positive	Zero	Small_Neg.

Figure 8. The Rule-matrix used for kitchen oven control.

Example: Let us assume that history is given in *n*-th step as *Decrease* and *SmallDecrease* (with numerically defined values ranging from 0 to 1) and let the temperature be *OK* and *High*. Then the following conclusion rules from the matrix in Figure 8 apply:

IF *History(n)* = *"Smal_Decrease"* AND *T(n)* = *"High"*

    THEN *Output(n)* = *"Small_Neg."*

IF *History(n)* = *"Small_Decrease"* AND *T(n)* = *"OK"*

    THEN *Output(n)* = *"Small_Pos."*

IF *History(n)* = *"Decrease"* AND *T(n)* = *"OK"*

    THEN *Output(n)* = *" Positive"*

IF *History(n)* = *"Decrease"* AND *T(n)* = *"High"*

    THEN *Output(n)* = *"Zero"*.   (3)

After processing the fuzzy rule and execution of defuzzification a numerical output value is obtained which is added to previous output value. It represents the portion of pre-defined time interval (0.5 seconds in this case) during which time the electrical supply to the selected heater is enabled.

3.3   Results of Fuzzy Control

A comparative analysis between the built-in thermostatical and fuzzy control is presented in this section. The quality of control as well as comparison of the energy consumption in the two alternative methods of control are of importance. All comparisons are based on experiments where the chosen temperature was 100°C. For other temperatures from 60 to 140°C the results were similar. The observations focus on two basic parameters for evaluation of controller performances:

- The first is the time interval in which the oven heats from room temperature to the chosen one and further to maximal temperature achieved. The criteria here is the length of the interval and height of first overshoot,
- The second is the observation of temperature oscillations around the chosen value over longer time period. The criteria used here is the average amplitude of deviations from chosen temperature.

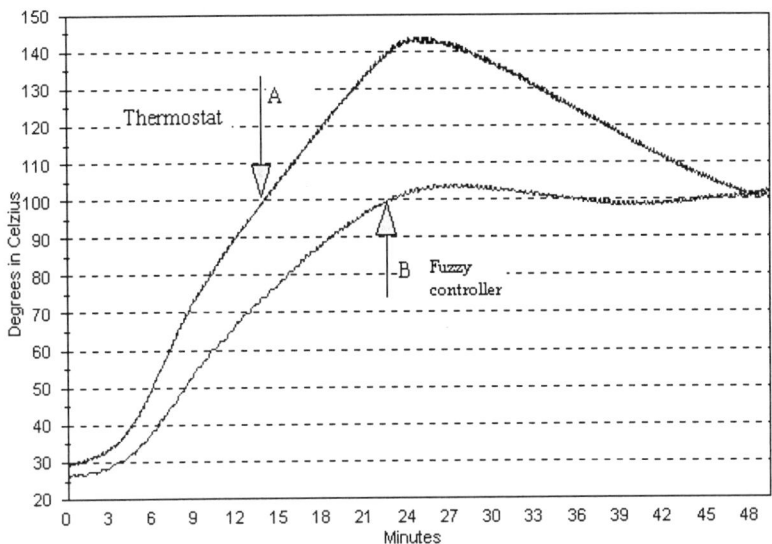

Figure 9. Comparing thermostatical and fuzzy control.

The first experiment consisted of running the 'cold' oven(s) for 48 minutes with chosen temperature at 100°C. From Figure 9 the following can be observed:

- The thermostatical control gives approximately 45 degrees of first overshoot while the fuzzy one gives only 4.
- The thermostatical control reaches the desired temperature T (point A) in 13 min while in the fuzzy case this time extends to 2 min point B). This could be improved if appropriate pre-control was applied, say a special control in first 5 min of 'cold' oven start). It can be seen that fuzzy control reaches 75°C in 13 minutes which is an acceptable approximation for the desired temperature so that corrections may not be required after all.
- The energy consumption from start to point A and B (the chosen temperature) was:
  - with thermostat control: 0.2144 kWh;
  - with fuzzy control: 0.2425 kWh;
- The energy consumption from points "A" and "B" to the end of experiment was:

- with thermostat control: 0.1719 kWh;
- with fuzzy control: 0.0830 kWh;
- The energy consumption over the entire experiment duration (48 minutes) was:
  - with thermostat control: 0.3863kWh;
  - with fuzzy control: 0.3255 kWh;

A rough estimate of the energy savings which can be achieved by the use of fuzzy control for these experiments chosen is 16 percent.

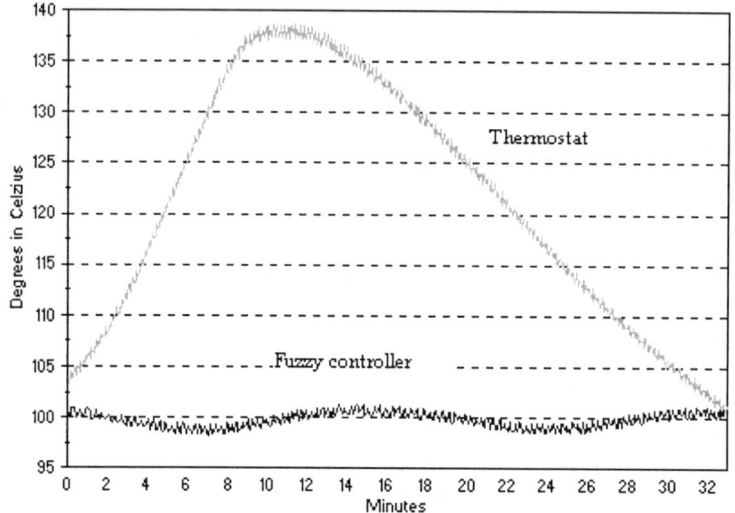

Figure 10. Comparison of sustain operation for thermostatic and fuzzy control.

The second series of experiments started from a warm oven(s). They were heated to the desired temperature and the ability of the control system to sustain and keep the oven at constant temperature was observed. The experiment duration was 32 min. The characteristics are as depicted in Figure 10:

- The energy consumption over the entire experiment duration:
  - with thermostat control: 0.1530 kWh;
  - with fuzzy control: 0.1285 kWh;
- Extremely high amplitude was observed, almost 40°C of deviation with thermostatic control.

A rough estimate of the energy savings using fuzzy control is again approximately 16%.

4   Refrigerator-Freezer Control Using Fuzzy Logic

In this experiment, Gorenje GA's two identical refrigerator-freezers were used with type identification HZOS 3361. This is a two-compartment appliance with two compressors, each used to cool its own compartment. Each compressor is controlled by its own thermostat. The first compartment is used to refrigerate food at a target temperature of +5° C. The second compartment is used for freezing at a temperature of -18°C regardless of the external heat conditions. The manufacturer of these appliances provided detailed information on standardised measuring procedures used in industry which specify long term tests with constant external thermal conditions. In the laboratory environment these requirements could not be met and that is why one device with fuzzy control and one with classical (built-in) control were used. As the two freezers were operating in the same environment comparative data was available. To ensure the quality of results both devices were tested using thermostatic control and proved to be functionally equal systems. The thermostatically controlled device *B* had an option for 7 manually set desired temperatures or operating modes for the refrigerating module. This guaranteed a temperature of +5°C depending on external temperatures. There were no such options for the freezing module.

The weakest elements of the devices analysed were the compressors. Due to economical considerations they are not continuously or gradually controllable, that is they are ON/OFF devices. Also they need to rest for 7 minutes after the working phase to reach stability of the cooling system. These conditions prevent fast response of control to exceptional events such as door opening. However this is not part of the standard tests. The temperature sensors were placed along the central vertical line in both compartments of each device for capturing ambient temperature. There were 4 in the refrigerating compartment and 2 in the freezing compartment. As an input to the fuzzy controller a reference temperature from a built-in sensor at the back of device was used.

Because the target temperature in thermostatic device B can be set by a continuous regulator it is hard to achieve the chosen temperature as opposed to the case of fuzzy control in device A which has exact given target temperature. Typically the chosen temperature is +/-2°C around the specified value of +5°C in device B. It was decided for this laboratory environment, that the fuzzy controlled device A should follow the temperature which is achieved by the thermostatically controlled device B. The target temperature for the fuzzy device A is computed as the average of the last 100 taken temperatures from thermostatic device B which represents approximately 10 minutes time window. In this manner an attempt was made to achieve a plausible and convincing relative difference in energy consumption for devices in an unstable environment.

In the first phase of the project the working environment for the software control was prepared by using a PC. The general characteristics of device functioning in the sense of internal temperature fluctuations and energy consumption was then analysed. The main purpose of the project was to decrease the consumption of electrical energy. It would improve the rating of devices with regard to 7 consumption classes of the European energy consumption classification. The European Community standards (EC) classify all home appliances into seven classes with respect to consumption of electrical energy. The class to which belong a device automatically determine the price of the device on the European market.

4.1   Refrigerating Operating Regime

Table 1 presents some characteristics of a refrigerating device. We can see that the device exhibits a relatively long cooling cycle. In case of regime 4 (medium) at 25°C external temperature this amounts to approximately 12 min of engaged and 36 min of disengaged compressor.

An attempt was made to shorten the cycle as much as possible. However, it was not possible to reduce the time below the limit of (x + 7 min) due to the compressor latency. The goal of fuzzy decision process is to calculate the time x, during which the compressor is operating. Such decisions are made at the end of each cycle i.

Table 1. Operational characteristics of refrigerating device, ambiental temper. 25°C.

Thermostat position	Medium	Maximum	Minimum
Average temp. [°C]	5.8	-0.6	7.2
Energy kWh/24h	0.4	0.8	0.3
Run time %	25.1	61.5	19.9
Running time [min]	12.2	118.7	7.3
Staying time [min]	36.3	74.1	29.5

4.2   Fuzzy Controller for the Refrigerating Device

The fuzzy controller software was implemented for the room temperature of between 23-27°C for the operating regime 4 (medium) and compared the results of the prototype device *A* directly with those obtained from the reference device *B*.

Let *e* be the average error or deviation from desired temperature in *i*-th cycle. This *e* is entered into fuzzification, where memberships of the 5 terms *Negative, SmallNegative, Zero, SmallPositive* and *Positive* are computed. An example of such fuzzification is given in Figure 2. The membership functions used here are once again triangular for the eventual simple implementation in microcontroller. This is because the triangular form requires less memory and is fast and accurate enough [5], [10].

Using the intersection of the temperature point with the triangles there are at the most two different triggering terms. High inertia of a system is reflected in delayed influence of the processed output on internal system temperature. This was the reason not to use secondary input variable Δe which is usually found in conventional fuzzy systems. The rule matrix is one-dimensional and contains at most 5 rules. In our case the matrix does not change for various desired temperatures. An example of rule matrix is given in expression (4).

```
if (Error=Negative)       then (ΔOutput=Positive)
if (Error=SmallNegative)  then (ΔOutput=SmallPositive)
if (Error=Zero)           then (ΔOutput=OK)
if (Error=SmallPositive)  then (ΔOutput=SmallNegative)
if (Error=Positive)       then (ΔOutput=Negative)        (4)
```

The descriptive values of deviation of average temperature throughout the time of past cycle are given on the left side of the rule matrix we have. On the right the descriptive output values for the correction of time interval length of ON cycle are defined. Figure 4 shows output classes in detail. The fuzzy output is then defuzzified which gives a crisp value representing the output ON length which is the time compressor operates for cycle $i+1$ with respect to output value in cycle i.

4.3 Results of Fuzzy Control of Refrigerating Device

The results of fuzzy control are given in Table 2. The temperature in fuzzy device A is kept equal to the thermostatically controlled reference device B. To improve the precision of results the freezing modules were disengaged completely.

Table 2. Comparison of results.

	Device A	Device B
Energy consumption [kwh]	4.20	4.39
Avg. internal temperature [°C]	7.29	7.31
Max. internal temperature [°C]	9.01	9.66
Min.internal temperature [°C]	1.91	1.67
External temperature [°C]	Max: 30.4, Avg: 28.7, Min: 26.0	
Test duration:	5 days, 21 hours, 25 minutes	

From the results obtained above it is deduced that fuzzy control has achieved 4.5% energy saving. This was verified by exchanging the roles of devices (fuzzy A and reference B) and practically the same results were obtained. Both cases also proved that the use of fuzzy control means smaller deviations from the chosen temperature (lower oscillation amplitude).

4.4 Results of Fuzzy Control of Freezing Device

When measuring the consumption of both devices' freezing modules the refrigerating parts were disengaged. In case of freezing operation even higher inertia was observed and the thermostatic control usually achieves the desired temperature. For this reason fuzzy device *A* was not set to follow the temperature of reference device *B*. Instead the fuzzy control of *A* was set to operate as a system which has to achieve and sustain some specified temperature.

From the results in table 3 it may be concluded that the fuzzy control for the desired temperature of –17°C with a slightly higher oscillating amplitude achieved 7.5% of energy savings compared to thermostatically controlled freezing device.

Table 3. Comparison of results.

	Device *A*	Device *B*
Desired temperature [°C]	–17.00	–17.00
Energy consumption [kwh]	6.59	7.14
Avg. internal temperature [°C]	–16.98	–17.15
Max. internal temperature [°C]	–12.42	–13.81
Min. internal temperature [°C]	–20.93	–20.26
External temperatures [°C]	min: 22.3, avg.: 24.8, max: 28.4	
Final ON/OFF cycle [sec]:	OFF: 1494, ON: 964	
Test duration:	6 days, 23 hours, 45 minutes	

4.5 Measuring Entire Appliance within Standard Test Environment

The final phase of the project was the measuring of the performance of entire appliance within standard environment. The device was first thermostatically stabilised and pre-tested. Later was used fuzzy control as well in order to eliminate the possible differences of both testing devices. The test lasted for several weeks under three different environment temperature conditions. Some surprising results were obtained:

- The energy consumption of the refrigerating module compressors was the same regardless of the control system used fuzzy control or thermostatic control
- The energy consumption of the freezing module compressors was 4.8% higher using thermostatic control compared to fuzzy control.

It can be seen that the fuzzy controlled appliance consumes 1.59 kwh/day while the thermostatically controlled appliance consumes 1.64 kwh/day. This represents 3.1% of savings in what could be seen as typical consumer environment. The surprising outcome of the results is hidden in the inter-relation of both compartments, the refrigerating and freezing modules.

5 Model and Simulation of Refrigerating-Freezing Appliance Using One Compressor

The results of the simulation of the refrigerating-freezing device controlled by a single fuzzy controller and using only one compressor are given in this section. The single compressor is used for both the refrigerating and freezing compartments. First it is necessary to design a simple simulation model of the device to show the physical and technological parameters influence in the temperature control process for refrigerating and freezing. For this purpose MATLAB, SIMULINK and the Fuzzy Logic Toolbox software tools are used. The following mathematical model of refrigerating-freezing device is developed [7],

$$T_z = \frac{k_z A_z T_{pz}}{K_S + k_z A_z T_{pz}} + \frac{k_0 A_0 T_0}{K_S + k_z A_z T_{pz}}$$

$$T_h = \frac{k_h A_h T_{ph}}{H_S + k_h A_h T_{ph}} + \frac{k_0 A_0 T_0}{H_S + k_h A_h T_{ph}} \qquad (5)$$

with model variables:

T_z: temp. of the freezing compartment [°C]
T_h: temp. of the refrigerating compartment [°C]
s: frequency parameter
T_0: device's environment temp. [°C]
T_{pz}: temp. of condensed gas in the freezing compartment [°C]
T_{ph}: temp. of condensed gas in the refrigerating compartment [°C]
k_z: heat transfer coefficient for the freezing compartment walls
A_z: reinforcing constant of heat transition through walls of freezer
A_h: reinforcing constant of heat transition through walls of refrigerator

k_h: heat transfer coefficient for the refrigerating compartment walls

K: coefficient used in calculations of heat with respect to freezing compartment temp.

H: coefficient used in calculations of heat with respect to refrigerating compartment temp.

The fuzzy controller provides two output variables: *CLK* and *RVKL*. From these (using simple temporal logic) we may calculate the length and time of engagement of the single compressor. *CLK* defines the cycle of compressor engagement and *RVKL* defines the length of relative engagement time within one cycle period. The fuzzy controller is based on 8 fuzzy rules. It is declared using a FIS-Editor which is a part of MATLAB & SIMULINK Fuzzy Logic Toolbox. The controller executes in accordance to Fuzzy Logic Toolbox software module of MATLAB.

5.1   Simulation Results

Figures 11 to 13 show the simulation results. Figure 11 was obtained for constant environment temperature $T_0 = 22°C$. The refrigerating and freezing compartment temperatures are around 6°C and 15°C, respectively. K_0 is the engagement function of the single device compressor. Simulation time is 5000 simulation seconds with 1 simulation second representing 1 minute of real device operation. The environment temperature is varied from 10 to 35°C (refer to Figure 12). As shown, the temperature disturbances are slow as well as fast. Figure 13 shows the conditions at the start, that is, up to 500 simulation seconds. It is concluded that the fuzzy controller performs better at levelling both temperatures T_h and T_z. This is especially relevant considering the need for low cost of implementation which is very important in refrigerating-freezing appliance design.

By changing the parameters of the model of the single compressor refrigerator-freezer the effects of control are improved or degraded. However, the simulation of fuzzy control to enforced conditions by using only a single compressor has shown to be effective. In this case the design of device drastically changes. In the classical two-compressor device any controller and control terms may be applied to an already existing device. As the simulation includes all necessary material data such as heat transfer coefficients, amplifying constants, etc. [7], it should not be difficult to develop a one-compressor device for the fuzzy control system. Despite the fact that this model is rather simplified, it is a good starting point for the further improvements in one-compressor refrigerating-freezing device analysis and design.

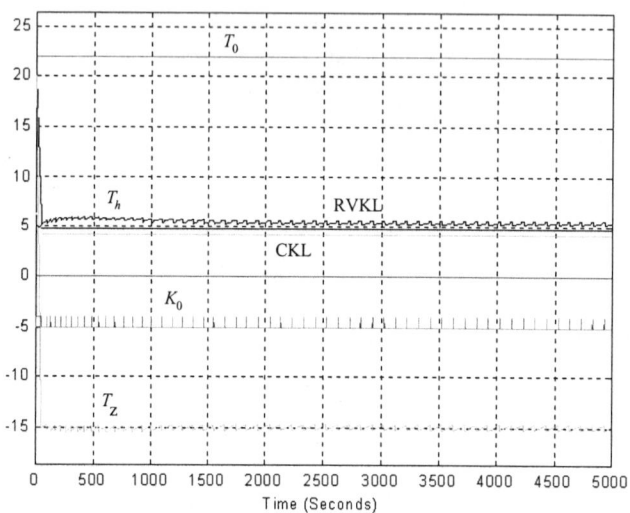

Figure 11. Fuzzy control of freezing and refrigerating, $T_0 = 22°C$.

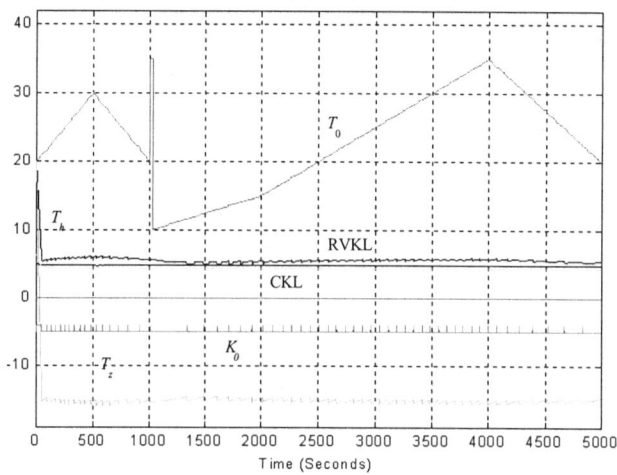

Figure 12. Fuzzy control of freezing and refrigerating, $T_0 = f(t)$.

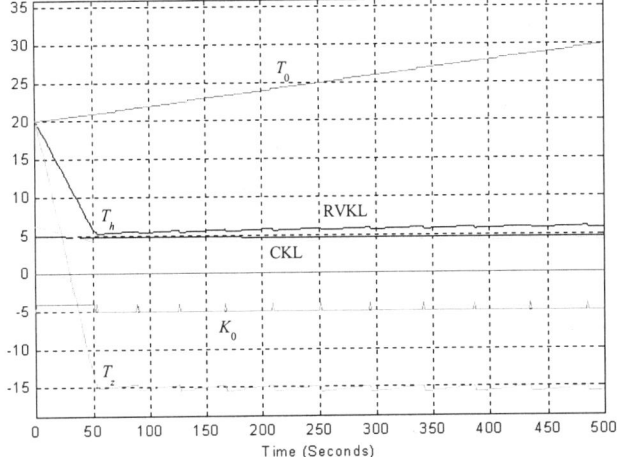

Figure 13. Beginning of fuzzy control of freezing and refrigerating ($y=T$).

6 Hardware Implementation

The laboratory prototype of a refrigerator/freezer fuzzy controller was tested without optimizing procedures in the sense of minimum electric energy consumption. Also we did not take into consideration the demands, which arise from mass production (1.5 million per a year) of appliance and the desired low cost of final product.

The controller is based on a micro-controller Intel 8052. Processor contains RAM (256 bytes), flash memory (8kb) and input/output connections. The main reason of the choice of the above mentioned processor is the minimal number of passive outer components. The LCD display is connected to micro-controller. Its function is to show the measured temperatures from cooling and freezing compartment and also to set a desired compartment temperatures, which is done by pressing four keys. Besides the display on the printed circuit there are also LED diodes. They are used as the indication of the device's status. The measurement of temperature is done with special digital sensors DS1620 built into integrated circuits. They communicate with micro-controller over serial lines. We use two outputs for compressor control which, connected with opto-couplers, represent the output. The control of freezing and cooling departments is divided into two independent processes. Figure 14 presents a basic scheme of a controller prototype.

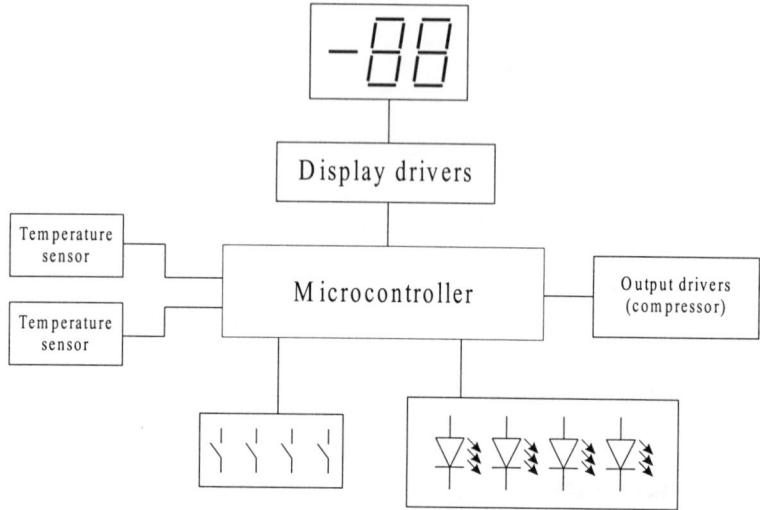

Figure 14. The general scheme of a fuzzy controller hardware prototype.

7 Decrease of Energy Consumption from National Point-of-View

In Tables 4 and 5 we show some results of the analysis of electric energy consumption in Austria. Results given in Table 4 are presented from national point of view while Table 5 provides data on household survey. In this table the electric energy used for heating is not included. It can be assumed that these results are similar for most European countries.

The thermostatically controlled freezing-refrigerating appliances produced by Gorenje GA, which were used in these experiments is ranked as third of seven European consumption classes. It is assumed that this class of appliances considering energy consumption is the most widely used in households in European countries. If all such devices became fuzzy controlled and it can be assumed that "intelligent" control causes similar savings of 3.1% as in the results, then energy would be:

- A 3.1% decrease of consumption of the 16% spent in house-keeping for freezing and refrigerating leads to a 0.496% savings in costs spent on electric energy from the family budget.
- A 0.496% saving of 27.8% of electric energy consumed by households from national point of view used for freezing and refrigerating leads to 0.138% savings in national energy consumption.

Table 4. Electric energy consumption in Austria from national point of view (source: VEW 1995).

Energy requirement	Energy Consumption
Industry [%]	29.0
Household [%]	27.8
Craft [%]	14.2
Public consumption [%]	11.0
Consumption in producing energy [%]	8.9
Accumulation [%]	3.3
Agriculture [%]	3.2
Traffic [%]	2.6

Table 5. Electric energy consumption in Austria from housekeeping point of view (source: VEW 1995).

Energy use	Energy Consumption [kWh/year]
Heating of water	2100 (= 30.6%)
Freezing, Refrigerating	1100 (= 16.0%)
Cooking	1000 (= 14.6%)
Cleaning of dishes	750 (= 10.9%)
Cleaning of clothes	450 (= 6.6%)
Drying of clothes	550 (= 8.0%)
Lighting	310 (= 4.5%)
Radio, TV	210 (= 3.1%)
Others	400 (= 5.8%)

8   Conclusion

In this chapter several concepts are described which involve fuzzy logic for controlling the household appliances produced by manufacturer Gorenje GA, Slovenia. Most of the control devices from the world leading manufacturers are mainly controlled by classical thermostatic systems. Incorporating fuzzy logic in the control process enables a decrease in electric energy consumption and also results in a higher degree of quality of control of improved functionality. The production costs of appliance does not change drastically due to the built-in digital system. Another advantage of fuzzy logic is that building uncertain knowledge applied into the control process is a relatively fast procedure when compared to classical PID controller development.

References

[1] Dubois, D. and Prade, H. (1980), *Fuzzy Set and Systems: Theory and Applications*, Academic Press, U.S.A.

[2] Mraz, M., Zimic, N., and Virant, J (1994), "Applying fuzzy logic in process of control with fuzzy history classification approach," *Proceedings of Mipro 94*, pp. 3/18-22, Opatija, Croatia.

[3] Mraz, M., Zimic, N., and Virant, J. (1995), "Time-dependent fuzzy control of a simple object," *Proceedings of VIth Intern. Fuzzy Systems Association World Congress IFSA*, SaoPaolo, Brasil, pp. 533-535.

[4] Mraz, M., Zimic, N., and Virant, J. (1997), "From conventional to fuzzy controlled electric kitchen," *Proceedings of Fifth European Congress on Intelligent Techniques and Soft Computing*, Aachen, Germany, pp. 2204-2209.

[5] Mraz, M., Zimic, N., and Virant, J. (1996), "Fuzzy arithmetic calculations and the presence of errors," *Proceedings of Mipro 96*, Opatija, Croatia, pp. 2.20-2.24.

[6] Badami, V.V. and Chbat, N.W. (1998), "Home appliances get smart," *IEEE Spectrum*, August, pp. 37-43.

[7] Stelzer, F. (1971), *Warmeübertragung und Stromung*, Verlag Karl Thiemig KG, Munchen, Germany.

[8]   Virant, J. (1992), *Using Fuzzy Logic in Modern Systems*, Didakta, Ljubljana, Slovenia. (In Slovene).

[9]   Virant, J. and Zimic, N. (1996), "Attention to time in fuzzy logic," *Fuzzy Sets and Systems*, vol.82, pp. 39-49, Elsevier Science.

[10] Virant, J. (2000), *Design Considerations of Time in Fuzzy Systems*, Kluwer Academic Publishers, U.S.A.

[11] Zimmerman, H.J. (1991), *Fuzzy Set Theory and Its Applications*, Kluwer Academic Publishers, the Netherlands.

Chapter 5

Neural Prediction in Industry: Increasing Reliability through Use of Confidence Measures and Model Combination

P.J. Edwards, G. Papadopoulos, and A.F. Murray

This chapter describes the application of neural networks to the prediction of "paper curl", an important quality metric in the papermaking industry. In particular we address the issue of reliability in neural network training and prediction. Model combination is used to compensate for the limitations of non-linear optimization algorithms used for neural network training. In addition, confidence measures are used to characterize prediction uncertainty. Reliability enhancement though model combination enables training to be automated. The provision of a confidence measure along with the prediction facilitates the user in knowing whether to trust the prediction or not.

1   Introduction

This chapter describes the application of neural network techniques to the papermaking industry, particularly for the prediction of paper "curl". Paper curl is an important and long-standing problem that is detrimental to the papermaking process at every stage of manufacture. In addition we consider neural network training, noting that it is frequently unreliable, finding poor solutions or failing. This is exacerbated by the fact that tasks taken from industrial environments are characterized by data that is noisy and sparse. Therefore for the prediction of paper curl, and by implication many other industrial tasks, accuracy and reliability are key issues. In this chapter we describe the development of a neural solution to curl prediction, addressing issues of accuracy and reliability through the use of model combination and confidence measures.

Individual models perform poorly because training data samples are always incomplete representations of the true distribution and because they are limited by the complexity available and by the efficiency of training algorithms. Models that are poor or erroneous may be used in real decision-support and may give rise to dangerous or expensive mistakes. A fully-automated training process is essential to real-time, in-house industrial use of neural technology, as most traditional industries cannot, and will not, employ an engineer with neural expertise. It is therefore vital that the influence of poor quality models be minimized. While it is not possible to overcome the limitations of a fixed data sample without the, often difficult and expensive, collection of more data, model combination allows us to improve upon the inadequacies of a training regime. In addition, as well as a prediction we provide the operator with a measure of confidence. With this information he/she knows whether the prediction should be acted upon or not. Therefore we have two levels of reliability enhancement. At the model development stage model combination is used to reduce the influence of poor or erroneous models. Then at the operational stage, confidence measures are provided to potentially increase the system reliability further.

In this chapter we describe the development of neural network models to encapsulate the non-linear processes underlying paper curl. By representing the task first as a classification and then as a regression problem, and also by calculating confidence intervals, we have made the tool (a neural network) fit the practical needs of the end-user. We have also defined a multiple-neural-network solution that is not a simple "black-box". We present parameters characterizing the current paper reel as inputs to a neural network and train the network to predict whether the resulting level of curl will be within a required specification (*i.e.* "in-specification") — a classification task. In parallel, we present these same data to another network and train it to predict the absolute level of curl, *i.e.* a regression task. Perhaps most importantly, we also put these two predictions in context by including confidence measures at every stage thus providing the machine operator with a powerful and insightful tool. The machine operator is then presented with a "red-light/green-light" indication of paper acceptability, a neural regression model on which the parameters can be altered to reduce curl if necessary and a clear indicator of the reliability of both diagnostics.

2   Paper Curl Prediction

Paper curl is an important measure of quality in papermaking, where paper exhibiting bad curl must be scrapped. Bad paper curl is the major cause of sheet-feeding problems in laser printers and photocopiers [27]. Curl is hard to control because it may only reliably be measured off-line. Thus the "control" of paper curl has traditionally been carried out using retrospective heuristic adjustments made to the paper machine. Therefore, paper curl is an important and long-standing problem that is detrimental to the papermaking process at every stage of manufacture.

Curl has received much interest in papermaking research, where the causes of paper curl have been studied [12], and where attempts have been made to control it using heuristic techniques and simple linear parametric models [14], [25], [26]. These attempts have been notoriously difficult due primarily to the difficulty in reliably measuring curl [25], [36], [51] and have proven to be unsatisfactory particularly for manufacturers like Tullis Russell who make quality paper for a wide variety of customers, with their own individual specifications and requirements. Paper curl is simply the tendency of paper to depart from the flat form and is affected by a number of complex, inter-related factors, including differing drying rates on the two sides of the paper during manufacture, relative variation in humidity or moisture within the sheet, and mechanical stresses within the fibers. Additionally there are a number of different recognized modes of curl, for example inherent curl and curl occurring as a result of changes in the surrounding environment. Traditionally paper curl has been controlled with limited success using variation in the drier temperatures or humidity levels. However, because it may only be measured off-line after an entire roll has been produced, its control is difficult and costly. Curl is therefore a significant problem to the papermaking industry. Although out-of-specification paper may be repulped and recycled in limited quantities or corrected using specially designed steam jet equipment [30], bad curl wastes plant time, engineering time and energy.

2.1   Data Collection

The data provided by Tullis Russell & Co. Ltd for the purpose of the work described here has a number of limitations, including missing records and measurement errors. Not least in this respect is the measurement of curl itself. Although curl is a simple quality measure, measuring curl is far from trivial. While it would seem naturally advantageous to measure curl continuously, to date suggested methods for achieving this aim, [25], [26], [51], have proven to be unreliable as standard techniques require the paper to be dried under controlled conditions before measurement [25]. In addition removing paper during manufacture is difficult due to the continuous nature of the task (drier timing is related to machine speed for example) and risky due to the danger of a paper break during the process. Therefore at Tullis Russell curl is measured after individual reels have been manufactured, leading to "out-of-specification" reels being scrapped (eventually to be re-cycled as "broke"), and a retrospective adjustment made to machine settings, according to unwritten heuristic rules developed by the skilled operators. The measurement is made using a sample of paper taken from the end of a reel and by cutting a cross-shape using a template. A glancing angle light source is then used to cast a shadow due to the curling paper at the centre of the cross. After a period of a few minutes has elapsed to allow the paper to relax the shadow is measured by hand, quantized to 5mm intervals. If the degree of curvature in the test sample is small, simple geometry indicates that the length of the shadow may be taken as being linearly proportional to curl. However, there may be error in the measurement due to quantization, operator error, paper misplacement, failure to allow for sufficient relaxation time, *etc*... Variability in the accuracy of measurements could lead to significant model error [48] and for the study reported in this paper, the limitations of the database are taken as an additional constraint to the modeling process.

Various parameters are measured prior to and during the manufacture of a reel of paper. Data collection is expensive in industry. At the time of this work only a limited subset of all possible parameters were available in sufficient numbers to allow investigation. One of the aspects of modeling that is typical of real-world tasks of this type is limitations in the data, whether in terms of noise, or number. In the event of there being a comprehensive automated data collection process, perhaps an optimal model may be produced. More commonly data collection is not a high priority for a production team working under tight deadlines, and it is because of this that more long-term development tasks such as the one described here must be carried out using imperfect data. Here, a model was produced using the following parameters, that were chosen for reasons of availability and where possible, through experimentation. These were:

1. Percentage softwood
2. Power used to process hardwood — refiner 1
3. Power used to process hardwood — refiner 2
4. Percentage broke
5. Caliper (paper thickness) at the paper-machine
6. Ash content
7. Moisture content at the paper machine
8. Porosity measure
9. Difference between surface moisture levels after top-coat application
10. Paper grade

These parameters were used to classify whether the current process settings and paper specification would lead to curl that was within a required specification and additionally the level of curl that would result. While at a later date it is hoped that other parameters may be collected, allowing improvement in the model, for the purpose of the experimentation described here only a limited number were available. In addition as neural network models are only semi-parametric, the assessment of the specific importance of these parameters in terms of curl is not trivial. A full assessment of the above and other parameters, as they become available, is an area for further work.

3   Neural Network Model Development

In this section the development of neural network models to provide solutions for this task are described. The data are preprocessed and models are trained.

3.1   Preprocessing

To preprocess data supplied directly from the papermaking plant a
number of operations were performed. After filtering the data to re-
move records with missing fields, the real and symbolic data within the
database were combined into a form that could be used for neural network training. In the case of symbolic data it is important that each field,
for example the grade of the paper — one-out-of-five for the purpose of
this task, is encoded to avoid creating an artificial "weighting" to any
case. Commonly a 1-out-of-N code is used. Here we note that this coding scheme contains redundant information and have used a more con-
cise scheme. Considering the encoding task geometrically a 1-out-of-N
code can be thought of as coordinates of unit points on the axes of N-
dimensional Euclidean space. In our coding scheme we use the vertices
of a hyper-tetrahedron where the coordinates of each vertex are incorporated into the code (*i.e.* the three vertices of an equilateral triangle in
our 3-D case). This transformation is shown diagrammatically in Figure 1 with the codes for three symbols encoded in 2 dimensions. More

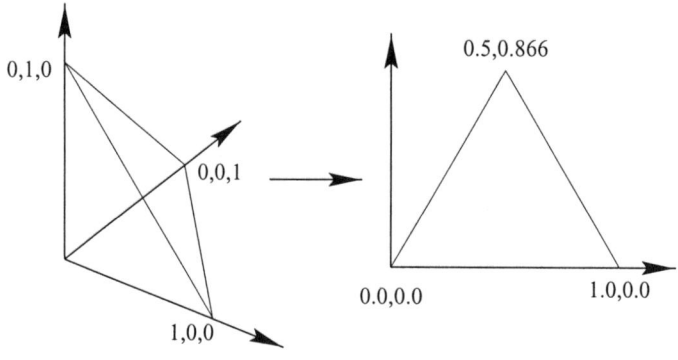

Code : 001, 010, 100 ⟶ Code : 0.0 0.0, 1.0 0.0, 0.5 0.866

Figure 1. Encoding three symbolic data fields, using a geometric transform of
the three dimensional binary codes into two dimensions.

generally, if we consider N symbolic parameters $(p_1, p_2, \ldots, p_N)$, then
a 1-out-of-N encoding would give N components $(c_1, c_2, \ldots, c_N)$ which

would be:

$$
\begin{array}{cccccc}
 & c_1 & c_2 & \cdots & c_N & \\
p_1 : & 1 & 0 & \cdots & 0 & \\
p_2 : & 0 & 1 & \cdots & 0 & \\
 & \vdots & \vdots & \vdots & \ddots & \vdots \\
p_N : & 0 & 0 & \cdots & 1 &
\end{array}
\tag{1}
$$

For the *reduced dimensionality code* we consider the following series of equations:

$$
x_1 = 1.0, \quad x_2 = x_1 cos(\pi/6), \quad \ldots, \quad x_N = x_{N-1} cos(\pi/6), \tag{2}
$$

where in terms of our geometric interpretation, component x_N defines the distance from the centroid of the $N - 1$ dimensional tetrahedron, perpendicularly to the extra vertex required to define an N dimensional tetrahedron.

Using the series of equations in (2) a general definition of the reduced dimensionality code is:

$$
\begin{array}{cccccc}
 & c_1 & c_2 & \cdots & c_{N-2} & c_{N-1} \\
p_1 : & 0.0 & 0.0 & \cdots & 0.0 & 0.0 \\
p_2 : & x_1 & 0.0 & \cdots & 0.0 & 0.0 \\
p_3 : & 0.5x_1 & x_2 & \cdots & 0.0 & 0.0 \\
 & \vdots & & & & \\
p_{N-1} : & 0.5x_1 & 0.5x_2 & \cdots & x_{N-2} & 0.0 \\
p_N : & 0.5x_1 & 0.5x_2 & \cdots & 0.5x_{N-2} & x_{N-1}
\end{array}
\tag{3}
$$

where the N symbolic parameters can be encoded using $N - 1$ components. This reduction in dimensionality is important in that it reduces the collinearity in the input vector, especially when there are significant numbers of symbolic parameters, which will greatly simplify the requirements of the training algorithm. As an aside, it should also be noted that sometimes a 2-out-of-N code can be used to encode symbolic fields, such as in the case of 15 symbolic parameters, where $N = 6$. In this case we can again reduce the dimensionality of the code by adding two of the codes defined in (3) to give a code with $N - 1$ components.

The second stage of preprocessing involved using the Karhunen-Loéve transformation [2]. This linear operation manipulates data such that they are aligned along their principal axes. The data are then normalized to have zero mean and unit variance in all dimensions. This operation removes correlation between parameters, thus greatly simplifying neural network training. Without this normalization stage the chosen neural network training scheme (described below) performed poorly. In this latest version of the curl estimator, therefore, preprocessing is data preserving.

3.2 Training

To develop models for the two tasks we use Multi-layer perceptron (MLP) neural networks with a sigmoidal output stage for the "in-specification" prediction and a linear output stage for the curl prediction task. The classification network contained ten hidden units and the regression network twelve. These architectures were chosen via extensive experimentation.

Network optimization was performed using MacKay's Bayesian Learning technique[28], [29], [46], using a simple Gaussian prior for the weights. Separate hyperparameters were used for all weights connected to each input (*i.e.* automatic relevance determination (ARD) [34][1]), hidden to output weights, hidden unit biases and output unit biases. An iterative re-estimation technique was used to update these parameters [28], [29]. Underlying the Bayesian formalism we used a standard steepest descent optimization scheme with incorporated line-search [40]. For the regression task a least squares cost function was used, while for the classification task we used cross entropy. While in theory the Bayesian framework mitigates against the need for a validation set in that the use of a Gaussian prior effectively introduces a regularization constraint into the learning [28], we found that over-fitting can still occur. In our previous work we showed that weight decay does not stop all over-fitting [6]. While there we suggest the use of a weight saliency regularizer,

---

[1] If the data could be used for training without the Karhunen-Loéve transformation then ARD could be used to assess the importance of each parameter to the prediction of curl. However, because of the correlations in the data the training process is greatly hindered if this normalization stage is not carried out. Detailed assessment of the importance of the various parameters described above, is an area for further work. The use of different regularization constraints for different network connections, however, increases the stability of the algorithm.

here for simplicity we combine weight decay with early stopping via the use of a validation data set. In [10] we consider the question of whether early stopping improves the solution enough to justify the inherent loss of training data when validating, and we find that it does.

The method of optimizing the neural network models described here was varied for the experiments carried out to consider different approaches to confidence measure estimation. The method described here is however, what we term our "baseline" approach to which all others are compared.

4   Model Combination

Model combination may be used to increase prediction accuracy and re-liability for real tasks (see [18], [19], [22], [54] for example), overcom-ing the limitations of an unreliable optimization method. The increase in performance over that of an individual model is most marked when the model output errors are uncorrelated (or "ambiguous") [24], [50]. Clearly, if the models are not different there is no benefit in consider-ing an ensemble. There are many approaches taken to achieve diversity, including: data partitioning[31], [38] or subsampling [3], the use of mul-tiple training algorithms [32], additional cost function terms to penalize collinearity in the model outputs [42] and through the use of competitive training rules and manipulation of the task [33].

Once a diverse pool of models exists, methods must be developed for op-timized model combination, or "committee formation". In the words of Merz and Pazzani [32], the task is to choose "which models are reliable and how much weight to give to each". Again, numerous methods have been proposed (see [21], [44], [54] for reviews) and many of these are founded on detailed theoretical analysis (see [23] for example). Perhaps most significantly the theory of "weak learning" [43] has been applied to make significant gains in classifier combination (see [13], [23]), al-though it is unclear about whether these approaches may be applied as successfully to non-classification applications. For regression tasks the most common form of committee formation is unweighted averaging [3]. The advantage of this approach is that no extra data is required and it is straightforward. More sophisticated approaches suggest methods for calculating non-equal model weights [15], [39], [49], although these methods suffer from the multicollinearity problem, as model outputs tend to be correlated. To overcome this, others apply principal component regression to summarize and extract relevant information from the model outputs [32], [54]. Heskes employs estimates of prediction error and a measure of output ambiguity to estimates model weightings [16]. Alternative approaches involve using nonlinear networks in a hierarchical structure to optimize the model weightings [4], [52], [53].

For our work we have developed a new technique, *cranking* (committee ranking), that builds on a number of approaches noted above. Full details of the algorithm development are described in [9].

4.1   Cranking

Cranking, or committee ranking, relies upon selecting the best from a pool of trained models using prediction error estimates. As shown in Figure 2, an initial population of M models is sorted (or ranked) and the best C models are used to form the final committee. The outputs of the C individual models are then aggregated (unweighted averaging) to form the committee output.

To calculate the prediction error estimates for the models we follow the approach of Heskes [16]. Each model is trained using a data set sampled via the bootstrap technique. If the original data set is of size P, then P patterns are sampled at random, *with replacement*, to form a training data set. Patterns that are not selected are referred to as out-of-sample and are used as a validation set to control early stopping. The validation error for model m is given by:

$$\epsilon_{\text{val}}(m) = \frac{1}{v_m} \sum_P (q_{mp} r_{mp}), \tag{4}$$

where q_{mp} indicates whether the validation data set includes pattern p ($q_{mp} = 1$) or not ($q_{mp} = 0$); r_{mp} is a conventional (squared) residual error for p and m, $r_{mp} = (y_{mp} - t_p)^2/2$, where y is an actual model output and t the associated target; and v_m is the number of out-of-sample patterns for m, used for model validation. This error measure is corrected for bias, introduced by bootstrap sampling, by using an expected validation

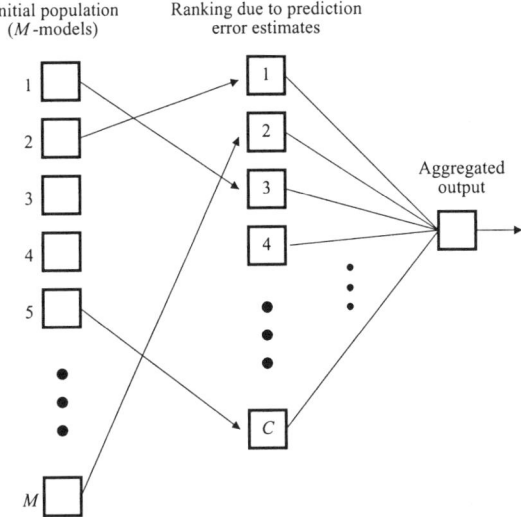

Figure 2. Diagram showing the *cranking* algorithm, where C models are se-lected from a population M, where $(M > C)$, due to a prediction error estimate. The C networks are aggregated to form the final committee.

error term. We define the average error for p as:

$$\epsilon_{\text{val},p} = \frac{1}{n_p} \sum_M \left(q_{mp} r_{mp} \right), \tag{5}$$

where M is the total number of models and n_p is the number of models for which p is in the validation set. The expected validation error is then given by:

$$\hat{\epsilon}_{\text{val}}(m) = \frac{1}{v_m} \sum_P \left(q_{mp} \epsilon_{\text{val},p} \right) \tag{6}$$

The ratio of the observed and expected validation error indicates whether the validation error for model m is relatively high or low. Therefore the prediction error (PE) estimate for model m is given by:

$$\text{PE}(m) = \frac{\epsilon_{\text{val}}(m)}{\hat{\epsilon}_{\text{val}}(m)} \frac{1}{P} \sum_P \epsilon_{\text{val},p} \tag{7}$$

Using this heuristic measure of the quality of an individual, all models in the initial population can be ranked in order of estimated quality. It is important to note that if too few models are trained a pattern p may

not be used for validation at all. In this case the mean residual error for
pattern p (*i.e.* $\epsilon_{\text{val},p}$) does not exist. Clearly, also, a reasonable number of
models is required before any accuracy in the estimate of prediction error
is achieved. However, the probability that p is never used for validation
decreases rapidly (0.368^M).

The procedure for the calculation of prediction error estimates is the same
as that for "balancing" [16]. In balancing the PE estimates are used in a
constrained optimization stage with an estimate of model ambiguity to
calculate model weightings. Here, we note that for many industrial applications training data are noisy and limited in number, leading to prediction error estimates that are inaccurate [7], [8]. For cranking therefore,
to reduce the influence of imprecise PE estimates we use them in a ranking process to select the best models and then use unweighted averaging.
Instead of optimizing the committee for model ambiguity, we rely on
bootstrap training data sampling to introduce ambiguity into the models.

5   Confidence Measures

Network reliability and usability are increased if network predictions are
supported by an associated confidence measure (CM). In this section we
present three confidence estimation techniques suitable for use in the
curl estimator. The currently used approach to confidence estimation (described below) has been chosen for reasons of simplicity and is probably
inferior to more sophisticated techniques. The methods investigated here
are *maximum likelihood* [35], the *approximate Bayesian* method [28],
[41] and the *bootstrap* technique [11], [17]. Since we are interested in a
multi-dimensional real application, we do not consider over-complicated
methods like the exact Bayesian approach [34]. This method requires
time-consuming Monte-Carlo integration over weight space and is therefore impractical for real applications. We only discuss the regression task
but the presented methods can be adapted for the classification tack as
well. Full details of this comparative study can be found in [37].

In this work, data noise is modeled as additive Gaussian i.i.d. errors with
zero mean and variance σ_ν^2. Data noise variance can be modeled as either constant over input space or as a function of the input. Data noise

is not the only source of uncertainty. The neural network introduces uncertainty due to model misspecification and inefficiencies of the training method. A network trained on a given data set forms a better representation of the data in regions of high input data density [1]. Moreover, the estimated weight values do not correspond to the global minimum of the error function (*i.e.* weight value misspecification). As explained in [1] weight value estimates are less accurate in low data density regions. We call the uncertainty due to the neural network and the training algorithm *model uncertainty*. This uncertainty means that the network output should be viewed as probabilistic. Here, the output distribution is considered Gaussian with mean equal to the most probable output and variance σ_m^2, referred to as *model uncertainty variance*.

These two sources of uncertainty (σ_ν^2 and σ_m^2) are assumed to be independent and the total prediction variance is thus the sum $\sigma_{\text{TOTAL}}^2 = \sigma_\nu^2 + \sigma_m^2$ (*e.g.* [17], [28], [47]). If a confidence estimation system is used to complement a real prediction system, it must provide estimates of both sources of uncertainty.

Our baseline approach to confidence estimation treats the noise inherent to the data as Gaussian with a constant variance σ_ν^2 and uses MacKay's approximate Bayesian method [28] and the *delta* method [47] to obtain estimates of the data noise variance σ_ν^2 and model uncertainty variance σ_m^2 respectively. There exist two popular variations of the delta method depending on whether the exact Hessian matrix of total error derivatives with respect to the weights **H** is used or the outer product approximation to the Hessian $\tilde{\mathbf{H}}$ (herein referred to as the "approximate Hessian") [5][47]. We refer to the delta method using the exact Hessian as the *exact delta* method and to the delta method using the approximate Hessian as the *approximate delta* method[2].

This approach can be extended for the case of input-dependent data noise variance using ML [35] or the Bayesian approach [41] to train the network. A new set of hidden units is used to compute $\sigma_\nu^2(\mathbf{x})$, the network estimate for input-dependent data noise variance. This leads to the "augmented" network architecture of Figure 3. The augmented network is

---

[2]The confidence intervals shown in Figures 4 and 5 use the exact Delta method since the exact Hessian is already calculated in the Bayesian training algorithm.

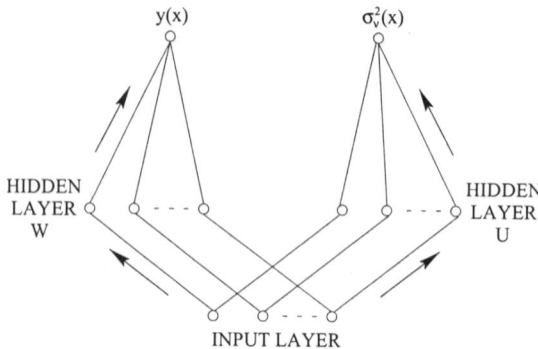

Figure 3. The augmented network architecture for obtaining an input dependent data noise variance estimate.

trained by optimizing the error with respect to weights $\mathbf{w}$ (the regression subnetwork weights) and $\mathbf{u}$ (the variance subnetwork weights). Compared to the Bayesian approach, ML is faster to train by a factor of about 30^3. However, unlike ML, the Bayesian estimate of σ_ν^2 is unbiased [2], [28]. Thus, when ML is used, data noise variance is likely to be under-estimated, particularly in regions of low data density where overfitting is more probable [41].

Finally, we consider the *bootstrap* technique [11]. The members of the bootstrap committee are trained on resampled versions of the training set as described in Section 4.1. The bootstrap regression estimate is obtained by averaging the predictions of the individual networks and a measure of model uncertainty is computed by calculating the variance of the individual network predictions around the average prediction. The predictions of each network on the corresponding out-of-sample set can be used to obtain an unbiased estimate of the overall model's residuals on the training data [17]. First, an unbiased regression estimate is computed by averaging the predictions of the individual networks on the out-of-sample sets. This unbiased regression estimate is then used to compute the model's residuals r^2. The set $(\mathbf{x}, r^2(\mathbf{x}))$ is used as the training set for a supplementary network that estimates data noise variance as a function of $\mathbf{x}$.

It is worth noting that in the bootstrap approach we implicitly adopt the

---

[3]Using the curl data and a Sun Ultra 10 workstation the average times for training one network was 1 and 30 hours for the ML and Bayesian methods respectively.

constant data noise variance assumption while fitting the regression networks and subsequently obtain an input-dependent data noise variance estimate. In contrast, both the ML and Bayesian methods use the augmented network architecture fitting the regression and noise variance functions at the same time. In the latter case, the data noise variance estimate enters the weight update equations for the regression weights, so that the regression subnetwork emphasizes regions of low data noise variance at the expense of high data noise variance regions [35]. Therefore, it can be expected that the regression function obtained using the augmented network methods (*i.e.* ML and Bayesian) will differ qualitatively from the one obtained using conventional architecture methods.

It is often useful to include a *novelty* detection scheme that highlights test inputs originating from regions of the input space that are underrepresented by the training set. This is essential when data preprocessing is lossy (*e.g.* when dimensionality reduction is used to ignore dimensions that carry little relevant information) [1]. Novelty detection must use the unprocessed input data to avoid misinterpreting a novel pattern as similar to the training data, due to information loss. Novelty detection is usually performed by thresholding the *probability density function* (PDF) of the training data [1]. Many standard methods for estimating the PDF are available (*e.g. Parzen windows*[45]).

The inverse PDF can be used as a measure of model uncertainty [1]. However, the inverse PDF is independent of the model at hand and does not scale well. Therefore, even if novelty detection precedes and "gates" the regression estimation, it is useful to obtain a measure of model variance using the delta or bootstrap methods. In theory, the model uncertainty estimate obtained by the delta or bootstrap methods should increase rapidly for regions unrepresented in the training set. However, in our experience [10], this is not always the case. Therefore, it is preferable to include a novelty detection stage so that novel inputs are omitted from the main estimation stage, even if data preprocessing is information preserving.

As far as curl estimation is concerned, data preprocessing is information preserving since it only involves the Karhunen-Loéve transformation However, the operator of the curl estimator, used for decision-support, may vary an input parameter manually to investigate its effect on curl. Therefore, the system may well be presented with inputs that differ significantly from the training data. Such inputs may not be identified by the delta or bootstrap estimates of model uncertainty but can be rejected by thresholding the PDF. The integration of a novelty detection stage into the curl predictors is an area of ongoing work.

6   Results

This section describes the results obtained from the neural network models trained to classify the current state of the manufacturing process as leading to paper "out-of-specification" in terms of curl and to predict the absolute level of that curl. In addition results are shown to indicate how accuracy and reliability enhancement are achieved through model combination and confidence measure estimation.

6.1   In-Specification/Out-of-Specification Classifier

In this experiment 40 networks were trained to classify the current paper characteristics as leading to paper "in-specification" or "out-of-specification" and combined using an unweighted average. For the purpose of this experiment a level of curl less than 20 was used as the limit of acceptable curl. This level was chosen as typical. Different grades of paper have different acceptable levels. We note that a multi-classifier system may be preferable in practice, giving an indication of the probability of curl being greater than 10, 20, 30, *etc*... The results of these experiments are shown graphically in Figure 4 and as a confusion matrix in Table 1. The overall classification error rate is 18.92% and this compares with an error rate of 24.32% achieved by a linear classifier trained to solve the same task (see confusion matrix in Table 2). The use of a non-linear neural network is thus beneficial for this task. The test cases are sorted for increasing levels of curl and Figure 4 shows that the majority of the errors (where the classification boundary is set at a probability of 0.5) occur at the centre of the graph, i.e. for cases where the measured curl is 20 or near 20. The graph also shows 68% confidence intervals calculated using our "baseline" approach.

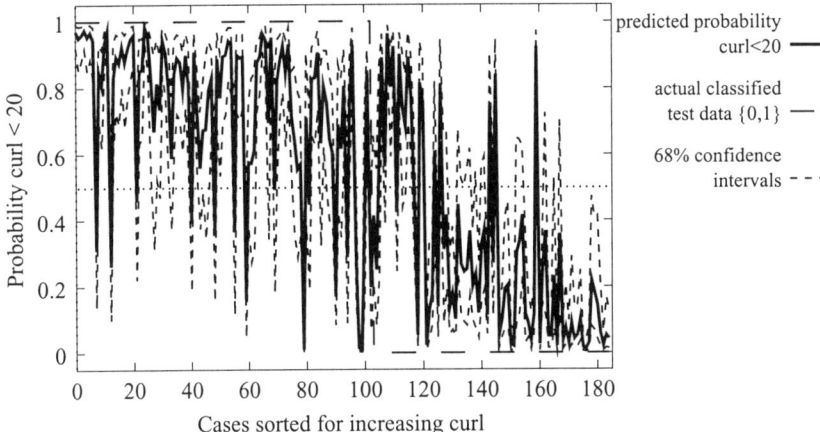

Figure 4. Graph showing variation in the predicted probability of curl < 20 for all cases in the test data set. The actual test data are sorted for increasing levels of curl and classified {0,1}, where "1" indicates "in-specification" (*i.e.* curl < 20).

Table 1. Confusion matrix of the test results shown in Figure 4 where the classification boundary is taken to be a probability of 0.5.

Test Class	Assignment Category	
	Curl < 20	Curl >= 20
Curl < 20	83.50%	16.50%
Curl >= 20	21.95%	78.05%

6.2   Curl Prediction

In the previous section the classification results were presented. Figure 5 shows the results for networks trained to predict the absolute level of curl and tested on the same data set as above. Clearly the model has encapsulated the trend underlying these measured data, although with some

Table 2. Confusion matrix of the test results of a linear classifier trained to solve the "out-of-specification" classification task.

Test Class	Assignment Category	
	Curl < 20	Curl >= 20
Curl < 20	84.47%	15.53%
Curl >= 20	35.37%	64.63%

imprecision. The results depicted in Figure 5 have a mean squared error (MSE) of 127.02, while a linear regression model, trained on the same data, achieves an MSE of 155.43. Again the non-linearity available in the neural network approach is beneficial for this task. The prediction of extremely high levels of measured curl is poor. This is perhaps due to this part of the model being under-represented in the database (note that the data are densest for low curl), or that the process is different for levels of curl greater than 40. In practice, however, some accuracy in the critical 10 < curl < 30 region is most important and the predictor achieves adequate accuracy in that critical regime. The prediction of curl is therefore possible to limited but usable accuracy using the database provided and a neural network model. In addition to the curl prediction, Figure 5 also shows 68% confidence intervals for those predictions again calculated using our "baseline" approach. These will be discussed in depth in following sections.

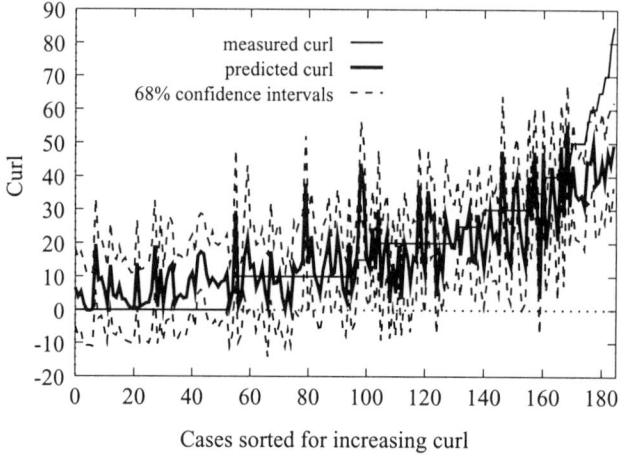

Figure 5. Graph showing the variation in the absolute value of curl as predicted by the committee of networks and additionally as measured at Tullis Russell and contained in the test data set.

6.3   Model Combination

To assess how the prediction accuracy and reliability may be improved for this task 800 neural network models were trained on bootstrap sub-samples of the curl dataset. In all of the experiments described in this

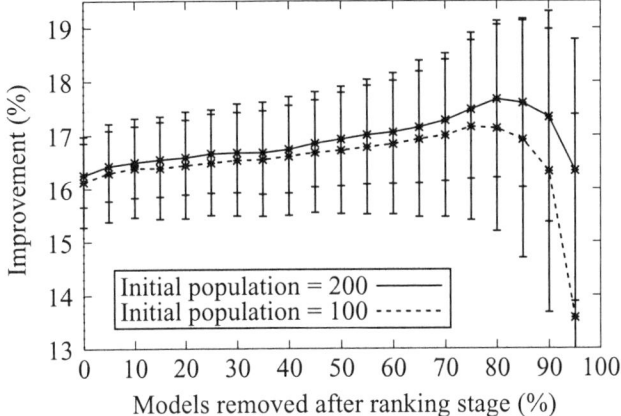

Figure 6. Graphs showing the effect on performance improvement of removing the models with highest prediction error from the committee. The error bars indicate plus/minus one standard deviation.

section performance improvement is expressed *as a percentage improvement over the average performance of the 800 individual models.*

Cranking ranks the committee models in order of their prediction error estimates. Figure 6 shows how removing the poorest models from randomly selected committees (M=100 and 200) affects performance. The results are averaged over 400 random committees. Clearly, by using a committee of networks significant performance enhancement can be obtained over the average model. In addition, as more of the poorer models are removed from the committee the percentage improvement increases, until eventually the committee becomes too small.

The optimal number of models in the initial population was then considered. For bagging, our experiments (not reported here due to lack of space) show that performance improvement reaches an asymptote at approximately M=100. Figure 7 shows the effect of the size of the initial population on the performance of a final committee of fixed size C. Each point is an average over 400 random committees. The results show that if there are sufficient models in the final committee (*i.e.* 20<=C<=30) then the initial population should be as large as is feasible. For the paper-curl prediction the trend shows that as more models are available to select from performance improves. In general we can say that larger initial

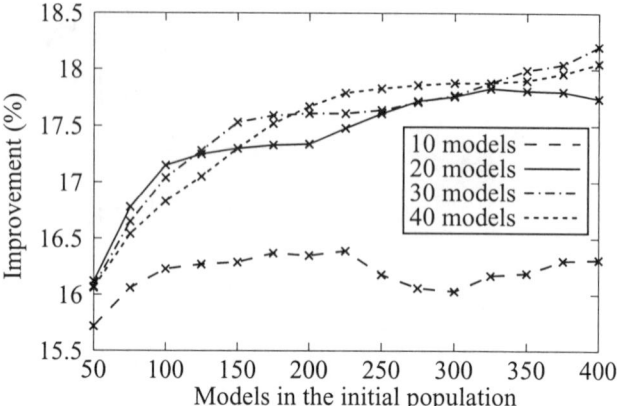

Figure 7. The effect of increasing the initial population of models on performance improvement. Committees with a final size of 10, 20 and 30 are shown. Error bars have not been included for clarity.

populations lead to better final committees. This is possibly due to the fact that there are more likely to be better models in the population to choose from. In addition it may be due to the fact that the accuracy of the prediction error estimates are related to average pattern validation error over the models which will potentially improve as the number of models increases.

We have also compared the performance of cranking with bagging [3] and balancing [16], [24] as these are related methods. For the paper-curl prediction task results of comparison experiments are shown in Table 3. Clearly, cranking achieves performance enhancement over bagging even under the difficult conditions of noisy and sparse data. The application of balancing to this task is more difficult as in a real situation additional data are not available for the estimation of ambiguity as data collection is difficult and expensive. However for the purposes of comparison here, the unlabeled test data (that in a real situation would not be available) are used to estimate ambiguity. In addition in a second experiment we also reuse the training data to estimate ambiguity (see Table 3). The results show that if unlabeled test data are available then balancing and cranking have approximately equivalent performance. If extra test data are not available and the training data must be used to estimate ambiguity then this has a detrimental effect on the performance of balancing. In addition,

Table 3. Comparative performance enhancement for paper-curl prediction where each method made use of a pool of 200 models.

Method	Final committee size	Improvement (%)	σ
Bagging	200	16.26	0.60
Cranking	20	17.34	1.92
Cranking	30	17.61	1.58
Balancing (test data)	17.29	16.90	2.62
Balancing (training data)	15.96	16.56	2.53

comparing the standard deviation of the results shows that balancing is less stable than cranking for this task. Therefore if an accurate measure of ambiguity is unavailable, the prediction error estimates are better used to rank the models, rather than in an optimization process where undue significance may be given to their precise values.

6.4   Confidence Measures

To assess the suitability of the alternative methods to the curl estimation task 100 networks were trained for each method. The results presented here are averaged over 100 committees chosen at random from the pool. The committee size was set to 20 networks for all methods.

Since Gaussian distributions are used, the width of the 68% confidence intervals[4] (CIs) is $y(\mathbf{x}) \pm \sigma_{\text{TOTAL}}$, 95% confidence intervals have width of $y(\mathbf{x}) \pm 2\sigma_{\text{TOTAL}}$ and so on. In theory, a $\alpha\%$ confidence interval should contain the true value of the parameter (*i.e.* in this case, the target) $\alpha\%$ of the time (*e.g.* [5]). We call this probability the *nominal coverage probability* of the interval. We can exploit this property of the CIs to evaluate the performance of a given confidence measure [5], [20]. By counting the number of targets that lie within the interval for a given test set we can get an estimate of the CI's *coverage probability* (CP) [5], [20]. Ideally, the estimated CP must be close to the nominal CP of the intervals *i.e.* 68% for intervals of $\pm\sigma_{\text{TOTAL}}$ width, 95% for intervals of $\pm 2\sigma_{\text{TOTAL}}$

---

[4] Actually, these are *prediction* intervals because they refer to target prediction, *i.e.* they are computed using the total standard deviation $\sigma_{\text{TOTAL}} = \sqrt{\sigma_\nu^2 + \sigma_m^2}$. The term confidence intervals refers to intervals for predicting the real regression. These are computed using the model uncertainty variance only.

Table 4. Comparative performance enhancement for the various CM methods.

Method	Number of models trained	Final committee size	Improvement (%)	σ
Baseline	100	20	-	2.32
Max. Lik.	100	20	3.74	1.83
Bayesian	100	20	4.32	1.37
Bootstrap	100	20	5.26	3.13

width and so on. Therefore, the optimum technique is the one that yields the estimated CP values with the smallest deviation from the nominal value.

We also compare the alternative methods in terms of generalization performance. As already mentioned, the augmented network methods will learn an inherently different representation of the data compared to the baseline method. In Table 4 the test MSE for the alternative methods is shown. The MSE is expressed as percentage of improvement over the baseline MSE. Evidently, the augmented network methods manage to improve the MSE over the baseline. This result implies that by assuming input-dependent data noise variance a better regression estimate can be obtained. The superiority of the bootstrap method over the augmented network methods should be viewed in terms of the bootstrap method's superior training regime rather than as an indication against the input-dependent noise variance assumption.

In Figure 8 the estimated coverage probability is shown for two different CI widths, one and two standard deviations. The CP is expressed as the difference from the nominal value (68% and 95% respectively). Clearly, the augmented network methods are superior exhibiting the smallest deviation from the nominal value in almost all cases. Only for the case of 68% intervals using the exact delta method (Figure 8(a)) does the baseline outperform an augmented network approach (namely the Bayesian approach).

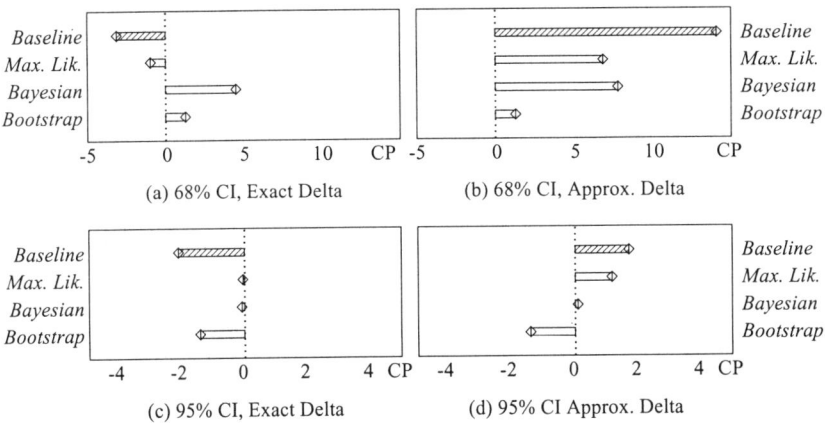

Figure 8. The estimated coverage probability for the curl task. The CP is expressed as the difference from the nominal value. Ideally, this difference should be zero. (a) 68% CI using the exact delta method, (b) 68% CI using the approximate delta method, (c) 95% CI using the exact delta method, and (d) 95% CI using the approximate delta method.

7   Discussion

This chapter has presented the application of neural network techniques to a complex non-linear task taken from the papermaking industry. The results show that paper curl can be predicted from parameters defining the current characteristics of a reel of paper and the plant machinery using neural network techniques. In addition we address issues of training unreliability and prediction uncertainty. We investigate how a reliable retraining strategy may be implemented through the use of model combination and compare techniques for confidence measure estimation.

Model combination and selection have been shown to increase the accuracy of the curl predictor. For the two predictors described here, and for neural networks in general, training is difficult often finding poor solutions or failing. Cranking may be used to discard poorer models and to combine the remainder into an accurate committee. In this application training must be automated and Cranking provides a means of achieving reliability in this context. In addition the use of a fixed small number of models in the final committee means that prediction speed is high.

In terms of confidence estimation the results demonstrate that the augmented network methods are more appropriate for the curl estimation task. These methods exhibit a lower MSE and better confidence estimation performance than the baseline method that assumes constant data noise variance. Although, clearly more work needs to be done the results presented here along with previous results using artificial data [37] strongly suggest the use of an input-dependent data noise variance approach for curl estimation. The bootstrap approach yields the best MSE and its confidence estimation performance is at least as good as the augmented network methods'. Therefore, it may be preferable to use the bootstrap technique in the curl estimator.

The results of this work are important because, firstly, predictions are accurate enough to be useful in the specific case of controlling curl at Tullis Russell. Secondly, tests using the models have indicated that current traditional corrective practice is effective and appropriate. Thirdly, the results have influenced data collection practice. Finally these results are important because they show how a combination of neural network techniques can be applied to complex and relevant practical problems in realistic working environments.

Acknowledgements

This work is supported by EPSRC grant number GR/L30930 and by Tullis Russell & Co. Ltd.

References

[1] Bishop, C.M. (1994), "Novelty detection and neural network validation," *IEE Proceedings in Vision, Image and Signal Processing*, vol. 141, pp. 217–222.

[2] Bishop, C.M. (1995), *Neural Networks for Pattern Recognition*, Oxford University Press.

[3] Breiman, L. (1996), "Bagging predictors," *Machine Learning*, vol. 26, no. 2, pp. 123–140.

[4] Breiman, L. (1999), "Using adaptive bagging to debias regressions," Technical Report 547, University of California at Berkeley. (Available at ftp://ftp.stat.berkeley.edu/pub/users/breiman).

[5] Donaldson, J.R. and Schnabel, R.B. (1987), "Computational experience with confidence regions and confidence intervals for nonlinear least squares," *Technometrics*, vol. 29, no. 1, pp. 67–82.

[6] Edwards, P.J. and Murray, A.F. (1998), "Toward optimally distributed computation," *Neural Computation*, vol. 10, pp. 997–1015.

[7] Edwards, P.J. and Murray, A.F. (2000), "Committee formation for reliable and accurate neural prediction in industry," *Proceedings of the European Symposium on Artificial Neural Networks*, pp. 141–146, Bruges, Belgium.

[8] Edwards, P.J. and Murray, A.F. (2000), "A study of early stopping and model selection applied to the papermaking industry," *International Journal of Neural Systems*. To appear.

[9] Edwards, P.J., Murray, A.F., and Papadopoulos, G. (1999), "Cranking : neural network committee formation in the context of high predictive loss," *IEEE Transactions on Pattern Analysis and Machine Intelligence*. Submitted.

[10] Edwards, P.J., Murray, A.F., Papadopoulos, G., Wallace, A.R., Barnard, J., and Smith, G. (1999), "The application of neural networks to the papermaking industry," *IEEE Transactions on neural networks*, vol. 10, no. 6, pp. 1456–1464.

[11] Efron, B. and Tibshirani, R.J. (1993), *An introduction to the bootstrap*, Chapman & Hall, New York.

[12] Eriksson, L-E., Cavlin, S., Fellers, C., and L.Carlsson (1987), "Curl and twist of paperboard — theory and measurement," *Nordic Pulp and Paper Research Journal*, vol. 2, no. 2, pp. 66–70.

[13] Freund, Y. (1995), "Boosting a weak learning algorithm by majority," *Inform. and Comput.*, vol. 21, pp. 256–285.

[14] Goldner, P. (1964), "Drying systems for curl control," *TAPPI Journal*, vol. 47, no. 7, pp. 168A–170A.

[15] Hashem, S. (1994), *Optimal Linear Combinations of neural networks*, Ph.D. thesis, Purdue University.

[16] Heskes, T. (1997), "Balancing between bumping and bagging," *Proc. Neural Information Processing Systems (NIPS) Conference*, pp. 466–472, Cambridge, Massachusetts. MIT Press.

[17] Heskes, T. (1997), "Practical confidence and prediction intervals," *Proc. Neural Information Processing Systems (NIPS) Conference*, pp. 176–182, Cambridge, Massachusetts. MIT Press.

[18] Ho, T.K., Hull, J.J., and S.N.Srihari (1992), "A computational model for recognition of multifont words images," *Machine vision and applications*, vol. 5, pp. 157–168.

[19] Huang, Y.S. and Suen, C.Y. (1995), "A method of combining multiple experts for the recognition of unconstrained handwritten numerals," *IEEE Trans. Pattern Analysis and Machine Intelligence*, vol. 17, no. 1, pp. 90–94.

[20] Hwang, J.T.G. and Ding, A.A. (1995), "Prediction intervals for artificial neural networks," *Journal of the American Statistical Association*, vol. 92, no. 438, pp. 748–757.

[21] Kittler, J., Hatef, M., Duin, R.P.W., and Matas, J. (1998), "On combining classifiers," *IEEE Trans. Pattern Analysis and Machine Intelligence*, vol. 20, no. 3, pp. 226–239.

[22] Kittler, J., Matas, J., Jonsson, K., and Sánchez, M.U. Ramos (1997), "Combining evidence in personal identity verification systems," *Pattern Recognition Letters*, vol. 18, no. 9, pp. 845–852.

[23] Kleinberg, E.M. (1996), "An overtraining-resistant stochastic modelling method for pattern recognition," *The Annals of Statistics*, vol. 24, no. 6, pp. 2319–2349.

[24] Krogh, A. and Vedelsby, J. (1995), "Neural network ensembles, cross validation and active learning," in Tesauro, G., Touretzky, D.S., and Leen, T.K. (Eds.), *Proc. Neural Information Processing Systems (NIPS) Conference*, pp. 231–238. MIT Press.

[25] Langevin, E.T. and Giguere, W. (1994), "Online curl measurement and control," *TAPPI Journal*, vol. 77, no. 8, pp. 105–110.

[26] Lebel, R. and Stadal, M. (1982), "Control of fine paper curl in papermaking," *Pulp and Paper-Canada*, vol. 83, no. 6, pp. 112–117.

[27] Lyne, M.B. (1988), "Paper requirements for non-impact," *International Printing and Graphic Arts Conference Proceedings*, pp. 89–97. TAPPI Press.

[28] MacKay, D.J.C. (1992), "Bayesian framework for backpropagation networks," *Neural Computation*, vol. 4, no. 3, pp. 448–472.

[29] MacKay, D.J.C. (1992), "Evidence framework applied to classification networks," *Neural Computation*, vol. 4, no. 5, pp. 720–736.

[30] Mann, K.C. and Huff, L.A. (1992), "Curl control with a Coanda actuator system," *TAPPI Journal*, vol. 75, no. 5, pp. 133–137.

[31] Meir, R. (1995), "Bias, variance and the combination of estimators," in Tesauro, G., Tourestzky, D., and Leen, T. (Eds.), *Proc. Neural Information Processing Systems (NIPS) Conference 7*, pp. 295–302. MIT Press.

[32] Merz, C.J. (1998), "Combining classifiers using correspondence analysis," in Jordan, M., Kearns, M.J., and Solla, S.A. (Eds.), *Proc. Neural Information Processing Systems (NIPS) Conference 10*, pp. 591–597. MIT.

[33] Munro, P.W. and Parmanto, B. (1997), "Competition among networks improves committee performance," *Proc. Neural Information Processing Systems (NIPS) Conference 9*, pp. 592–598, Cambridge, Massachusetts. MIT Press.

[34] Neal, R.M. (1996), *Bayesian learning for neural networks*, Springer–Verlag, New York.

[35] Nix, D.A. and Weigend, A.S. (1995), "Learning local error bars for nonlinear regression," in Tesauro, G., Tourestzky, D., and Leen, T. (Eds.), *Proc. Neural Information Processing Systems (NIPS) Conference*, pp. 489–496. MIT Press.

[36] Nordstrom, A., Carlsson, L.A., and Hagglund, J.E. (1997), "Measuring curl of thin papers," *TAPPI Journal*, vol. 80, no. 1, pp. 238–244.

[37] Papadopoulos, G., Edwards, P.J., and Murray, A.F. (2000), "Confidence estimation methods for neural networks: a practical comparison," *Proceedings of the European Symposium on Artificial Neural Networks*, pp. 75–80, Bruges, Belgium.

[38] Parmanto, B., Munro, P.W., and Doyle, H.R. (1996), "Improving committee diagnosis with resampling techniques," in Tourestzky, D.S., Mozer, M.C., and Hasselmo, M.E. (Eds.), *Proc. Neural Information Processing Systems (NIPS) Conference 8*, pp. 882–888. MIT Press.

[39] Perrone, M.P. and Cooper, L.N. (1993), *When networks disagree: ensemble methods for hybrid neural networks*, pp. 126–142. Chapman & Hall, London, UK.

[40] Press, W.H., Teukolsky, S.A., Vetterling, W.T., and Flannery, B.P (1992), *Numerical Recipes in C*, Cambridge University Press.

[41] Qazaz, C. (1996), *Bayesian Error Bars for Regression*, PhD thesis, Aston University.

[42] Rosen, B.E. (1996), "Ensemble learning using decorrelated neural networks," *Connection Science*, vol. 8, no. 3, pp. 373–383.

[43] Schapire, R.E. (1990), "The strength of weak learnability," *Machine learning*, vol. 5, pp. 197–227.

[44] Sharkey, A.J.C. (1996), "On combining artificial neural nets," *Connection Science*, vol. 8, no. 3, pp. 299–313.

[45] Silverman, B.W. (1986), *Density Estimation for Statistics and Data Analysis*, Chapman & Hall, London.

[46] Thodberg, H.H. (1996), "A review of Bayesian neural networks with an application to near infrared spectroscopy," *IEEE Trans. Neural Networks*, vol. 7, no. 1, pp. 56–72.

[47] Tibshirani, R.J. (1996), "A comparison of some error estimates for neural network models," *Neural Computation*, vol. 8, no. 1, pp. 152–163.

[48] Tresp, V., Ahmad, S., and Neuneier, R. (1994), "Training neural networks with deficient data," *Proc. Neural Information Processing Systems (NIPS) Conference 6*, pp. 128–135. Morgan Kaufmann.

[49] Tresp, V. and Taniguchi, M. (1995), "Combining estimators using non-constant weighting functions," *Proc. Neural Information Processing Systems (NIPS) Conference 7*, pp. 419–426, Cambridge, Massachusetts. MIT Press.

[50] Ueda, N. and Nakano, R. (1996), "Generalisation error of ensemble estimators," *Proc. International Conference on Neural Networks*, vol. 1, pp. 90–95, Washington D.C.

[51] Viitaharju, P., Kajanto, I., and Niskanen, K. (1997), "Heavy papers and curl measurement," *Paper and Timber*, vol. 79, no. 2, pp. 115–120.

[52] Wolpert, D. and Macready, W. (1996), "Combining stacking with bagging to improve a learning algorithm," Technical Report SF1-TR-96-03-123, Santa Fe Institute.

[53] Wolpert, D.H. (1992), "Stacked generalisation," *Neural Networks*, vol. 5, no. 2, pp. 241–259.

[54] Zhang, J. (1999), "Inferential estimation of polymer quality using bootstrap aggregated networks," *Neural Networks*, vol. 12, pp. 927–938.

Chapter 6

Handling the Back Calculation Problem in Aerial Spray Models Using a Genetic Algorithm

W.D. Potter, W. Bi, D. Twardus, H. Thistle,
M.J. Twery, J. Ghent, and M. Teske

The United States Department of Agriculture – Forest Service (USDA-FS) has been involved in the development of computer models to simulate deposition from aerial pesticide spraying since the early 1970s. Originally, this work was driven by the need to improve the percentage of aerially sprayed material that actually deposited on a target area. The amount of on-target deposition is a primary factor in determining the level of pest control achieved. A second focus of this modeling work that has become the objective in much of the recent work is to use modeling to determine the amount of sprayed material that does not land on the target area and is defined as "drift". It is assumed that drift causes unintended environmental consequences and is a form of environmental pollution.

Current model operation requires that the user input a mechanical system including aircraft type and spray system, a volatile fraction, release height and forward speed, meteorological data and other parameter values. The model will then output the deposition prediction across an area including the spray block. It would be more useful from an operational standpoint to input a desired measure of deposition, constrain the model with things that are fixed (an applicator may be limited to one aircraft type for instance) and then let the model optimize on, for example, nozzle type or release height. This desired facility is called back calculation and is the focus of our research efforts using the genetic algorithm. This chapter covers our Spray Advisor Genetic Algorithm system for optimizing aerial spray parameters.

1 Introduction

The Spray Advisor Genetic Algorithm (SAGA) project is a cooperative effort between the USDA-FS and the Artificial Intelligence Center at the University of Georgia. This chapter combines, enhances, and extends previous works describing the various stages of development of the SAGA project [1], [23], [24]. The goal of this project is to provide Forest Service managers with a tool to predict forest aerial spray performance and dynamically optimize the spray parameters to save substantial effort, time, and cost in practical spray tasks.

It has always been a difficult problem to identify ideal spray parameters to achieve a desired deposition, reduce spray material evaporation or drift, and save time and money devoted to the spray process [25], [31]. The difficult part of the problem is that there are many spray parameters in spray practice and each of them has many possible values. The total number of possible spray parameter value settings equates to a huge search space that is not searchable using traditional techniques [28]. For example, 20 parameters each with 20 possible values will lead to a total of 20^{20} possible settings. It is indeed beyond current computing technological capacity to find the best solution using an approach such as exhaustive search. In this project we use the Genetic Algorithm (GA) to reduce this workload to a large extent by searching for optimal or near-optimal solutions based on Darwin's theory of evolution and the survival of the fittest. We report on the efforts and achievements of the SAGA project in this chapter.

In the following sections, we present detailed background discussions of aerial spray practice and simulation, and briefly cover GA fundamentals and applications. This is followed by a review of our development work on SAGA, each section focusing on a version of SAGA at a different development stage, namely Fortran-SAGA, VB-SAGA1.0, and VB-SAGA2.0. The experimental results and discussions from these versions are also included. Finally, we conclude with a summary of this project and some future expectations.

2   Early Spray Models

The earliest models used to simulate aerial spraying of pesticides were based on dispersion models developed by the United States Army and generally of the type referred to as Gaussian models [2], [31]. These models were not sufficient to describe aerial spraying because the initial dispersion of the spray (and the entire dispersion field in many applications where the aircraft is near the ground or near the foliage surface) is controlled by vortices created by the aircraft. The description of the near-field wake dynamics critical to accurately simulate the dispersion of the spray released into this wake is not easily incorporated into a Gaussian modeling framework. This then drove the development of a simplified wake model that was computationally manageable given the constraints of the day but accurately simulated observed deposition. The result was a simplified near-wake algorithm that describes the dominant wing-tip vortices on fixed wing aircraft and corresponding vortices shed as 'rotor-wash' from helicopters [28].

This algorithm was developed in the late 1980s and has been incorporated into a large family of spray deposition models as well as other applications. It has been extensively validated and is elegant in its simplicity considering that it fulfills the requirement of being reasonably physically accurate. The algorithm is currently being used by the United States Environmental Protection Agency (EPA) and chemical manufacturers to evaluate drift from pesticide applications. It is also being used by applicators and land managers as a component of larger decision support systems to design and understand pesticide application projects.

2.1   FSCBG

The first model the FS used was known as the FS Cramer-Barry-Grim (FSCBG) spray dispersion model. This model is based on Gaussian modeling techniques developed by the United States Army. Gaussian models are so named because they assume the downwind distribution of a dispersing gas from a point source is normal or Gaussian when considering the concentration profile in the crosswind and vertical direction. The centerline is directly downwind and concentration decreases laterally. This type of model has become the workhorse regulatory air pollution model. It has evolved a "Christmas tree" architecture as other modules, such as handling dispersion near non-uniform terrain and dispersion near buildings, were hung on it to compensate for the a priori assumption of a Gaussian concentration distribution in the lateral and vertical directions [25].

FSCBG could not describe the intricacies of the near-wake flow field that is critical to accurately describing pesticide deposition from aerial spraying. Therefore, AGDISP was developed. The original AGDISP stood alone and did not have many of the features of FSCBG such as extensive aircraft libraries, nozzle libraries (which associated spray nozzles with droplet size distributions), a canopy penetration algorithm and various other features. It was therefore decided to merge FSCBG and AGDISP. The resulting model retained the FSCBG name but now included the near-field algorithm [26].

The combined model interfaced the near-wake algorithm with the Gaussian algorithm to model dispersion from the near field to the far field. The interface required an arbitrary matching of the two approaches at some downwind distance. This resulted in what the development team referred to as the "seam" that manifested itself as a discontinuity in the deposition field. The model allows the user to turn off either algorithm and in time the AGDISP near-wake algorithm became the approach of choice for the entire deposition field. So, though FSCBG remained as the only readily available applied model to simulate pesticide deposition from aerial spraying, the core calculation typically being implemented had changed from Gaussian to the AGDISP near-wake computation in the majority of applications (Figure 1).

2.2 AGDISP

The near wake algorithm used in SAGA is known as AGDISP and uses aircraft weight and wing or rotor span to determine the strength, size and position of the vortices that dominate the wake flow field behind a flying aircraft. The model mathematics are described in detail elsewhere [3], [28]. The model describes the wing tip vortex strength (Γ) as:

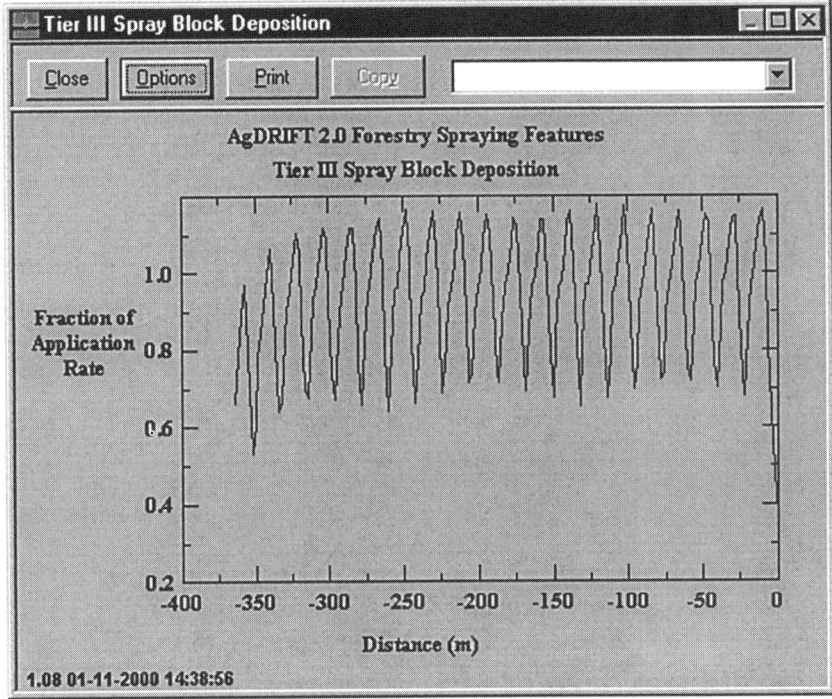

Figure 1. This screen from AgDRIFT/FS shows deposition across a target area with peaks roughly corresponding to the aircraft flight path. This screen is a staple for forestry spraying and was originally in FSCBG with the deposition calculation performed by the AGDISP near-wake algorithm.

$$\Gamma = \frac{W}{2\,\rho_a\,s\,U_\infty} \qquad (1)$$

and the swirl velocity around each vortex as:

$$V_s = \frac{\Gamma}{2\pi}\,\frac{r}{(r+r_c)^2} \qquad (2)$$

where W is aircraft weight, s is aircraft semi-span, U_∞ is aircraft speed, r is radius from the vortex center, r_c is the core radius and ρ_a is the density of air. The trajectory of the droplet after it leaves the spray nozzle and enters the vortex velocity field (where X_i indicates position) is given by:

$$\frac{d^2X_i}{dt^2} = [U_i - V_i]\tau_p^{-1} + g_i \tag{3a}$$

and

$$\frac{dX_i}{dt} = V_i \tag{3b}$$

where t is time, U_i is the mean local velocity at drop location, V_i is the mean drop velocity, τ_p is the relaxation time (dependent on droplet diameter and drag) and g_i is gravity and only operates in the vertical direction (Figure 2). The model is much more involved than shown above but these are some of the basic principles. The algorithm also incorporates local meteorology and evaporation. After extensive validation and sensitivity work, of the dozens of input variables the model can consider, the three most important influences on the location of deposition of an individual droplet are the droplet size, wind speed and release height.

2.3   AgDRIFT

At the beginning of the 1990s, a pesticide drift modeling effort developed between a consortium of pesticide manufacturers and the EPA. The EPA requires new pesticide formulations to be registered. As part of this registration process, field trials are run to measure the drift propensity of a given formulation. According to the EPA, if drift propensity could be anticipated for a given chemical based on physical chemistry, atomization parameters, applicator practice, meteorology and other items, then the expensive step of performing field trials could be foregone. The leading pesticide manufacturers formed an association called the Spray Drift Task Force (SDTF) and began investigating modeling approaches. The SDTF adopted the AGDISP near-wake algorithm as the core algorithm for a drift simulation model for aerial pesticide spraying [29]. The most important contribution of the SDTF was a database derived from over 180 field deposition studies. The SDTF also undertook an overhaul of the model interface and gave the model a modern "look and feel" which is critical to user acceptance. The model that this effort produced is known as AgDRIFT and though it is specifically designed to calculate drift from agricultural spraying, it also calculates pathways to water, long term depositional loading and many other items of interest to the EPA. Currently, some of the features of FSCBG are being added into AgDRIFT with the resulting module named AgDRIFT/FS. These features include canopy penetration, on block deposition analyses, the FS aircraft and nozzle libraries, and other features. AgDRIFT/FS will also have new features such as a simple slope correction and a library of deciduous canopy architecture to be used in an updated canopy penetration scheme.

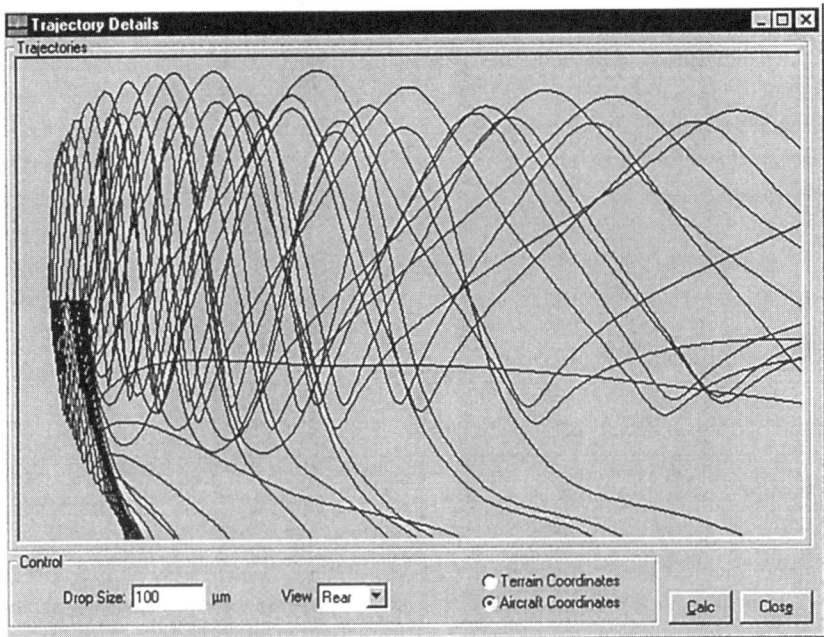

Figure 2. This output screen from the AgDRIFT model plots trajectories as calculated by the AGDISP near-wake algorithm. Droplets originate from boom-mounted nozzles under the aircraft wing (small black edge, center-left) and are influenced by the wing tip vortex. Note that this only shows the 100 μm drops. Droplets of this size are considered fine and are prone to drift.

2.4   Computer Simulation Models in Common

Both FSCBG and AGDISP predict the movement of the spray material above the forest canopy, the movement among the trees, and the amount of material that actually reaches the ground [25], [26], [30]. The routines within these programs estimate the droplets leaving the aircraft and the events encountered by the droplets as they make their way through the aircraft wake and descend onto the spray block (forest or crop area).

Spray deposition and drift are dependent upon: (1) the altitude of the aircraft when the material is released, (2) the speed of the aircraft, (3) whether the aircraft is an airplane or a helicopter, (4) the type of boom and nozzle system used to discharge the spray material, (5) the swath width of each pass of the aircraft, (6) the type and density of the forest, (7) wind speed and direction, (8) relative humidity, and (9) spray material characteristics. These parameters need to be specified before using FSCBG or AGDISP in order to obtain the deposition and dispersion of the spray materials. However, optimal values for these parameters cannot be obtained using FSCBG or AGDISP directly since both programs use these parameters as inputs to carry out batch operations. The SAGA project is designed to help optimize these spray factors using the genetic algorithm in order to achieve maximal deposition and even distribution of spray materials.

The output of the various computer simulation models typically includes several important values that describe deposition: the volume median diameter (VMD), the drift fraction, and the coefficient of variance (COV). VMD is a measure of spray material droplet size distribution. It is the drop diameter (in microns) that divides the spray volume into two equal parts. For example, a VMD of 150 microns means that 50 percent of the spray volume is in droplets smaller than 150 microns, and the remaining 50 percent is in droplets larger than 150 microns. It is important to know the expected droplet size of the spray material as it leaves the aircraft nozzle, and also to know the droplet size that hits the ground. Variations in these two values are due to a number of factors including evaporation. Some of the spray material is likely to drift away from the target area onto adjacent lands due to wind. Drift is defined here as a measure of the amount of spray material deposited outside the spray block (smaller drift is better since that means the spray material stays within the spray block or has evaporated). On block deposition is the amount of spray material that is deposited on the spray block. On block deposition and drift fraction are inversely related to each other, we can either take the approach to maximize the former or minimize the latter. The COV is used to indicate the uniformity of the deposited spray material. Ideally, the spray material should be evenly distributed over the entire spray block. The calculation of COV is based on the standard deviation of the deposition and the average deposition [26].

3   Genetic Algorithms

Modern Genetic Algorithms (GAs) were introduced by John Holland in the early 1970's to solve problems using the processes of natural evolution [9], [10]. Holland was inspired by Darwin's theory about evolution and constructed GAs based upon the fundamental principle of the theory: survival of the fittest. The theoretical basis for the GA is the Schema Theorem, which states that individual chromosomes with short, low-order, highly fit schemata or building blocks receive an exponentially increasing number of trials in successive generations [9].

A typical GA is started with a set of solutions (chromosomes) called a population [16]. A chromosome in the GA is a legal solution to the problem and has the form of a string of genes that can take on some value from a specified finite range or alphabet. An initial population of chromosomes is then constructed at random. All the chromosomes in the population are evaluated using a fitness function. The chromosomes from one population are selected and used to form a new population according to certain selection methods. The common selection schemes are roulette wheel selection and tournament selection. Several further operations such as crossover and mutation are then applied on the newly selected individuals to mimic inheritance and mutation in natural evolution. Crossover is a key operator in the GA, and is used to exchange the main characteristics of parent individuals and pass them on to the children. Mutation is applied after crossover to maintain the genetic diversity of the population and recover possible lost characteristics during crossover. This process is repeated again and again until some terminating condition is met (for example, the desired number of generations is reached).

3.1   Main GA Components and How the GA Works

Typically, genetic algorithms consist of the following main components [7]:

- Chromosomal representation
- Initial population
- Fitness evaluation
- Selection
- Crossover and mutation

How to represent a valid solution to the given problem is an important step when initializing the GA. The concept of a chromosome is normally used in the GA to stand for a valid solution to the problem. The chromosome consists of a string of genes just as the human chromosome does. The specific chromosome representation varies based on the particular problem properties and requirements. In fact, almost any representation can be used as long as it enables a solution to be encoded as a finite length string. A binary representation based on bits [1, 0] is commonly used due to its convenient features such as easy coding and decoding. Integers or real numbers are also frequently used in certain applications.

Once a suitable representation has been decided upon for the chromosomes, an initial population is created randomly or by using specialized and problem specific techniques. This initial population is the starting point for a GA to evolve to desired solutions. The individuals in the initial population are generally assigned random values within their valid ranges and the GA evolves new individuals and populations in the subsequent generations until convergence stability is reached.

Fitness evaluation is used throughout the GA evolutionary process. Every chromosome is tested by the fitness function to acquire its fitness value. This fitness value is a measure of whether the chromosome is suited for the environment under consideration. Chromosomes with higher fitness will receive larger probabilities of inheritance in subsequent generations, while chromosomes with low fitness will more likely be eliminated. As the GA proceeds we would expect the individual fitness of the "best" chromosome to increase as well as the average fitness of the population. The selection of a good and accurate fitness function is thus key to the success of solving any problem quickly. Only those fitness functions that truly map the problem properties should be used. In some cases it might be extremely hard to find an appropriate fitness function to accurately reflect the complex problem properties. Sometimes a single fitness is not sufficient in cases such as multi-objective problems and complex inputs problems. Some advanced GA implementations are needed under these circumstances to handle the complexity [20].

Selection in the GA is a scheme used to select mating pairs for reproduction. Pairs of individuals in the current generation are selected as parents to reproduce offspring. Roulette wheel selection is a simple selection scheme that weights the probability of selecting an individual based on its fitness value. Tournament selection picks parent individuals by choosing the best one from a group of randomly selected individuals.

Crossover is applied on the individuals after selection so that the children inherit partial characteristics from each parent respectively. The crossover probability is introduced here to stipulate the chance that a mating pair is going to be crossed. A crossover probability of 1.0 indicates that all the selected chromosomes are used in the next generation. Empirical studies have shown that better results are achieved by a crossover probability between 0.65 and 0.85 [7], which implies that the probability of a selected chromosome surviving to the next generation being unchanged (excluding any changes arising from mutation) ranges from 0.35 to 0.15. The common crossover approaches are 1-point, 2-point, uniform, and average crossover. 1-point crossover involves taking the two selected parents and crossing them at a randomly chosen point to produce two children. 2-point crossover is similar to single-point crossover, but swapping applies to the two segments between the two randomly chosen points. In uniform crossover, each child gene is randomly selected from either one parent or the other. Average crossover differs from other schemes in that for each child gene, the average of the corresponding genes from both parents is used.

The mutation operation is needed after the crossover operation to maintain population diversity and recover possible loss of some good characteristics. An example of the necessity of the mutation operator is that if all the chromosomes in the initial population have the same value at a particular position, then all future offspring will have this same value at this position. Mutation is introduced in order to generate some random alteration of the genes, e.g. 0 becomes 1 and vice versa in a binary representation. If mutation occurs frequently, then the population is quite unsteady. The mutation probability is normally on the order of one thousandth.

3.2   Sample GA Applications

The GA has been widely used in many fields such as scheduling, telecommunication, engineering simulation modeling, and various optimization fields. A good example of scheduling using GAs is the optimization of airline crew scheduling [15]. The multi-fault diagnosis system (MFD) is an automated approach to diagnosing multiple simultaneous problems [21]. IDA-NET, which is a battlefield communication network system to support specific military missions, configures a "shopping list" for type and number of communication equipment components [22]. The optimization of irregular computer architecture using GAs is another promising application to optimize the interconnections between processors for modern computers [4]. Using GAs for intelligent internet search is a new application and has shown good performance to search and retrieve documents from the enormous number of documents on the internet [17]. A particular GA research project of interest to us is the decision support system developed by Pabico [18] that determines simulation inputs. Genetic Algorithms were used in this project to help determine the cultivar coefficients in crop models. Similarly, SAGA determines simulation input parameter values that produce optimal or near-optimal aerial spray predictions.

4   Development of Fortran-SAGA

Our first step in developing SAGA was to test the feasibility of combining the spray simulation model with a GA. For the spray simulation model part, we experimented with both FSCBG and AGDISP DOS 7.0 to compare their performance and I/O features. Both programs work well as individual spraying simulation programs in specific applications. However, for our initial SAGA efforts, we found AGDISP DOS 7.0 more suitable due to its convenient I/O features (described below) that made it easier to modify and connect with the

GA. For the GA part, we started with a Fortran GA for compatibility and connection convenience with the simulation models.

4.1   AGDISP DOS Version 7.0

AGDISP DOS 7.0 is an MS-DOS program that simulates the behavior of spray material released from aircraft to predict the spray deposition result. Figure 3 shows the AGDISP main interface when it's running. The program calculates the mean position of the spray particles and the position variance about the mean as a result of turbulent fluctuations. It reads inputs from an ASCII data file to get a well-defined set of input values in a specific order. The results are written to another ASCII file when the run ends. The program also displays current computing information to the screen during the process. Two separate library files are called during the computation to provide aircraft and drop size distribution information.

The convenient I/O features of AGDISP DOS Version 7.0 enabled us to develop the methodology to make full use of it in order to establish the interconnections between SAGA and the model, which was one of the most important aspects of the SAGA project at this preliminary stage.

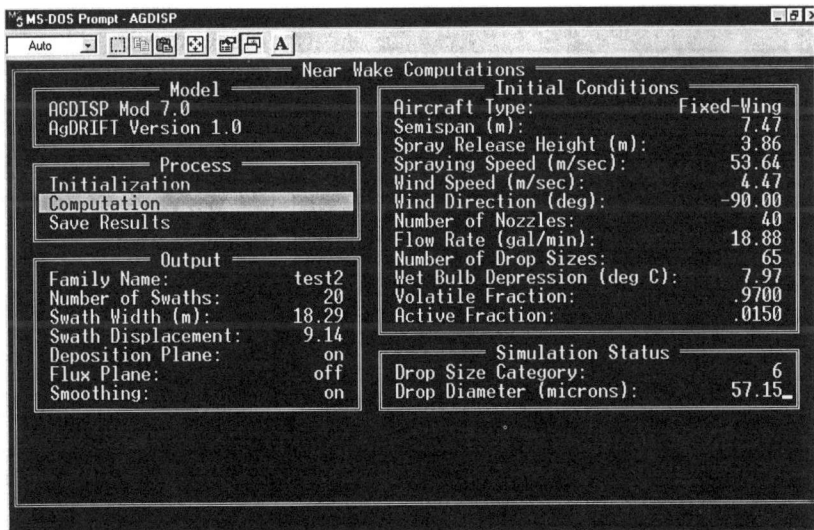

Figure 3. AGDISP DOS 7.0 main interface when running.

4.2   The Fortran GA

The GA we used in this initial stage of SAGA was a modified Fortran version of the Simple GA described by Goldberg [9]. The reason we started with a Fortran GA was because AGDISP DOS 7.0 is implemented in Fortran. We searched the web and found a shareware Fortran GA at http://www.staff.uiuc.edu/~carroll/ga.html developed by David L. Carroll. This GA uses a binary representation. Each individual corresponds to a set of AGDISP parameters in our problem. The selection scheme is tournament selection with a shuffling technique for choosing random pairs for mating. We have the option of using jump mutation or creep mutation, and the option for single-point or uniform crossover. For our SAGA project, we added roulette wheel selection as another selection scheme option, added two-point crossover, used intermediate output file generation for AGDISP input, and changed the standard I/O formats to meet our project requirements.

4.3   Preliminary Fortran-Based SAGA

Figure 4 shows the basic architecture of our SAGA system. The GA sends a set of spray parameters to the AGDISP simulation model. The AGDISP model calculates and sends back the spray results for each parameter set. AGDISP DOS 7.0 returns the expected deposition only. This computation process may take between 5 and 45 seconds depending on the input parameter characteristics and the working platform properties. Based on the fitness function values mapped from the spray results (deposition and the COV for this Fortran based SAGA), the GA attempts to evolve an improved set of parameters. This process is repeated from generation to generation for each individual in the population. The best parameter set is returned as the proposed set-up to achieve the desired deposition.

AGDISP DOS Version 7.0 is implemented using Fortran, and reads input and writes output through text files. The Fortran GA driver also relies on text files for GA parameter input and final results output. We thus combined their I/O features to establish the inter-connections. In our approach, the GA characteristics are first specified in the GA input file (saga.inp). The simple GA has been modified in order to generate a text file containing the twelve key parameters and all other necessary

AGDISP parameters in the format of the input file for AGDISP. This file is named 'agdisp.inp'. AGDISP is then initialized by the GA main routine to compute the deposition. Since the GA and AGDISP are two separate programs that run as separate processes, the GA program halts until AGDISP generates and saves the deposition results in an output file, 'agdisp.dep'. This file contains two columns of data, one for downwind distance and the other for deposition. Then the GA continues execution. It reads in the deposition values from 'agdisp.dep'. The COV of depositions would be computed (we have it set to a constant value in these initial experiments) and combined with the deposition to map the objective function to form the fitness function. The ultimate goal of our SAGA experiments and the fundamental principle of our fitness function are to maximize the deposition and minimize the COV. Based on the fitness value, the GA evolves an improved set of parameters to send back to AGDISP. We focused on twelve specific parameters in this initial stage. Each of them has a lower and upper bound as shown in Table 1.

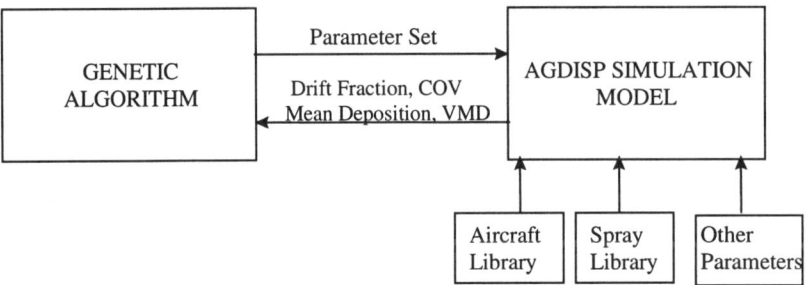

Figure 4. SAGA Architecture.

The GA parameters used in this preliminary study are: population size between 40 and 100, generations between 50 and 200, crossover probability between 0.6 and 0.9, jump mutation probability between 0.005 and 0.05, and creep mutation probability between 0.002 and 0.05.

4.4   Results and Discussion of Fortran-Based SAGA

At this stage, we focused on the determination of (hopefully) optimal spray parameter settings. Some preliminary results are shown in Table 2. It should be noted that we are dealing with two sets of parameters:

one set for the Fortran GA driver which includes population size, generations, crossover and mutation probability, and one set for the spray simulation model which includes release height, drop size, and the other spray parameters. From the evolution of the fitness values, we can see that the Fortran-SAGA has done a good job of improving the parameter values in order to obtain better depositions. For example, comparing the depositions at the edge of the spray block, we can see that the deposition has improved from 98.34 mg/m^2 in the first generation to 146.53 mg/m^2 after 70 generations. This was exactly what the FS wanted to see in this preliminary proof of concept stage.

Table 1. Fortran-SAGA parameters and their ranges.

PARAMETER	LOWER	UPPER
Release Height (m)	1	100
Wind Speed (m/s)	0.5	10.0
Drop Size Distribution (μm)	100	200
Wind Direction (deg)	-360	360
Number of Nozzles	1	60
Total Flow Rate (gal/min)	0.1	1000.0
Volatile Fraction	0.0	1.0
Flight Speed (m/s)	10	200
Dry Bulb Temperature (degC)	1.0	51.67
Relative Humidity (%)	5.0	100.0
Number of Swaths	1	20
Width of Swath (m)	5	300

Table 2. An example of preliminary results of Fortran-SAGA.

GENERATION	DEPOSITION (mg/m^2)
1	98.34
5	99.46
10	102.56
20	108.25
30	116.84
40	119.25
50	124.29
60	137.58
70	146.53

There were a few simplifications that we embedded during this testing stage such as setting the COV to a constant value of 0.3, and restricting the droplet size range. The primary reason for these simplifications was that it allowed us to begin the spray parameter optimization process and test the feasibility of the project fairly quickly after setting up the genetic algorithm and its connection with the spray model. The computation of the COV was not incorporated in the original AGDISP DOS 7.0 and thus COV was not directly available to be mapped into the fitness function. We felt it might require implementing another routine to determine the COV. This problem was solved in a later stage by incorporating the computation of COV within a new AGDISP DLL file created from AGDISP DOS 7.0. The other simplification at this stage dealt with droplet size distribution. Here we set the range for droplet size to be between 100μm and 200μm. This range was subdivided into ten droplet size categories with an increment of 10μm. Each droplet size category was assigned a mass fraction of 0.1. In spray practice the droplet size distribution may be dependent on certain factors such as nozzle specifications and spray speed. This simplification was also used to speed the spray parameter optimization process. Later, these simplifications are removed in order to obtain more accurate and reliable results.

We also ran numerous experiments to determine which GA parameters produced the best results. The selection of GA parameters such as population size, number of generations, crossover type and probability, and mutation probability is a key facet of the speed and success of the evolutionary process. These parameters are typically domain dependent. One big problem with this initial SAGA was that it was to a certain extent limited by the runtime of AGDISP DOS 7.0. The runtime of the main GA program was negligible compared to the AGDISP runtime. Thus for example if we set the population size to 50 and number of generations to 100, then assume an average AGDISP runtime length of 15 seconds, it would take about 20 hours to complete the SAGA run. During these initial experiments, we usually let SAGA run overnight and collected data the next morning. Therefore, the number of generations was accordingly set to around 50 and the population size was set to between 50 and 100. Table 3 shows some comparisons of the results obtained with different GA population sizes. Similar experiments were run to help determine appropriate values for other GA parameters.

Table 3. Fortran-SAGA results at different population size.

GEN	DEPOSITION (mg/m^2)		
	Pop size = 50	Pop size = 40	Pop size = 20
1	98.34	98.34	98.34
20	108.25	107.36	105.42
50	124.29	122.68	116.35

Another key issue in the initial development of SAGA was the mapping of the deposition and the COV onto the fitness function. It is highly desired to get the exact amount of spray material evenly distributed over the spray block. We followed the rule of thumb suggested in [19] and set the COV to 0.3 temporarily. We tested and compared different mapping functions having linear and exponential characteristics, and decided to use the exponential function formulated below and the graph shown in Figure 5 for our initial experiments.

$$Fitness = 3.0 - \exp(1 - 0.04(a \times Deposition - b \times COV))$$

It should be noted that COV is dependent on swath width in most cases, but in the above formulation, we temporarily fixed the COV and set the goal to maximize deposition only. Later on we removed this simplification and modified our fitness formulation accordingly.

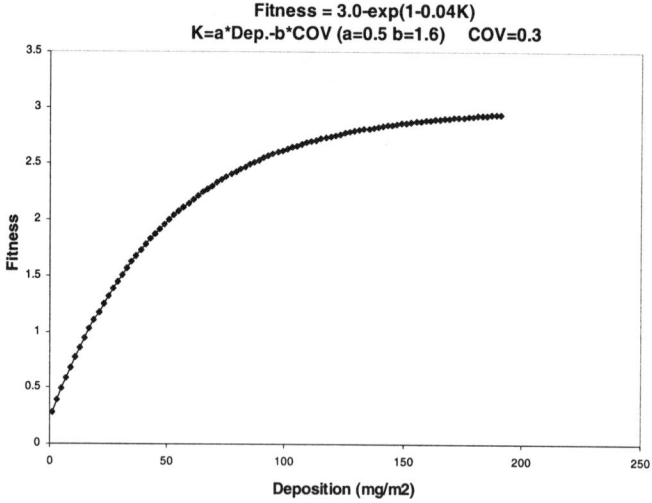

Figure 5. Fortran-SAGA fitness function graph.

In addition, some other work we did in the initial stage was to test the parameter sensitivity of AGDISP. The approach we took was to set one of the twelve SAGA parameters constant and test the impact of this change on the deposition evolution. Release height, wind direction, and wind speed are the three main parameters we focused on. The results are presented in Table 4. It is indicated that setting the release height has a large impact on the deposition evolution. Likewise, keeping the wind parameters constant also has a considerable impact on SAGA results. The trend is consistent with the results obtained by Teske and Barry [27], namely that the input parameters for aerial spray can be ranked in order of importance. The approach they took to measure the relative importance was to change an input variable linearly and measure the corresponding relative sensitivity of the results. Two parameter values, Figure of Merit and Mean Horizontal Position were used to measure the effectiveness of swath width deposition and the level of off-target drift, respectively. Our results need further technical verification compared to their approach. But the similar trend indicated by our results provided support for the important roles of these key parameters and their relative importance.

Table 4. Testing of the Fortran-SAGA parameter importance.

GEN	DEPOS (mg/m^2)	DEPOS (mg/m^2)	DEPOS (mg/m^2)	DEPOS (mg/m^2)
		Release Height = 75m	Wind Dir = 150 degree	Wind Speed = 5.0m/s
1	98.34	97.38	96.52	96.82
10	102.56	100.25	100.34	101.25
20	108.25	104.39	103.95	103.49
40	119.25	112.65	115.87	114.58
60	137.58	120.87	125.75	124.68

Based upon the results and experience from these initial SAGA experiments, we successfully showed the feasibility of the SAGA project, and the preliminary results helped us to make necessary modifications to improve the system. The main areas for improvements needed for the Fortran-SAGA included the user-interface, the computation of COV, and the run time. The user interface was not friendly enough mainly due to the Fortran/DOS implementation. The user has to specify all GA parameters in a text file before the run and the SAGA results are stored in a text file after the run. Some significant changes were needed to solve these problems.

We decided to change our GA driver to a new GA implemented with Microsoft Visual Basic 5.0 (VB) to take advantage of the language's nice interface development features. The new interface would facilitate the use of SAGA. AGDISP 7.0 was also replaced with a new AGDISP DLL that returns deposition, COV and the resulting VMD. The new SAGA was expected to speed up significantly based on the improved DLL and VB-GA. We also expected to incorporate AGDISP parameter dependencies and practical application considerations (spray knowledge) into a revised fitness measure.

5   Development of VB-SAGA 1.0

After the initial testing stage with Fortran-SAGA, some significant changes were made to improve the SAGA user-friendliness and overall performance. We implemented a new SAGA GA with Microsoft VB 5.0. This new VB-GA features highly user-friendly interfaces. A new AGDISP DLL created from the AGDISP model was used as the spray simulation engine. The new SAGA program was named VB-SAGA 1.0. The inter-connection between the Fortran GA and AGDISP DOS Version 7.0 was based on the reading and writing of intermediate files. In VB-SAGA 1.0 these files were replaced by inter-program calls that speed up execution significantly. As requested by the Forest Service, an exhaustive search scheme was set up to validate the GA and test/compare the performance of our VB-SAGA. These items are discussed in the following sections.

5.1   VB-SAGA 1.0

Figures 6 to 9 show the interfaces of the new VB-SAGA. These interface windows are designed to provide user convenience and high flexibility to specify GA parameters, preset necessary spray parameters, chart ongoing SAGA evolution, view the dynamic evolution of the spray parameters, and view final SAGA results. The top half of the main interface is primarily for GA control parameters and the bottom half is mainly for spray parameters and results.

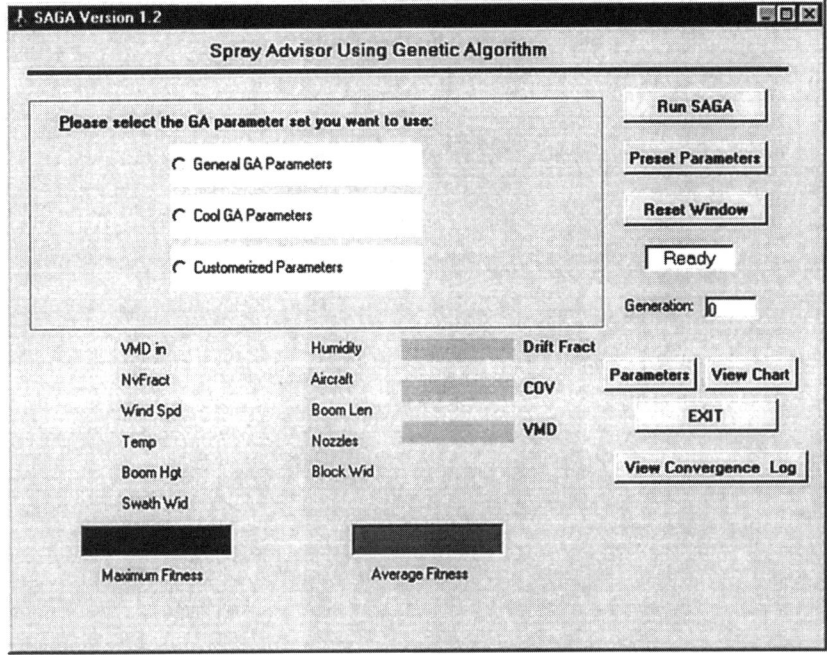

Figure 6. Main interface of VB-SAGA1.0.

As shown in Figure 6, depending on the user's knowledge of the GA and the application purpose, the user can select either [General GA Parameters] which is a set of recommended GA parameters for gypsy moth spray, [Cool GA Parameters] which is a set of recommended GA parameters for regular spray, or the advanced [Customized GA Parameters]. If the user selects the [Customized GA Parameters], groups of GA parameters will appear (shown in Figure 7) and the user can modify the default settings as desired.

The new VB-GA driver originated from the Simple Genetic Algorithm (SGA) described by Goldberg [9]. We made use of one of the convenient features of VB, the "Type" statement, to define a new data structure that consists of the spray parameters (defined as a *Single* array), return values from the DLL, and the fitness. This new data type is named "individual". This "individual" corresponds to the chromosome string representation in the traditional GA. We use a real number representation for the parameters and the individuals.

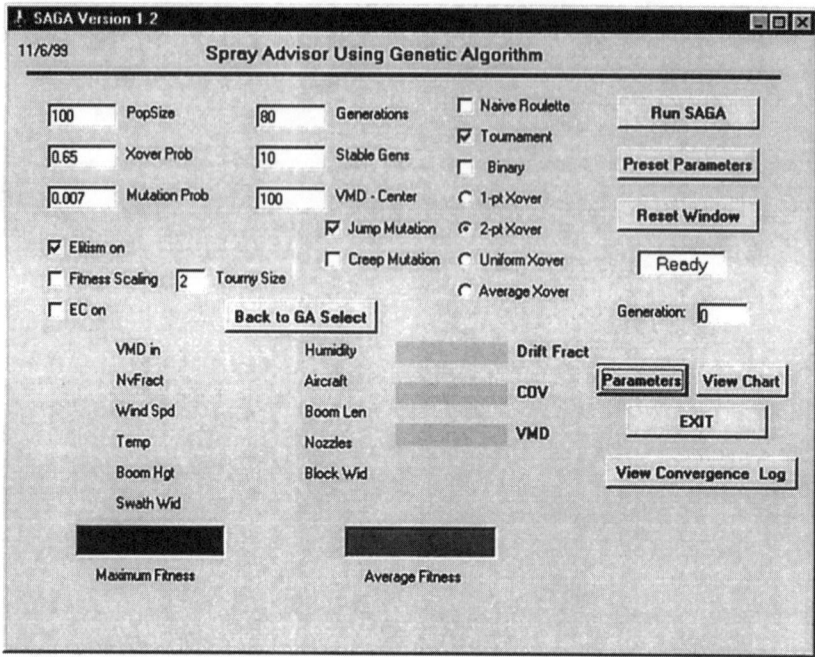

Figure 7. Main interface of VB-SAGA1.0 with user-specified GA parameters.

For the GA parameters, as shown in Figure 7, we have various GA options that users can select to group a set of GA parameters for SAGA. The user can enter population size, generations, crossover probability, and mutation probability. Each of these parameters is provided a default value, e.g., 100 for population size, 80 for generations, 0.65 for crossover probability, and 0.007 for mutation probability. For the GA operators, we provide several options for each. For the selection scheme, users can choose among [Naive Roulette Wheel] selection, [Tournament] selection and [Binary] selection. For the crossover operation, users have the options of [1-point], [2-point], [uniform], and [average] crossover. We have [Jump Mutation] and [Creep Mutation] for mutation options. The former is used to randomly select a new value for a parameter within its valid range. The latter is used to change the old parameter by a small increment (error checking is added to make sure the new value is valid). Besides these basic GA parameters, we also added some new features such as [Elitism], which enables the GA to inherit the best individual from the previous generation. Another useful option is [Fitness Scaling] which is an advanced GA feature that is used to overcome the "local maximum" problem. With [Elitism] and [Fitness Scaling] turned on, SAGA normally converges in less than 30 generations. The GA population becomes basically homogenous after that. We thus provide a [Stable Generations] option for the user to specify how many stable generations (no changes in maximum fitness) are allowed before stopping SAGA. The current default value is 12. The user can also specify the size used in the tournament selection scheme. The recommended value is 2 for selection in pairs.

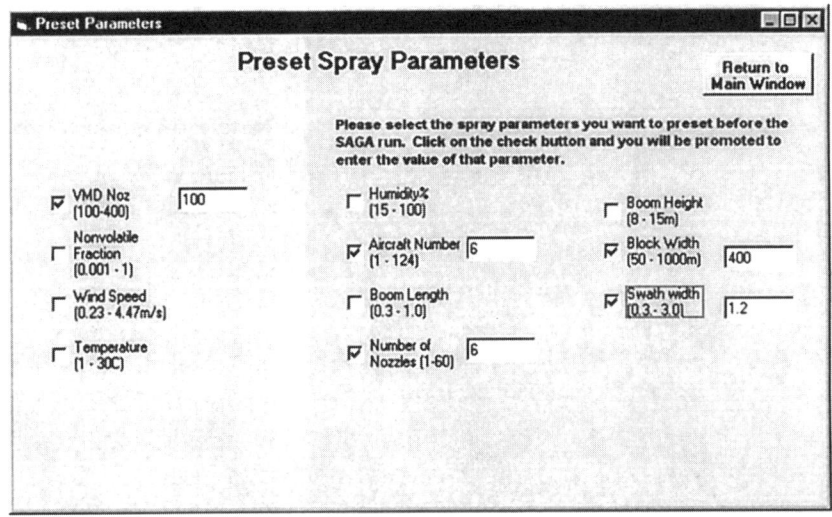

Figure 8. Secondary interface of VB-SAGA1.0 to preset spray parameters.

In practical spray applications, it is quite common that some spray parameters can and should be fixed according to the spray requirements. We thus provide the option to preset certain spray parameters by selecting [Preset Parameters]. A new interface window appears with the spray parameters listed (shown in Figure 8). The user can select the ones to preset and fill in appropriate values. The rest of the parameters are left open to evolution by SAGA.

The bottom half of the main interface is designed to display intermediate results with two options provided. The first option allows the user to view the dynamic values of the spray parameters and the spray results (shown in Figures 6 and 7). These values are associated with the best individual so far as the program evolves from generation to generation. This option is set as the default output mode. The user can also click on the [View Chart] button to switch the bottom half to a fitness growth chart with the maximum and average fitness values displayed dynamically (shown in Figure 9). The user can click on the [View Parameters] to return to the parameters option.

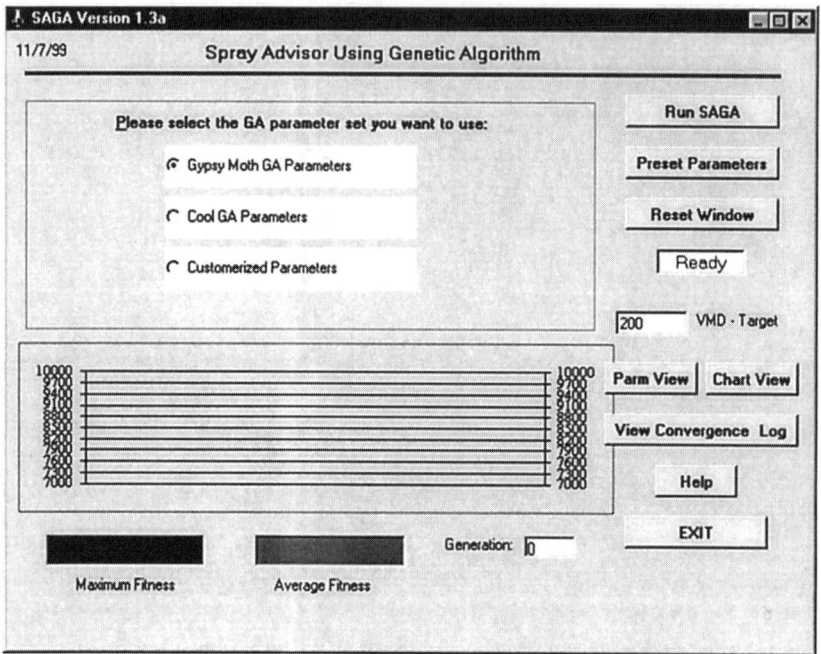

Figure 9. VB-SAGA1.0 main interface, chart view option turned on.

After the user finishes setting the GA and spray parameters, clicking on the [Run SAGA] button starts the run, or clicking [Reset Window] resets the parameters to their default values. Besides the spray parameters and results being displayed dynamically in the main interface, the user can also click on [View Convergence Log] after the program stops to look at a detailed report.

The spray parameters to be optimized by VB-SAGA1.0 are not the same as those used in Fortran-SAGA as shown in Table 1. As suggested by Forest Service experts, we introduced several more representative spray parameters such as VMD Input, Aircraft ID

Number, and Block Size. We removed some old parameters such as
Number of Swaths, Drop Size Distribution, and Total Flow Rate. Table
5 shows the eleven spray parameters that are used by VB-SAGA 1.0.
These eleven spray parameters are also the input parameters for the new
AGDISP DLL. Other less important or more static parameters are kept
constant during our experiments. However, they can become part of the
variable parameter set (i.e., we can easily include additional parameters
to the parameter set we are searching for) by specifying them at the
beginning of each SAGA run.

Table 5. VB-SAGA1.0 spray parameters and their ranges.

PARAMETER	LOWER	UPPER
VMD Input (μm)	100	400
Nonvolatile Fraction	0.001	1.0
Wind Speed (m/s)	0.23	4.47
Temperature (degree C)	1	30
Boom Height (m)	3	30
Swath Width (fraction of wingspan)	0.3	3.0
Humidity (%)	0.0	100.0
Aircraft ID Number	1	124
Boom Length (fraction of wingspan)	0.3	1.0
Number of Nozzles	1	60
Block Size (m)	50	1000

VB-SAGA1.0 has a very similar architecture to that of Fortran-SAGA
shown in Figure 4. However, there are two major differences. One is
that instead of Deposition and COV, the new AGDISP DLL returns
three outputs, Drift Fraction, COV, and VMD Output. We adopt a new
fitness function (shown below) suggested by the Forest Service that
incorporates these three outputs with different weights. VMDCenter is
the desired VMD value specified by the user before the run. The other
difference is that the connection between the VB-GA and the AGDISP
simulation model is now based on inter-program calls instead of the I/O
intermediate files for the Fortran-SAGA. This improvement greatly
speeds up the SAGA to a large extent.

$$Fitness = 100 \times \left\{ 50 \times (1.0 - DriftFrac) + 25 \times (1 - COV) + 25 \times \exp\left(-8.0 \times \left(\tfrac{VMD}{VMDCenter} - 1\right)^2\right)\right\}$$

5.2   Exhaustive Search Test

The exhaustive search test (actually pseudo-exhaustive) was requested by the Forest Service to validate the SAGA results. The goal of the test was to compare SAGA results with exhaustive search results to make sure that SAGA was able to find optimal or near-optimal solutions. The exhaustive test program interface is shown in Figure 10. Because of the extremely large search space for the eleven parameters, it was desired to finish the exhaustive test with reasonable time and economical efforts. We thus needed to reduce the huge search space to run the exhaustive search within several days. The approach we took to reduce the huge search space was to fix eight spray parameters as shown in Table 6 and test the remaining combinations of the other three parameters as shown in Table 7.

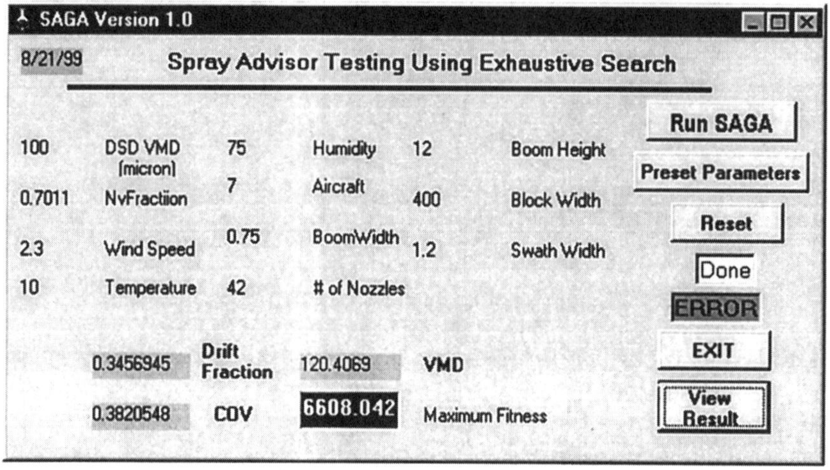

Figure 10. Exhaustive search test main interface.

Another effort to reduce the search space was to use narrower ranges (reduce upper bound and increase lower bound) of these three parameters. Our earlier test runs gave us some idea of good ranges for these three parameters. We therefore used these smaller ranges (also shown in Table 7) instead of the full range as shown in Table 5. These spray parameters were imported into the AGDISP DLL to produce batch results and we used the same fitness function in SAGA to obtain their fitness value. The total combination of all parameter sets is about

15*12*100=18,000. If we estimate an average running time to be about 20 seconds for each run, the total running time for the exhaustive test would be approximately 4 days. The actual exhaustive search experiment took about three and one half days and the top ten solutions are listed in Table 8.

Table 6. Fixed spray parameters in exhaustive search test.

DSD-VMD	100.0 μm
Temp	10.0 degC
Humidity	75.0
Aircraft ID	7
BoomWidth	0.75
NumNoz	42
BlockWidth	400.0 m
SwathWidth	1.2 m

Table 7. Changing spray parameters in exhaustive search test.

	Lower Bound	Upper Bound	Increment Step
NvFrac	0.75	0.9	0.01
Wind Speed	0.23m/s	0.35m/s	0.01
BoomHeight	6.0m	7.0m	0.01

Table 8. Fitness results from exhaustive experiment (8 fixed parameters).

No.	Best Fitness
1	9428.176
2	9427.911
3	9427.605
4	9427.257
5	9426.553
6	9426.434
7	9425.577
8	9425.041
9	9423.479
10	9422.863

It should be noted that the exhaustive experiment results are dependent on the increment step adopted. The exhaustive test scheme being used here is in fact a pseudo exhaustive search, because we are actually selecting very closely spaced points in the search space, though the difference between the points is very small to match as closely as possible to a real exhaustive search. However, the problem does exist that using this pseudo exhaustive search could possibly leave out some good points and reduce the certainty of finding the best individual. We therefore need to minimize the steps in order to approach closely enough to a continuous search to obtain the best results. However, the smaller the steps are, the longer time it will take to finish the exhaustive search. We want to complete the experiment within a reasonable time and have the results satisfy the precision requirements.

5.2.1 VB-SAGA1.0 Test

We then ran VB-SAGA1.0 with the same eight spray parameters fixed with the same values, and let VB-SAGA1.0 evolve Non-Volatile Fraction, Wind Speed, and Boom Height to obtain the best spray results. The results are displayed in Table 9. It only took 1.5 hours to finish and the best result from SAGA was found among the top 0.1% of the exhaustive results. Table 10 shows a side-by-side comparison of best exhaustive with best SAGA results. This is a good indication that SAGA is capable of finding near-optimal solutions for our spray application in a relatively short time.

Table 9. Fitness results from VB-SAGA 1.0 experiment
(8 fixed parameters, Mutation 0.003 and 0.02 not shown).

Mut.→ Xover	0.001	0.007	0.01	0.03	Row Average
0.60	9384.353	9354.186	9345.647	9416.380	9372.866
0.65	9322.343	9399.037	9416.356	9324.382	9376.907
0.70	9402.429	9406.360	9387.536	9351.680	9390.633
0.75	9403.283	9401.998	9364.615	9398.096	9388.505
0.80	9423.766	9393.582	9413.530	9397.563	9409.692
0.85	9321.064	**9427.255**	9414.876	9358.782	9375.631
Average	9376.206	9397.07	9390.427	9374.481	

Table 10. The maximum fitness for exhaustive and VB-SAGA 1.0 tests.

	EXHAUSTIVE TEST	GA TEST
Maximum Fitness	9428.176	9427.255
Non-Volatile Fraction	0.780	0.789
Wind Speed (m/s)	0.280	0.282
Boom Height (m)	6.100	5.777
Drift Fraction	0.0309	0.0297
COV	0.165	0.167
VMD (μm)	101.625	104.223

5.3   VB-SAGA1.0 Experiments and Results

After the exhaustive test, we began to use VB-SAGA 1.0 under different circumstances, mainly with and without spray parameter restrictions. We ran many experiments based on Forest Service applications. With no spray parameters fixed, SAGA is expected to generate better results compared to those with certain spray parameter restrictions. The best fitness and the corresponding spray parameters are listed in Table 11.

Table 11. The maximum fitness from VB-SAGA 1.0 without restrictions on spray parameters (GA crossover rate=0.65 and mutation rate=0.007).

Item	Best Result
Maximum Fitness	9924.08
DSD-VMD (μm)	100
Non-Volatile Fraction	0.788
Wind Speed (m/s)	0.264
Temperature (degC)	4.941
Humidity (%)	62.715
Aircraft ID	110
Boom Length (fraction of wingspan)	0.529
Nozzles	9
Boom Height (m)	7.086
Block Size (m)	964.9
Swath Width (fraction of wingspan)	0.543
Drift Fraction	0.00301
COV	0.0242
VMD (μm)	99.58

We ran many experiments based on the practical spray parameter specifications provided by Forest Service managers. In total there are six groups of experiments that belong to two categories. The first category includes two groups of experiments of which four and seven spray parameters are fixed respectively. The second category includes the other four groups of experiments that focus on investigating the roles of aircraft and swath width. For each group, we ran 10 experiments with the combination of crossover rate 0.65, 0.7, 0.75, 0.8, 0.85 and mutation rate 0.007 and 0.012. The population size was 100 and the generation was 70 for all experiments.

The maximum fitness obtained based on the first group of specifications was 9710.885 and the spray parameters corresponding to this maximum fitness are listed in Table 12. Detailed results are listed in Table 13. DSD-VMD, Aircraft, Block Size and Swath Width were fixed in this case. The second group of experiments has the highest fitness of 9750.743 and its corresponding spray parameters are listed in Table 14. Detailed results are listed in Table 15. DSD-VMD, Non-Volatile Fraction, Wind Speed, Humidity, Temperature, Number of Nozzles, and, Block Size were fixed in this case.

Besides the above two groups of experiments, we also ran four groups of experiments with different configurations of fixed aircraft and swath width. Tables 16 to 19 show the results from these four groups of experiments. There is often only one type of aircraft available on an operational spray project. It is therefore of great practical importance to determine what optimal or near optimal values for the other parameters should be used when the aircraft and swath width are fixed. These four groups of experiments were expected to give forest managers this benefit.

Table 12. Experiment 1: practical settings with maximum fitness=9710.885.

Evolve Parms	NvFrac	Wind Spd	Temp	Humid	Boom Len	Noz	Boom Height
	0.853	0.252	24.651	66.689	0.604	29	3.219
Fixed Parms	DSD-VMD		Aircraft	Block Size		Swath Width	
	100		6	400		1.2	

Table 13. Experiment 1: practical settings details.

Xover	Mutation	Max Fit.	Drift Frac.	COV	VMD
0.65	0.007	9681.238	0.024264	0.075137	102.1929
0.65	0.02	9521.977	0.043287	0.104543	100.3391
0.7	0.007	9669.712	0.028459	0.082142	100.5931
0.7	0.02	**9710.885**	0.022601	0.064461	102.2375
0.75	0.007	9651.816	0.028348	0.080639	101.5573
0.75	0.02	9574.656	0.026514	0.117077	100.2037
0.8	0.007	9609.734	0.033884	0.086671	101.4445
0.8	0.02	9691.885	0.021839	0.07862	100.6032
0.85	0.007	9694.788	0.025709	0.070153	100.8018
0.85	0.02	9639.846	0.033055	0.077817	99.59053

Table 14. Experiment 2: practical settings with maximum fitness=9750.743.

Fixed Parms	DSD-VMD	NvFrac	Wind Speed	Humid	Temp	Noz	Block Size
	200	0.45	0.5	75	10	6	400
Evolve Parms	Boom Height	Boom Length		Aircraft		Swath Width	
	4.233	0.322		7		0.698	

Table 15. Experiment 2: practical settings details.

Xover	Mutation	Max Fit	Drift Frac	COV	VMD
0.65	0.007	9643.027	0.010314	0.065517	217.0752
0.65	0.02	9021.266	0.040434	0.096909	234.6726
0.7	0.007	9623.292	0.014304	0.063063	217.4393
0.7	0.02	9250.907	0.028835	0.098085	227.8699
0.75	0.007	**9750.743**	0.074591	0.030589	216.6913
0.75	0.02	9642.524	0.011115	0.06124	217.5165
0.8	0.007	9642.524	0.011115	0.06124	217.5165
0.8	0.02	9263.371	0.029171	0.099688	227.102
0.85	0.007	9263.371	0.029171	0.099688	227.102
0.85	0.007	9551.506	0.011498	0.071744	221.0305

Table 16. Experiment 3: practical settings details (Aircraft: 100, swath width: 2.5).

Xover	Mutation	Max Fit	COV	VMD	Drift Frac
0.65	0.02	8478.91	0.4693	206.174	0.06913
0.65	0.007	8489.09	0.4747	200.007	0.0648
0.7	0.02	8492.34	0.468	200.104	0.0674
0.7	0.007	8486.53	0.442	199.89	0.0812
0.75	0.02	8476.31	0.4767	199.95	0.0663
0.75	0.007	8489.97	0.468	199.79	0.0673
0.8	0.02	8488.16	0.4343	199.23	0.0833
0.8	0.007	8492.53	0.4625	199.769	0.06967
0.85	0.02	**8494.48**	0.4673	199.91	0.06722
0.85	0.007	8477.4	0.4476	199.708	0.07998

Table 17. Experiment 4: practical settings details (Aircraft: 106, swath width: 2.25)

Xover	Mutation	Max Fit	COV	VMD	Drift Frac
0.65	0.02	8717.63	0.384	199.37	0.063
0.65	0.007	8728.38	0.3798	199.434	0.063
0.7	0.02	8716.13	0.385	199.292	0.0625
0.7	0.007	8730.29	0.378	199.44	0.0634
0.75	0.02	8730.85	0.378	199.42	0.0636
0.75	0.007	8719.01	0.381	199.244	0.064
0.8	0.02	**8738.82**	0.377	200.112	0.06346
0.8	0.007	8729.69	0.375	199.565	0.654
0.85	0.02	8720.44	0.383	199.39	0.063
0.85	0.007	8730.54	0.375	199.52	0.0652

Table 18. Experiment 5: practical setting details (Aircraft: 5, swath width: 2.3).

Xover	Mutation	Max Fit	COV	VMD	Drift Frac
0.65	0.02	8345.86	0.362	200.33	0.198
0.65	0.007	8345.78	0.36	200.77	0.149
0.7	0.02	8346.01	0.362	201.18	0.147
0.7	0.007	8351.57	0.362	200.2	0.15
0.75	0.02	8351.86	0.363	200.404	0.147
0.75	0.007	8353.54	0.36	200.17	0.148
0.8	0.02	8349.72	0.36	200.3	0.149
0.8	0.007	8353.4	0.361	200.3	0.148
0.85	0.02	**8357.87**	0.3607	200.27	0.1474
0.85	0.007	8354.77	0.3608	200.078	0.14844

Table 19. Experiment 6: practical setting details (Aircraft: 10, swath width: 2.2).

Xover	Mutation	Max Fit	COV	VMD	Drift Frac
0.65	0.02	8441.29	0.3694	200.12	0.127
0.65	0.007	8353.47	0.357	199.08	0.148
0.7	0.02	8405.04	0.378	201.7	0.13
0.7	0.007	8432.74	0.368	199.777	0.1291
0.75	0.02	8438.65	0.3689	199.655	0.1269
0.75	0.007	8434.23	0.37	200.5	0.128
0.8	0.02	**8444.23**	0.37	199.93	0.126
0.8	0.007	8440.37	0.365	199.989	0.129
0.85	0.02	8436.35	0.371	200.82	0.125
0.85	0.007	8439.76	0.362	200.111	0.13

These results were evaluated by spray experts and regarded as excellent predictions with significant practical importance.

6   Development of VB-SAGA 2.0

VB-SAGA2.0 inherits most of the important features of VB-SAGA1.0 and adds some significant new features. The two most important new features are the menu and the self-adaptive GA. Figure 11 shows a typical VB-SAGA 2.0 interface with these two new features. In addition, VB-SAGA2.0 uses a slightly modified fitness function listed below.

$$Fitness = 100 \times \{[50 \times (1.0 - DriftFrac)] + [25 \times (1 - COV)] + [25 * VMDTerm]\}$$

$$VMDTerm = 1.0 - Abs\left(1.0 - \frac{VMD}{VMDCenter}\right)$$

6.1   VB-SAGA2.0 Menu Items

VB-SAGA2.0 replaced the buttons of VB-SAGA1.0 with a menu bar as shown in Figure 11. All the functionality of the buttons on the VB-SAGA1.0 main interface is now replaced by the handy menu bar. The menu bar is added onto the top-left corner of the VB-SAGA 2.0 main interface. The four main menus on the menu bar are [Command], [Configuration], [View], and [Help]. Each main menu has certain sub-items. For example, under the [Command] item, there are [Run

SAGA], [Preset Spray Parameters], [Reset Parameters], [View and Print SAGA Results], and [Exit].

Under the [Configuration] item, there are [Enable Adaptive GA], [Disable Adaptive GA], [Change Frame Color], and [Change Window Size]. The adaptive GA feature can be enabled and disabled by selecting the first or second item. Details of the adaptive GA feature will be introduced in the following section. The item [Change Frame Color] has several sub-items that lead to different color combinations of the frame. [Change Window Size] has [Small], [Medium], or [Large] to choose to modify the size of the program window.

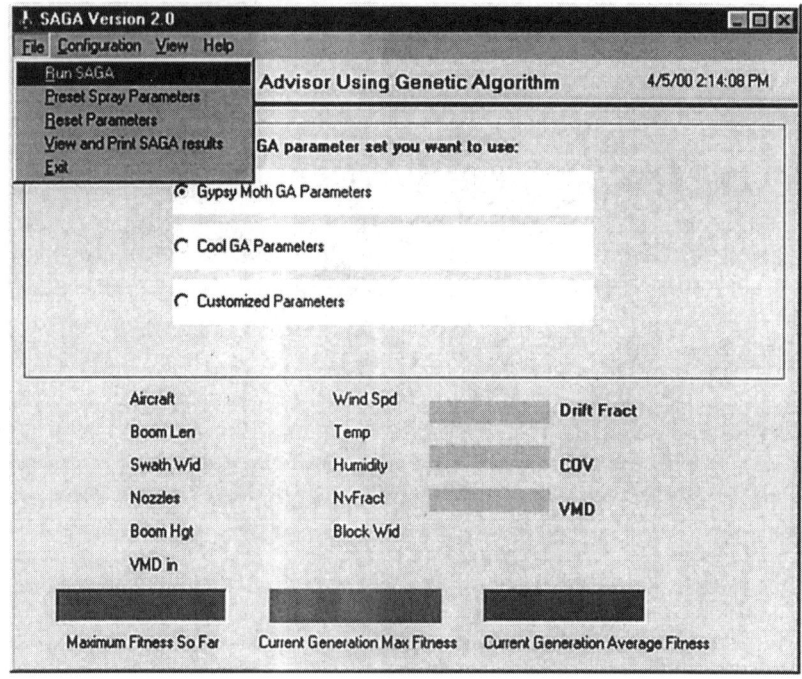

Figure 11. The main interface of VB-SAGA2.0.

The item [View] has sub-items such as [View Default GA Settings], [View Customized Settings], [View SAGA Progress Chart], [View SAGA Parameters List], [AGDISP DLL Information], and [Spray Advisor Information]. [View Default GA Settings] provides two default GA Settings [Gypsy Moth Parameters] and [Cool GA Parameters].

[View Customized Settings] displays the specific GA parameters for the user to specify. [View SAGA Progress Chart] shows the evolving curves of the maximum and average fitness of the generations in the bottom half of the window and [View SAGA Parameters List] changes the bottom half back to evolving spray parameters. [AGDISP DLL Information] gives some introduction of the AGDISP model and its DLL version.

The item [Help] has sub-items such as [View Help File], [View Recent SAGA Paper], [View General GA Tutorial], and [Contact Information]. [View Help File] enables the user to view an introduction document about SAGA. [View Recent SAGA Paper] presents the user with the latest published SAGA paper so that the user can have comprehensive access to the development and achievements of SAGA. [View General GA Tutorial] provides a quick tutorial about basic concepts and working principles of the GA. [Contact Information] provides the authors information for comments or inquiries.

6.2   The Self-Adaptive GA

VB-SAGA1.0 has performed satisfactorily. However, we believe it can be improved. For example, the program requires the user to be familiar with the GA technique. The user must set appropriate GA parameters before the run. The rule of thumb for the best values of the GA parameters is 0.65 - 0.85 for crossover rate, 0.005 - 0.01 for mutation rate. Our previous experiments showed that for SAGA, a crossover rate between 0.75 and 0.85 and mutation rate between 0.005 and 0.012 usually produced good results, but the specific values may differ with different problems. For an inexperienced user, it may take many tests before determining the appropriate range and exact values of these GA parameters. This is not always possible given time constraints. It is also not easy for a novice user to understand the GA concepts such as crossover and mutation. One main goal of our project is that the user with little GA knowledge can start to use SAGA quickly and correctly. We thus continued our efforts to develop an improved SAGA with a self-adaptive GA feature so that users with little GA knowledge or even little computer knowledge are able to use SAGA easily. We name the new program VB-SAGA2.0. With this self-adaptive GA feature on, the new VB-SAGA2.0 can actually start at any random valid GA operator values (crossover and mutation only at this stage). The program tends to evolve to the best GA values as well as the best spray parameters.

6.2.1 Fuzzy Logic Control

The use of Fuzzy Logic Control is pivotal for the self-adaptive feature in VB-SAGA2.0. Fuzzy Logic is basically a multi-valued logic that is used to handle the concept of partial truth instead of "completely true" and "completely false" notions such as yes/no, true/false, and black/white [13]. By using fuzzy logic, notions like small, big, warm, or pretty cold can be formulated mathematically and processed by computers. Fuzzy logic was first introduced by Dr. Lotfi Zadeh at UC Berkeley in the 1960's as a means to model the uncertainty of natural language [12]. It has emerged as a powerful tool for the control of transportation systems and industrial processes, as well as for household and entertainment electronics, diagnosis systems and other expert systems.

The membership function is one of the important concepts in fuzzy logic. It is used to convert an input to be anywhere in the range of [0, 1] [12]. Triangular or Gaussian functions are commonly used representations of membership functions. A set of IF-THEN rules is used in fuzzy logic to indicate what actions should be taken under certain conditions. Fuzzification is the process used to convert crisp inputs to values in the range of [0, 1] (degree of the membership) based on the membership functions. If the fuzzified values match the conditions of one or more rules, the consequents of these rules will be used to produce outputs. If more than one rule is fired, the outputs need to be aggregated together to generate an output region. Defuzzification is the last process in fuzzy control to deduce the crisp output from the output region. Centroid, maximizer, and weighted average are the three commonly used approaches to locate crisp output.

6.2.2 Development of Self-Adaptive GA in VB-SAGA2.0

The idea for the self-adaptive GA came from the work of Lee and Takagi [14]. They use fuzzy logic techniques to dynamically control parameter settings for their GA. We simplified their approach and designed our adaptive scheme based on similar principles. For our self-adaptive SAGA, there are three inputs and two outputs. The three inputs are:

      A1: (average fitness)/(best fitness)
      A2: (worst fitness)/(average fitness)
      A3: change in fitness since last generation

The two outputs are:

      B1: the crossover probability change
      B2: the mutation probability change

Each input or output has three membership values: small, medium and large. Triangular membership functions are used for this fuzzy control (the membership functions are shown in Figures 12 to 16). There are altogether 27 control rules for our self-adaptive GA. Some examples of the rules are as follows:

    IF A1 is small, A2 is small, and A3 is small,
      THEN B1 is small and B2 is small.
    IF A1 is small, A2 is medium, and A3 is medium,
      THEN B1 is large and B2 is medium.
    IF A1 is medium, A2 is small, and A3 is medium,
      THEN B1 is medium and B2 is large.

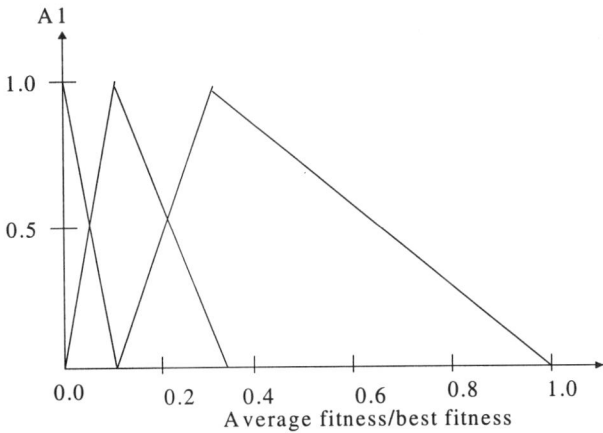

Figure 12. A1 membership function used in VB-SAGA 2.0

When the self-adaptive feature is turned on, the GA watches the changes of A1, A2 and A3, and makes modifications to B1 and B2 when one or more rules are fired. We use triangular membership functions in fuzzification and defuzzification to obtain crisp outputs. The goal is to force the GA to evolve to the GA parameters that maximize the fitness based on the underlying rules. The new crossover and mutation parameters are restricted such that they can at most change half of their previous values at any time. The valid range for both crossover and mutation rates is [0, 1].

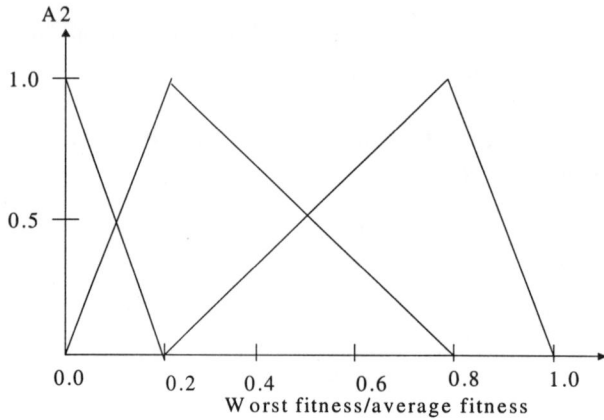

Figure 13. A2 membership function used in VB-SAGA 2.0

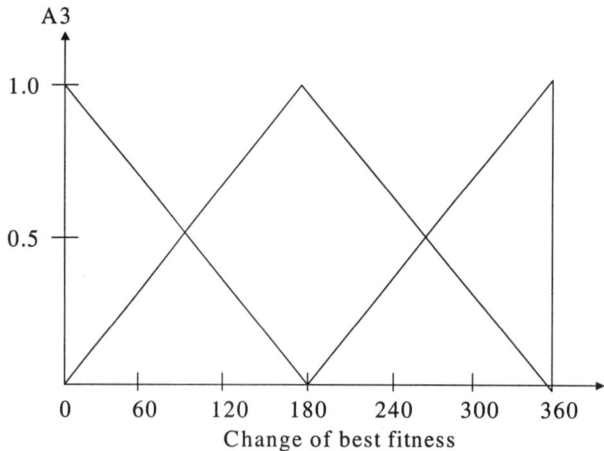

Figure 14. A3 membership function used in VB-SAGA 2.0

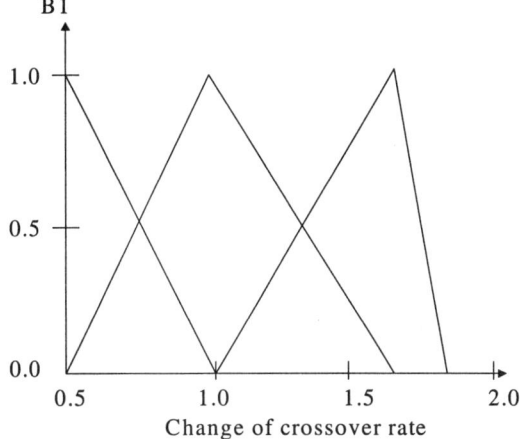

Figure 15. B1 membership function used in VB-SAGA 2.0

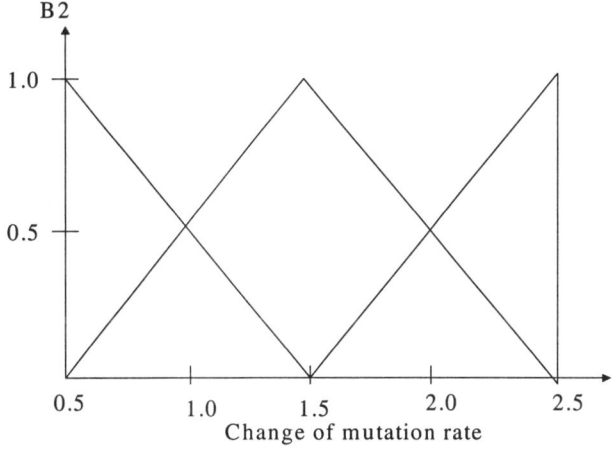

Figure 16. B2 membership function used in VB-SAGA 2.0

6.3   Results of VB-SAGA2.0

Table 20 gives an example of the best fitness from VB-SAGA2.0 with the self-adaptive GA feature on. In this example, the VB-SAGA2.0 started with population size 100, generations 70, crossover 0.75, mutation 0.012, and VMD-target 100. No spray parameter restrictions were set. The best fitness obtained was 9935.24 and the corresponding best spray parameters are also shown in the table. The final crossover rate was 0.9203 and mutation rate was 0.0125 due to self-adaptive change. The best fitness from VB-SAGA1.0 with the same initial conditions is also listed in the table for comparison.

We then ran two experiments to test VB-SAGA2.0 performance with the same initial spray conditions of experiments 1 and 2. That is, for experiment 1, we fixed DSD-VMD, Aircraft Number, Block Size, and Swath Width, while the other spray parameters were left to evolve. Experiment 2 was repeated for VB-SAGA2.0 with its same initial conditions as well. The results are shown in Table 21. The best fitness results from the two experiments of VB-SAGA1.0 are also listed for comparison.

We further ran several more tests with VB-SAGA2.0 repeating the conditions of experiments 3 to 6 to compare the performance of VB-SAGA1.0 and VB-SAGA2.0. Table 22 gives the details of the results.

Table 20. Results from VB-SAGA 1.0 and VB-SAGA 2.0.

	Max Fit	COV	VMD	Drift Frac
VB-SAGA1.0	9924.08	0.0242	99.58	0.00301
VB-SAGA2.0	9935.24	0.0215	100.73	0.00312

Table 21. VB-SAGA 2.0 results for experiment 1 and 2.

Exp	Max Fit	COV	VMD	Drift Frac	VB-SAGA1.0 Best Fit
1	9788.236	0.0632	102.132	0.0223	9710.885
2	9802.384	0.0312	205.434	0.0651	9750.743

Table 22. VB-SAGA2.0 results for experiment 3-6 (Aircraft ID omitted in this table).

Exp	Swath Width	Max Fit	COV	VMD	Drift Frac	SAGA1.0 Best Fit
3	2.25	9500.97	0.141	200.56	0.02797	8738.82
4	2.5	9327.26	0.265	200.25	0.03841	8494.48
5	2.2	9405.37	0.149	200.17	0.1536	8444.23
6	2.3	8386.54	0.213	199.85	0.01257	8357.87

As we can see from Tables 21 and 22, the self-adaptive VB-SAGA2.0 has obtained much better results than the regular VB-SAGA1.0 for experiment 3 to 6. However, in Table 21, the results of the self-adaptive VB-SAGA2.0 are only a little better than those of the VB-SAGA1.0 for experiments 1 and 2. One of the reasons for this difference is the degree of spray parameter restriction. Experiments 1 and 2 fixed four and seven spray parameters, respectively, while experiments 3 to 6 fixed only two parameters. As we know, the crossover and mutation operators apply on individuals to exchange their characteristics and maintain certain diversity. If many spray parameters are already fixed, the effect of crossover and mutation is reduced by a large extent. The self-adaptive GA in particular relies more on the proper functioning of crossover and mutation operators to optimize crossover and mutation as well as optimize spray parameters.

The self-adaptive GA is the latest addition to our SAGA project. We are still working on it, running more experiments to verify the results and attempting to improve the system based on the results and feedback. The adaptive GA has already been shown to be a feasible way to improve GA performance [14]. However, the implementation approach for different problems may differ greatly. The results of dynamic control of GA parameters in SAGA have indicated that this new feature can improve SAGA performance under our circumstances. We are expecting to add new dynamic control features in future improvements.

7   Summary and Conclusions

The development of SAGA consists of three stages as discussed in earlier sections, Fortran-SAGA, VB-SAGA1.0, and VB-SAGA2.0. The experimental results from these different versions of SAGA were evaluated by spray experts and regarded as extremely useful for practical applications. By using SAGA, the user is able to find optimal or near-optimal spray parameters in order to achieve minimal drift loss, uniform deposition and desired droplet size. SAGA can usually find the optimal or near optimal spray parameters in a short time. If the user presets one or more of the spray parameters, SAGA will spend even less time to find the optimal/near-optimal values due to the reduced complexity of the problem. The user is also able to use SAGA as a regular spray simulation program by specifying some or all spray parameters to obtain spray results, such as drift fraction, VMD and COV. Based on user feedback, we will be able to make further modifications to the user interface and the program operation. In addition, a new fitness formula is currently being developed. We are making continuous efforts to improve the GA as well as the overall user friendliness.

One new goal of interest is to apply SAGA to optimize additional practical factors in spray practice. An example of important factors affecting the spray time and cost is the flight path of the spraying aircraft. We currently assume the number of flight lines is determined by dividing the block width by the swath width and the aircraft follows these flight lines. However, many blocks have irregular shapes. Some blocks could be combined to reduce spray time. The problem of flying these blocks is a more complicated version of the famous traveling salesperson problem where a salesperson is expected to visit a group of cities in such an order that the total traveling distance is minimized [11]. We expect to add this new functionality to SAGA so that it will be able to find the optimal or near-optimal flight path.

It is also one of our future expectations to incorporate a multi-objective GA into the SAGA project. Our current work focuses on optimizing spray parameters to achieve maximal spray deposition, minimal evaporation loss, and even spray distribution. We combine all these objectives into one single fitness function. This approach may inadequately model the original problem [5]. Approaches such as min-max optimum [5], combination of the Pareto method with weights [6], and ranked solutions based on Pareto-optimal theory [8] have been taken to tackle such multi-objective problems. We plan to consider the feasibility of including multi-objective optimization in our future SAGA work.

References

[1] Bi, W., Potter, W.D., Twardus, D., Thistle, H., Ghent, J., Twery, M., and Teske, M. (2000), "Aerial spray optimization," *Proceedings of the International Conference on Artificial Intelligence*, (IC-AI'2000), Las Vegas, NV, pp.743-480.

[2] Bilanin, A.J. and Teske, M.E. (1984), "Numerical Studies of the Deposition of Material Released from Fixed and Rotary Wing Aircraft," NASA Contractor Report 3779.

[3] Bilanin, A.J., Teske, M.E., Barry, J.W., and Ekblad, R.B. (1989), "AGDISP: the aircraft spray dispersion model, code development and experimental validation," *Transactions of the ASAE*, 32(1): 327-334.

[4] Burgess, C.J., Chalmers, A.J. (1999), "The optimization of irregular computer architectures using genetic agorithms," *Baltzer Journals*.

[5] Coello, C.A.C., Dudnick, M., and Christiansen, A.D. (1994), "Using genetic algorithms for optimal design of trusses," *Proceedings of the Sixth International Conference on Tools with Artificial Intelligence*, pp. 88-94, New Orleans, LA, November.

[6] Cvetkovic, D. and Parmee, I.C., (1999), "Genetic algorithm-based multi-objective optimization and conceptual engineering design," *Proceedings of the 1999 Congress on Evolutionary Computation*, CEC'99, Washington D.C., July.

[7] Davis, L. (Ed.) (1991), *Handbook of Genetic Algorithms*, Van Nostrand Reinhold, New York.

[8] Dick, R.P., and Jha, N.K. (1997), "MOGAC: a multiobjective genetic algorithm for the co-synthesis of hardware-software embedded systems," *Proceedings of the 1997 International Conference on Computer-Aided Design* (ICCAD '97), San Jose, CA.

[9] Goldberg, D.E. (1989), *Genetic Algorithms in Search, Optimization, and Machine Learning*, Addison-Wesley Publishing Co.

[10] Holland, J.H. (1975), *Adaptation in Natural and Artificial Systems*, Ann Arbor: The University of Michigan Press.

[11] Jog, P., Suh, J.Y., and Gucht, D.V. (1989), "The effects of population size, heuristic crossover and local improvement on a genetic algorithm for the traveling salesman problem," *Proceedings of the Third International Conference on Genetic Algorithms*, Morgan Kaufmann Publishing, San Mateo, CA.

[12] Klir, G.J., St.Clair, U.H., and Yuan, B. (1997), *Fuzzy Set Theory: Foundations and Applications*, Prentice Hall, Englewood Cliffs, NJ.

[13] Kosko, B. (1991), *Neural Networks and Fuzzy Systems, a Dynamical Systems Approach to Machine Intelligence*, Prentice Hall, Englewood Cliffs, NJ.

[14] Lee, M.A. and Takagi, H. (1993), "Dynamic control of genetic algorithms using fuzzy logic techniques," *Proceeding of Fifth International Conference on Genetic Algorithms* (ICGA'93), Urbana-Champaign, IL, pp.76-83, July.

[15] Levine, D.M. (1993), "A genetic algorithm for the set partitioning problem," *Proceedings of the Fifth International Conference on Genetic Algorithms*, July, pp. 481-487.

[16] Michalewicz, Z. (1992), *Genetic Algorithms + Data Structures = Evolution Programs*, Spinger-Verlag, New York, NY.

[17] Mirkovic, J., Cvetkovic, D., Tomca, N., Cveticanin, S., Slijepcevic, S., Obradovic, V., Mrkic, M., Cakulev, I., Kraus, L., and Milutinovic, V. (1999), "Genetic algorithms for intelligent internet search: a survey and a package for experimenting with various locality types," *IEEE TCCA Newsletter*, pp. 77-87.

[18] Pabico, J.P. (1996), *A Genetic Algorithm Approach for the Determination of Cultivar Coefficients in Crop Models*, Master's Thesis, University of Georgia, Athens, GA.

[19] Parkin, C.S. and Wyatt, J.C. (1982), "The determination of flight-lane separations for the aerial application of herbicides," *Crop Protection*, vol. 1, no. 3, pp. 309-321.

[20] Parmee, I.C. and Dragan, C. (1999), "Use of preferences for GA-based multi-objective optimization", in Banzhaf, W. *et al.* (Eds.), *Proceedings of the Genetic and Evolutionary Computation Conference* (GECCO '99), Orlando, FL.

[21] Potter, W.D., Miller, J.A., Tonn, B.E., Gandham, R.V., and Lapena, C.N. (1991), "Improving the reliability of heuristic multiple fault diagnosis via the environmental conditioning operator," *International Journal of Applied Intelligence*, vol. 2, pp. 5-23.

[22] Potter, W.D., Pitts, R., Gillis, P., Young, J., and Caramadre, J. (1992), "IDA-NET: an intelligent decision aid for battlefield communication network configuration," *Proceedings of the Eighth IEEE Conference on Artificial Intelligence for Applications* (CAIA'92), pp. 247-253.

[23] Potter, W.D., Bi, W., Twardus, D., Thistle, H., Ghent, J., Twery, M., and Teske, M. (1999), "Intelligent decision support for US forest aerial spray," *Papers from the AAAI Workshop on Environmental Decision Support Systems and Artificial Intelligence*, Orlando, FL.

[24] Potter, W.D., Bi, W., Twardus, D., Thistle, H., Ghent, J., Twery, M., and Teske, M. (2000), "Aerial spray management using the genetic algorithm," *Proceedings of the 13th International Conference on Industrial and Engineering Applications of Artificial Intelligence and Expert Systems*, (IEA-AIE'2000) New Orleans, LA, pp. 210-219.

[25] Teske, M.E. and Curbishley, T.B. (1990), "Forest Service Aerial Spray Computer Model FSCBG 4.0 User Manual," Technical Report Number 90-06, Continuum Dynamics, Inc., Princeton, NJ.

[26] Teske, M.E., Bowers, J.F., Rafferty, J.E., and Barry, J.W. (1993), "FSCBG: an aerial spray dispersion model for predicting the fate of released material behind aircraft," *Environmental Toxicology and Chemistry*, vol. 12, pp. 453-464.

[27] Teske, M.E. and Barry, J.W. (1993), "Parameter sensitivity in aerial application," *Transactions of the American Society of Agricultural Engineers*, vol. 36, no. 1, pp. 27-33.

[28] Teske M.E., Barry, J.W., and Thistle, H.W. (1994), "Aerial spray drift modeling," *Environmental Modeling Vol. II.* P. Zanetti (Ed.), Computational Mechanics Publications, Boston.

[29] Teske, M.E., Bird, S.L., Esterly, D.M., Ray, S.L., and Perry, S.G. (1997), "A User's Guide for AgDRIFT 1.0: a Tiered Approach for the Assessment of Spray Drift of Pesticides," Technical Note Number 95-10. Continuum Dynamics, Inc., Princeton, NJ.

[30] Teske, M.E. (1998), "AGDISP DOS Version 7.0 User Manual," Technical Note Number 98-13, Continuum Dynamics, Inc., Princeton, NJ.

[31] Teske, M.E., Thistle, H.W., and Eav, B. (1998), "New ways to predict aerial spray deposition and drift," *Journal of Forestry*, vol. 96, no. 6, pp. 25-31.

Chapter 7

Genetic Algorithm Optimization of a Filament Winding Process Modeled in WITNESS

E. Wilson, C.L. Karr, and S. Messimer

With the advent of smaller, less expensive, and generally more effective computers, simulation models have become increasingly popular tools for solving engineering problems. More and more, engineers are turning to simulation environments in order to achieve increased system performance at a reduced cost. One such environment found to be very effective is WITNESS, a modeling program developed by AT&T and Istel. This chapter describes an effort to link a genetic algorithm with WITNESS in order to optimize a model of a manufacturing process called filament winding. Results show the genetic algorithm to be an effective, optimization tool for use with WITNESS models.

1   Introduction

As computer technology continues to improve at an increasingly rapid pace, engineers in all fields are expanding their use of computer simulation in research, design, and real-time analysis. Industrial engineering, in particular, has made heavy use of this new technology in the area of manufacturing systems analysis. Manufacturing plants and businesses are continually looking for cost-effective ways to improve performance and enhance research concerning the interactions between competing objectives. Simulation software is now available for companies to use in designing efficient production lines to implement at lower development costs.

Several simulation packages have suitable optimization programs, available for use either within their respective modeling environments or as add-on programs, which are closely linked with their respective simulation environments. One of the better known simulation environments, ProModel, has an add-on module called SimRunner that can be plugged directly into ProModel and used for multi-variable optimization. OptQuest, a program designed by Glover et al. [1], performs multivariable optimization in the Micro Saint 2.0 modeling environment via a scatter search. In addition, several researchers have successfully designed stand-alone optimization routines for use with specific simulation models. Brennan and Rogers [2] used infinitesimal perturbation analysis (implemented in the C++ programming language) to optimize a single assembly line model (designed in SLAM IV) that contained both manual and automatic stations connected by conveyors. Sammons and Cochran [3] optimized a cellular manufacturing system (simulated in SLAM II) by using a pattern search methodology. Morito et al. [4] applied simulated annealing with a simulation model to find an appropriate dispatching priority of operations to minimize the total tardiness for a commercial flexible manufacturing system.

Clearly, optimization programs that are developed outside of simulation environments can also serve as effective means of performing optimization. It is also clear that several different optimization methods can be used in combination with various modeling environments to solve a wide range of problems. However, as is typically the case in the field of optimization, several of the methods listed above are problem specific – that is, effective on that particular problem, but much more difficult to use on a different problem. Ideally, a developer would like to have available an optimization routine that would work equally as well on a wide range of simulation models – a robust algorithm. Genetic algorithms seem ideal for this task.

Genetic algorithms and evolution strategies have both been suggested and implemented for optimizing the performance of simulation models [5]-[7]. In particular, it seems that a genetic algorithm would be a very useful tool for performing optimization within the WITNESS modeling environment. WITNESS is a well-known, highly advanced simulation environment designed by AT&T and Istel that is generally geared toward manufacturing applications. Its interface is graphical in nature – production lines can be drawn from scratch by pointing and clicking with a mouse. Defining the specifics of production lines becomes a more complex task (several options exist for every component in WITNESS), but the data output that WITNESS gives from simulation runs is quite comprehensive. Genetic algorithms appear to be tailor-made for handling the complexity of WITNESS models while using the immense information WITNESS provides about a model for optimization.

Given this optimism for using genetic algorithms within WITNESS, a general research goal is formulated: use a genetic algorithm to optimize a filament winding model in WITNESS by adjusting specific model parameters in order to reach a minimum cost while also achieving a minimum production level. The genetic algorithm will modify eight parameters within the filament winding model, run the model for a preset amount of simulation time, and then evaluate the model's performance based on the number of filaments produced and the amount of cost incurred by the setup of the model. Results show that the genetic algorithm is consistent in finding good filament winding setups with low overall cost that meet the minimum production requirement.

2   Filament Winding Model

Filament winding is a manufacturing process used by the advanced composites industry to make polymer-based composite products. This process involves winding resin-coated fibers around mandrels, hardening (curing) the fiber/resin matrix, and then cutting and assembling the winding into a finished product. Products produced using the filament winding process include rocket-motor cases, helicopter blades, piping, tubing, and drive shafts.

A model of filament winding has already been fully developed in WITNESS [8]. This model is very complex, involving a great deal of accounting for scrap materials, material balance, and environmental effects within the model. However, when scaled down to its most essential components, the filament winding model in WITNESS is made up of six main components: (1) mandrel preparation, (2) filament winding, (3) curing, (4) mandrel removal, (5) finishing, and (6) quality inspection. A schematic of this model is shown in Figure 1.

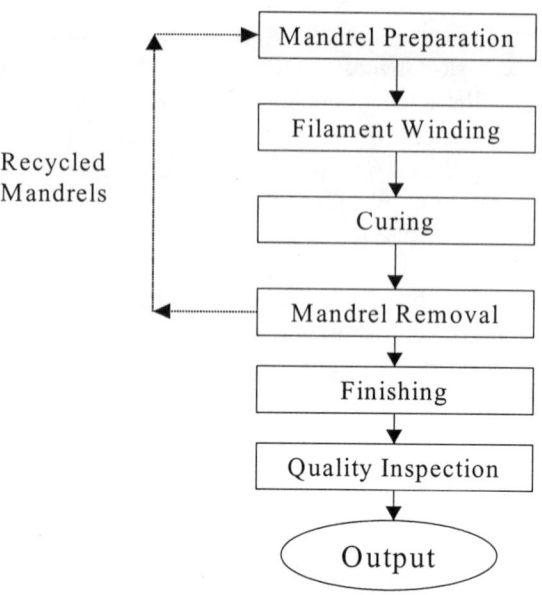

Figure 1. Filament winding model.

The first stage of the filament winding process is mandrel preparation. This stage involves taking a one-use mandrel or reusable mandrel and doing one or more of the following: applying a mold release to the mandrel, cleaning the mandrel, applying a liner to the mandrel, or attaching parts to the mandrel. The main role of mandrel preparation is to prepare the one-use or reusable mandrels for the second stage of filament winding, which is the filament winding itself. Filament winding involves winding a resin-coated fiber around a prepared mandrel. Most filament winding operations include a stage where fibers such as glass or graphite are covered in a mixture of resin, a curing agent, and various other additives. In other cases, the composite material already has the resin mixture impregnated in the filament fibers. In both cases, the wound filaments are then passed on to the third stage of filament winding, which is curing. Curing takes the wound filaments and hardens them. The curing itself may be done at room temperature, in an oven, or in an autoclave, and may be done either continuously or in batches. Once the filaments are baked and hardened, they are shipped forward to mandrel removal. In mandrel removal, the mandrels are removed from the cured filaments. One-use mandrels are discarded, while reusable mandrels are sent back to mandrel preparation for reuse. From here, the cured filaments are sent to finishing equipment. Finishing involves three different types of operations: (1) machining, (2) cutting, and (3) assembling. Machining involves removing material from the wound filament. Cutting divides the filament into equally sized parts. Assembly takes filaments and attaches them together. These three processes may be completed in any order. The final stage of filament winding is quality inspection. Throughout the filament winding process, poor quality filaments that can be visually inspected during processing are scrapped. Those filaments that have not been visually inspected are inspected at this stage. Quality inspection makes sure that poor quality filaments not already removed from the assembly are scrapped here.

Given the size and complexity of this manufacturing model, there are several parameters that the genetic algorithm could adjust in order to improve the filament winding model's performance. However, it is decided to limit the genetic algorithm to adjusting only eight parameters of the filament winding model. Specifically, it is decided that the genetic algorithm will adjust two parameters on each of four stages of the filament winding model – the mandrel preparation, the filament winding, the curing, and the mandrel removal. So, the genetic algorithm is given eight different parameters to adjust: two each for mandrel preparation, filament winding, curing, and mandrel removal.

For each workstation, it is decided that the genetic algorithm will adjust the number of machines at the workstation. Also, a certain amount of cost will be associated with each additional machine. This is implemented such that if the genetic algorithm uses too many machines in the filament winding model, the requisite number of parts will be produced within the time limit but at too high of a setup cost. This high cost will drive the genetic algorithm towards line setups that use fewer machines but still achieve the required throughput of filament wound parts. In addition, each workstation is given a second parameter for adjustment by the genetic algorithm. At mandrel preparation, the genetic algorithm adjusts the setup used by each machine before mandrel preparation is performed. At filament winding and curing, the genetic algorithm adjusts the batch size used by each machine. This batch size, like the machines themselves, also has an associated cost with it — larger batch sizes can work with more parts at a time, but at a more expensive cost to the filament winding manufacturing line. At mandrel removal, the genetic algorithm adjusts the cycle time (the time taken to complete the process done at the workstation) of each machine. This too has a cost associated with it – smaller cycle times will process filaments faster, but at a more expensive setup cost to the filament winding assembly line.

In summary, the genetic algorithm adjusts eight parameters within the filament winding model in order to achieve a minimum cost while also reaching a minimum throughput of parts. These goals are reached through a defined cost function. This cost function adds cost for each additional machine to the filament winding manufacturing line, as well as each increase in batch size, and each reduction in cycle time. Cost is also added for each filament below one thousand that is not produced within the time limit. In effect, this production requirement adds a "soft" constraint to the cost function – if the assembly line setup does not produce one thousand filament wound parts, the setup is not tossed out, but is given a penalty for each filament not produced. This penalty drives the genetic algorithm towards solutions that reach the throughput requirement. The specific cost function used by the genetic algorithm is shown below.

Workstations Costs:
Mandrel Preparation:
        $6,000 per machine per year
Filament Winding
        ($32,000 + $8,000 *(batch size -1)) per machine per year
Cure Oven
        ($16,000 +$3,200 *(batch size -1))/ per machine per year
Mandrel Removal
        ($2,000 + $325 *(3 - cycle time))/ per machine per year

Cost Formula:

Let A = No. of Mandrel Preparation machines
Let B = No. of Filament Winding machines
Let C = Batch size of Filament Winding machines
Let D = No. of Curing machines
Let E = Batch size of Curing machines
Let F = No. of Mandrel Removal machines
Let G = Cycle Time of Mandrel Removal machines

Let H = The number of parts produced by the filament winding model
Let I = 0   if H ≥ 1000
        = 1000–H if H < 1000

Total cost per part produced of filament winding model (in dollars) =

$$Cost = \frac{6000A + 32000B + 8000(C-1) + 16000D + 3200(E-1) + [325*(3-G) + 2000]F + 250I}{H}$$

3   Genetic Algorithm Interface

In genetic algorithms, the optimization input parameters are represented as strings of characters. The genetic algorithm manipulates these strings using three genetic operators – reproduction, crossover, and mutation. Reproduction takes the current population of strings (that have already been evaluated and given a fitness value), makes copies of the strings with better fitness values, and places these strings in a "mating pool." Here, the strings are paired up, and a percentage of the pairs trade parts of their strings. This is known as crossover. These newly formed strings are then subjected to a random screening, where a percentage of the population strings are selected and modified. This is known as mutation. The resulting strings from mutation form the next generation of population strings. These strings are then evaluated on their performance, given a fitness value, and again subjected to reproduction, crossover, and mutation. This combined process of exploiting knowledge about a search space (reproduction) and exploring a search space (crossover, mutation) is what drives the performance of a genetic algorithm. Over time, bad population strings disappear from a population, while good population strings live on and recombine with other good strings to form even better strings.

In most genetic algorithm research, bit-string (or binary) genetic algorithms are used. These "binary" genetic algorithms use a coding scheme of zeroes and ones to represent input parameters in the particular problem being examined. The effectiveness of such binary genetic algorithms has been documented in theoretical papers, as well as in real-world research. However, real-value genetic algorithms are becoming increasingly popular within the evolutionary computing community. Real-value genetic algorithms differ from their bit-string cousins in one major way: instead of using a coding scheme to represent parameters, the actual parameters themselves are used as inputs. This use of actual parameters within the real-value genetic algorithm provides it one major advantage over more traditional bit-string genetic algorithms. Bit-string genetic algorithms require that the user parameterize the input space. For example, given that there are thirty-two possible answers to a given problem, bit-strings with five bits in each string must be assigned to represent each parameter. For many research problems, this does not present any difficulty. However, in some cases, the number of input parameters is so large that the length of the bit string needed to represent all input parameters becomes huge and unwieldy. Also, the solution space is often not discrete, but rather continuous. Parameterization of the input space requires that possible solutions be excluded from the input space. Real-value genetic algorithms circumvent both of these problems by allowing the use of real numbers as search parameters. Because of the need to search a continuous input space, as well as the desire to avoid problems with bit strings of excessive length, a real-value genetic algorithm is applied to all of the research presented in this chapter.

The real-value genetic algorithm used in this research uses the same genetic operators as the bit-string genetic algorithm, but with slight modifications. For this case, tournament selection is used as the reproduction operator. Tournament selection selects a certain number of population strings at random (called the poolsize), and compares their fitness values. The population string with the best fitness value is then selected and placed in the mating pool. This process is repeated until the mating pool has the same number of strings as the original population. Standard single-point crossover [9] is then applied to the mating pool. Here, the eight parameters adjusted by the genetic algorithm are placed in sequence in the population string (parameter 1, parameter 2, etc.). Single-point crossover selects a point between two parameters, cuts two mating pool strings at that point, and swaps between the two strings the parameters to the right of the crossover point. Figure 2 below illustrates how this single-point crossover works in the real-value genetic algorithm used on the filament winding problem.

Figure 2 shows two parameter strings (String A and String B), both with eight parameters. These strings are randomly crossed between parameters three and four. Parameters four through eight then switch strings to form new strings, which are then subjected to mutation. Note that the crossover point is chosen randomly each time two strings are subjected to crossover.

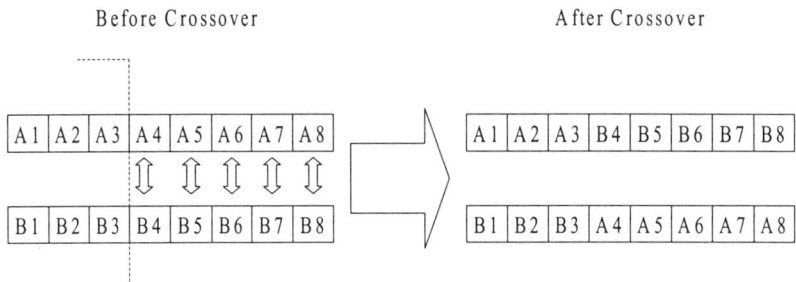

Figure 2. Single-point crossover, real-value genetic algorithm.

The new strings that are formed from crossover are then run through the mutation operator. Each parameter in each string is randomly assigned a mutation value between zero and one. If this value is lower than the specified mutation percentage, the parameter is changed to a random value between the specified minimum and maximum values for the parameter. Once the populations of strings have been subjected to tournament selection, single-point crossover, and mutation, the strings are then tested in the filament winding model in order to assign each string a cost value. These strings are then sent back into tournament selection, and the process of reproduction, crossover, and mutation is repeated.

Figure 3 shows how the genetic algorithm interacts with the filament winding model. The genetic algorithm here takes the model parameters determined in the population string and sends them to the model. The model is then run for two thousand five hundred minutes of simulation time. Once this task is completed, the genetic algorithm evaluates the model parameters by calculating its cost function value. This associated cost serves as the fitness of the genetic algorithm parameter string. The genetic algorithm evaluates each parameter string in a population of strings, adjusts the strings accordingly, and then repeats the process with the goal of finding the set of model parameters that produce the least amount of cost.

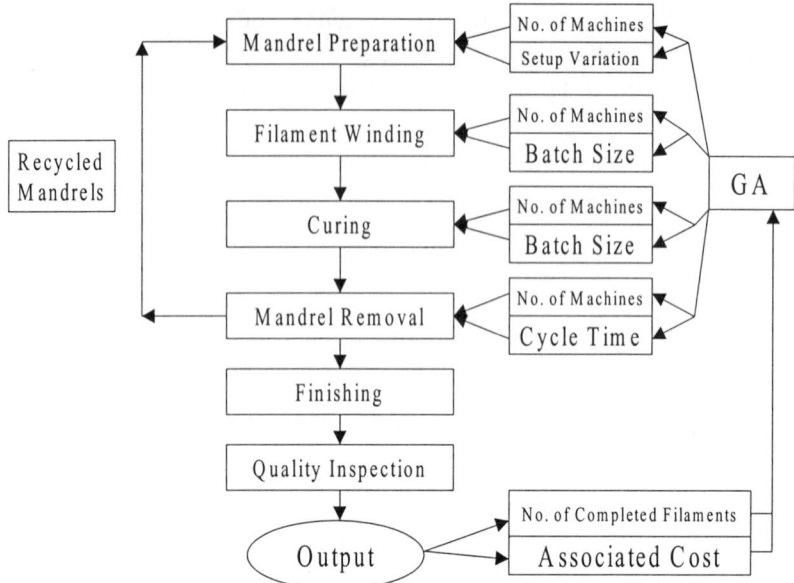

Figure 3. Genetic algorithm interaction.

As mentioned above, the cost function is calculated by assigning a cost for each additional machine on the filament winding model, as well as adding cost for increased batch size, for lowered cycle time, and for not completing a minimum number of filaments within the prescribed time. This cost function entices the genetic algorithm to find a solution that strikes a balance between efficiency and throughput. For example, if the genetic algorithm uses too many machines, enough filaments are made but the cost is overly high due to extra machines. However, if the genetic algorithm uses too few machines, it saves on the cost of the assembly line but adds cost for not completing the number of desired filaments.

Before running the genetic algorithm, each workstation is provided with minimum and maximum values for each adjustable parameter. These values are shown in Table 1. These minimum and maximum parameters served as boundaries for the actual cycle times sent to WITNESS by the genetic algorithm. Note that in this problem, each parameter sent to WITNESS is an integer ranging from the minimum to the maximum values. Also note that for Setup Variation under Mandrel

Preparation, this integer represents an overall setup scheme for the machines and is not a parameter itself in the model.

Table 1. Variable Ranges – Filament Winding Problem.

	Minimum	Maximum
Mandrel Preparation		
No. of Machines	1	5
Setup Variation	1	2
Filament Winding		
No. of Machines	1	5
Batch Size	1	5
Curing		
No. of Machines	1	5
Batch Size	1	20
Mandrel Removal		
No. of Machines	1	5
Cycle Time	1	3

Tournament selection is used by the genetic algorithm with a poolsize of four. Other important parameters used by the genetic algorithm include the number of generations run (100), the population size (40), the probability of mutation (0.1), and the probability of crossover (0.85).

4   Results

For the purposes of this research, only two methods of determining the lowest cost per part are used. The first is the genetic algorithm, and the second is a random search. For the random search, 2000 sets of parameters (the same total used by the genetic algorithm) are randomly generated and tested. Their best and average performances are calculated. Below are two charts portraying this performance. The first chart shows the average performance of the genetic algorithm and the random search. The second chart shows the minimum performance of the genetic algorithm and the random search.

Analysis of Figures 4 and 5 clearly demonstrates that the genetic algorithm outperformed the random search algorithm. The genetic algorithm is able to quickly find a result that did better than any result found by the random search. This improved performance is very important here. The biggest difference in the random search's result and the genetic algorithm's result is about $15 per part. Multiplying $15 by 1,000 (the number of required filament wound parts), we find that the total difference between the two assembly lines comes out to $15,000 – a substantial difference, especially when one considers the tremendous volume of parts typically passing through such an operation. By using a cost function as the fitness function of the genetic algorithm, the output given by the genetic algorithm can be easily interpreted for further use.

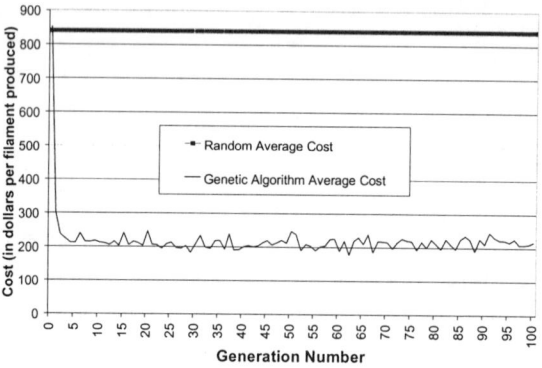

Figure 4. Average performance, optimization methods.

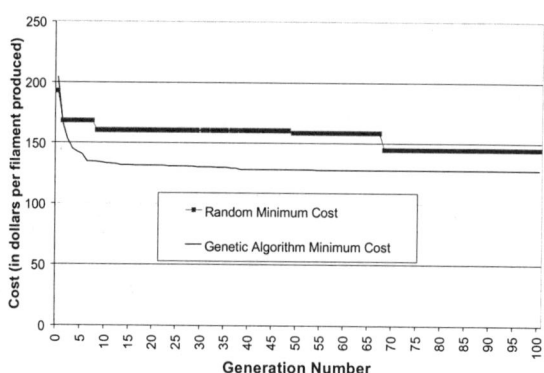

Figure 5. Minimum performance, optimization methods.

One of the concerns of this research is whether there is more than one "good" setup for the filament winding model. Below is a table comparing the best results of several genetic algorithm runs along with the best result found during the random search. Note that the genetic algorithm often repeated its own results on this problem – runs one and five resulted in the exact same answer, as did runs three and six, and runs four and seven.

Table 2. Best performances, filament winding problem.

	Genetic Algorithm Runs 1, 5	Genetic Algorithm Runs 3, 6	Genetic Algorithm Runs 4, 7	Random Search
Total Cost	$127.93	$123.44	$127.97	$144.82
Mandrel Preparation:				
No. of Machines	5	5	4	4
Setup Variation	2	2	2	2
Filament Winding:				
No. of Machines	1	1	1	1
Batch Size	4	5	4	3
Curing:				
No. of Machines	4	3	3	4
Batch Size	4	5	4	7
Mandrel Removal:				
No. of Machines	2	2	1	1
Cycle Time	1	1	1	1

All eight runs of the genetic algorithm did better than the random search algorithm (the other two runs achieved minimum costs of around $132). The three answers shown by the genetic algorithm above are very similar. They all use the maximum or near the maximum number of machines for mandrel preparation, only one machine for filament winding, a moderate to high number of machines for curing, and only one or two machines for mandrel removal. Each setup given by the genetic algorithm reached the goal of one thousand produced parts. In addition, each genetic algorithm output uses the same setup for mandrel preparation, nearly the same batch sizes for both filament winding and curing, and the minimum cycle time for mandrel removal. So, although the genetic algorithm's answers differ slightly, they all have the same basic trend.

5   Conclusions

Clearly, the genetic algorithm has proven to be an effective optimization method for use with the filament winding model. In experiments on other assembly line models, the genetic algorithm proved to be just as effective [10]. However, for the filament winding model as well as for the other models, the genetic algorithm was case-specific. In other words, the genetic algorithm was modified for each model in order to accommodate difference between models and optimization goals. For each trial case, changes in the genetic algorithm code were made in order to adjust the fitness function, the variables values (and the number of variables to be considered), and the genetic algorithm running parameters (maximum generation number, probability of crossover, probability of mutation, etc.). Thus, the genetic algorithm as it is presented here could be used on other WITNESS models, but the user would have to program the necessary changes to the genetic algorithm in Visual Basic.

Ideally, a genetic algorithm could be developed for general use on WITNESS models. This, in fact, is one of the main reasons for pursuing the research presented here. The University of Alabama in Huntsville (UAH) has been developing a user interface for WITNESS in the Visual Basic programming language. Their user interface would allow users to construct their assembly lines without having to "click-and-drag" components in WITNESS. In addition, the user interface would allow users to define parameters associated with each assembly line component (such as cycle time, conveyor length, etc.). Including an optimization tool in this user interface would add to the value of such an interface – the user would not only be able to completely define an assembly line, but be able to optimize it as well. Thus, a well-developed general genetic algorithm could be added as a component to a WITNESS user interface for use in optimizing constructed assembly lines for various goals.

However, this general genetic algorithm would require quite a bit of re-coding. The first part of this process would be to adjust the way in which variables are handled in the genetic algorithm. Currently, the number of model parameters to be adjusted in WITNESS is hard-coded into the genetic algorithm. Obviously, different users will want to adjust different model parameters when performing an optimization. Thus, the general genetic algorithm must allow users to specify the number of parameters to in the optimization process. Understand, however, that the user would be limited to a maximum number of parameters to adjust (because of memory constraints in Visual Basic). Also, each parameter must be assigned a minimum and maximum value, in order to specify the range of values available for each parameter. Again, this is currently hard-coded into the genetic algorithm. The general genetic algorithm would have to allow this information to be input into the genetic algorithm. Both of the tasks presented above would most likely be done via the Visual Basic interface designed by UAH.

Another important aspect of the genetic algorithm that would need to be re-coded would be the algorithm-specific parameters. A genetic algorithm has several parameters that "control" various aspects of its optimization. These parameters include the maximum number of generations, the probability of crossover, and the probability of mutation. Allowing the user to adjust these parameters would give the general genetic algorithm more flexibility for use on various problems.

The most important aspect of the genetic algorithm that would need to be re-coded to general use is the fitness function. In the research presented above, a cost fitness function was used. Clearly, different users might consider different optimization goals for their respective models. Thus, it would be necessary to allow the user to choose between several different fitness functions. Ideally, users would have the option of designing their own fitness function. However, given programming constraints, this is not practical. Again, users would choose an appropriate fitness functions from a Visual Basic interface.

In addition to the components listed above, it would be desirable to add the capability in the genetic algorithm of adjusting breakdowns in WITNESS models. This was a component of the research presented here that was not fully pursued because of problems in development. Assuming that these problems could be worked out, the user would then be allowed to adjust the breakdown/repair schedules of each machine. This would open up a new area of optimization for the genetic algorithm. Adjustments in possible fitness functions would need to be made in order to account for breakdowns in the model.

WITNESS itself has recently released its own optimization program for use. However, a comparison between the genetic algorithm and the WITNESS optimization package was not researched. This stemmed from one main reason – the groups that the authors have been doing WITNESS-related work with do not have the WITNESS optimization package because it is too expensive. Given its effectiveness, the genetic algorithm serves as a viable low-cost alternative for those WITNESS programmers looking to optimization of their simulation models.

6   Summary

This chapter examines the possibility of applying a genetic algorithm to optimize a filament winding process modeled in the WITNESS simulation environment. A real-value genetic algorithm is given the task of adjusting eight assembly line parameters within the filament winding model in order to achieve a minimum production throughput of filaments while reaching a minimum overall cost. The genetic algorithm proves to be an effective optimization method on the filament winding model. Extensions to the genetic algorithm can be made to allow the genetic algorithm to be applied to any model designed in WITNESS.

Acknowledgments

Many thanks to those at The University of Alabama, The University of Alabama in Huntsville, and MICOM who made this research possible. Specials thanks to Dawn Russell for her help in working with the filament winding model.

References

[1] Glover, F. (1996), "New advances and applications of combining simulation and optimization," *Proceedings of the 1996 Winter Simulation Conference*, pp. 144-152.

[2] Brennan, R. and Rogers, P. (1995), "Stochastic optimization applied to a manufacturing system operation problem,"

Proceedings of the 1995 Winter Simulation Conference, pp. 857-864.

[3] Sammons, S. and Cochran, J. (1996), "The use of simulation in the optimization of a cellular manufacturing system," *Proceedings of the 1996 Winter Simulation Conference*, pp. 1129-1134.

[4] Morito, S., Lee, K.H., Mizoguchi, K., and Awane, H. (1993), "Exploration of a minimum tardiness dispatching priority for a flexible manufacturing system – a combined simulation/optimization approach," *Proceedings of the 1993 Winter Simulation Conference*, pp. 829-837.

[5] Stuckman, B., Evans, G., and Mollaghasemi, M. (1991), "Comparison of global search methods for design optimization using simulation," *Proceedings of the 1991 Winter Simulation Conference*, pp. 937-943.

[6] Azadivar, F. and Tompkins, G. (1995), "Genetic algorithms in optimizing simulated systems," *Proceedings of the 1995 Winter Simulation Conference*, pp. 757-762.

[7] Hall, J. and Bowden, R. (1996), "Simulation optimization for a manufacturing problem," *Proceedings of the 1996 Southeastern Simulation Conference*, pp. 135-140.

[8] Russell, D. (1997), "A Methodology for Designing Modular Multi-Criteria Discrete Event Simulations," Dissertation published by The University of Alabama in Huntsville.

[9] Goldberg, D.E. (1989), *Genetic Algorithms in Search, Optimization, and Machine Learning*, Addison-Wesley Publishing Company, Inc.

[10] Wilson, E. (2000), "Genetic Algorithm Optimization of Assembly Lines Modeled in the WITNESS Simulation Environment," Thesis to be published by The University of Alabama.

Chapter 8

Genetic Algorithm for Optimizing the Gust Loads for Predicting Aircraft Loads and Dynamic Response

R. Mehrotra, C.L. Karr, and T.A. Zeiler

A genetic algorithm (GA) is shown to be a feasible approach to determining worst-case gust loads in aircraft structures. The mathematical modeling of extreme turbulence is discussed, as well as methods of analysis and prediction of aircraft response to atmospheric turbulence in the context of aircraft design. A representation of this problem for a GA approach is proposed and results of the technique are presented.

1   Introduction

Atmospheric turbulence can be dangerous for aircraft occupants and can present tremendous challenges for aircraft designers. Of course, pilots are faced with having to maintain control of their aircraft during turbulence that can be extremely violent. Aircraft designers, on the other hand, are faced with the problem of ensuring that the aircraft they design are capable of withstanding the severe dynamic loads that can be associated with turbulence. Digital flight records of turbulence encounters during scheduled airline flights have been found to contain cases of severe atmospheric turbulence occurring near mountains and thunderstorms. These instances of turbulence typically appear as sharp changes in the response of the aircraft. Other cases of severe turbulence are found in strong updrafts above thunderstorm buildups that may be undetected by onboard weather radar [14].

Pilots must be sufficiently trained to deal with such instances of turbulence. Further, they must spend a substantial amount of time in

flight simulators to ensure that they are prepared to react appropriately
to such turbulence encounters. Even with ample training, turbulence
presents pilots with perhaps their most challenging endeavor.

Aircraft designers are just as important as pilots when it comes to
dealing with turbulence. Designers must ensure that aircraft structures
can adequately support the loading that results from these natural
occurrences. Aircraft designs must be strong enough structurally to
survive extreme cases of such encounters. To assist designers in this
effort, it is highly desirable to (1) formulate a mathematical model of
extreme turbulence to be used in the design process [8] and (2) develop
methods of analysis and prediction of aircraft response to atmospheric
turbulence so that adequate designs can be developed.

In the past, a substantial amount of work has been done to model
turbulence using mathematical and statistical methods. Specifically [1]
has reviewed alternative theoretical descriptions of the turbulent wind
developed for the prediction of the response of aircraft. He states that
"the models of the wind have to accommodate those events that are
perceived as discrete, and described as gusts, as well as the
phenomenon described as continuous turbulence." Etkin outlined
current models for discrete gusts and random turbulence. He describes
discrete events as isolated encounters with steep gradients in the speed
of the air, typically occurring at the edges of thermals and downdraft, in
the wakes of mountains. These discrete gusts may also appear as rare
extremes of turbulence in clouds and storms. Etkin further argues that
these rare extremes are not adequately represented in the usual
Gaussian probability models of continuous random turbulence [1].

The statistical nature of turbulence has typically been treated in the past
by means of power-spectral-density (PSD) methods [1], based on the
theory of stationary Gaussian random processes. Such methods are
generally employed to meet current airworthiness requirements for
aircraft flight in "continuous turbulence," although it is generally
accepted that, even in continuous turbulence, the more extreme
fluctuations are not adequately represented by Gaussian models [1].

It has been shown that the PSD model [4] of turbulence in the current
airworthiness requirements leads to results that implicitly assume the

phase components of the turbulence to be uncorrelated, and achieves the exponential tails on the distribution of loads through the composition of a succession of "Gaussian patches." These results are mathematically expedient, but bear no relation to physical reality. Thus, it is imperative that new approaches be developed.

On the other hand methods of analysis and prediction of aircraft response to atmospheric turbulence have been developed in the past and are being used by different airframe manufacturers to compute gust loads to satisfy certification requirements. These methods include both stochastic and deterministic techniques.

One approach which is used for the calculation of time-correlated gust loads is matched filter theory (MFT) [10]. It has been shown, for instance by Zeiler [17], that the concept of MFT could be applied to the analysis of time-correlated, worst-case gust loads for linear aircraft models. This gust-load analysis method works well only for linear systems by taking advantage of the principle of superposition. Thus, the method is not applicable to nonlinear systems. Though it is important to note that computing time-correlated gust loads for linear systems in this manner has the twin advantages of being computationally fast and solves the problem directly. Historically, MFT was first utilized in the detection of returning radar signals. It was shown by Papoulis [11] that the principles of MFT can be used to obtain maximized responses without the use of calculus of variations and in applications other than signal detection.

Another recent method is based on Statistical Discrete Gust (SDG) method and its implementation for calculating the gust response in conjunction with new techniques, is primarily associated with wavelet analysis [6], [7]. This approach employs the SDG method [3], [4] as a gust-loads analysis method and calculates the gust response by convolving the linear aircraft impulse response function with single ramp inputs of varying length and a modified von Karman gust pre-filter to determine the wavelet surface from which the worst case input pattern is calculated. It is a time-domain approach and yields time-correlated gust loads. Again this approach is applicable only to linear aircraft models and is not robust enough to be applicable to non-linear systems.

The next section of this chapter describes in detail the particulars of the problem considered here, including the mathematical modeling technique used and the associated aircraft model. Section 3 provides the GA approach and representation scheme used. Section 4 presents the results from the application of this approach to the problem and Section 5 discusses the conclusions and research directions.

2   Problem Statement and Related Mathematical Underpinnings

The primary structure of an aircraft must be designed to withstand all of the static and dynamic loads it is expected to encounter during the life of the aircraft. In general, the dynamic loads are more difficult to determine. The dynamic loads that must be accounted for include landing and taxi loads, maneuver loads, and gust loads. Of these, gust loads are by far the most difficult to account for due to the stochastic nature of gust loads as compared to other dynamic loads that at least tend to be deterministic.

To assist designers in the effort to account for gust loading, it is highly desirable to formulate a mathematical model of extreme turbulence to be used in the design process. Further, methods of analysis and prediction of aircraft response to atmospheric turbulence are required for use in the aircraft design process. This chapter describes an effort to develop such a method of analysis based on the search characteristics of a genetic algorithm (GA). Specifically, the focus of the chapter is on applying a GA to the problem of identifying worst-case gust on the loading of a particular aircraft.

The problem as discussed above is to employ a GA to determine the worst-case or critical gust profile for specific loading conditions on a specific aircraft model. This worst-case gust is one that results in maximum values of one of four responses (wing root bending moment, wing root torque, engine lateral acceleration, or c.g. normal acceleration) in a given aircraft model when the given gust is used as the forcing function in the model. This problem statement gives rise to three fundamental questions that must be addressed: (1) what model of the atmospheric turbulence should be used in the study? (2) what methodology should be utilized to search for that "critical gust?" and (3) what aircraft model, or models, should be used to compute the peak response quantities?

2.1   Statistical Discrete Gust (SDG) Model

This study uses the SDG representation of atmospheric turbulence. This model of turbulence is preferred over the Power Spectral Density (PSD) model primarily because the PSD model assumes the phase components of the turbulence to be uncorrelated, and achieves the exponential tails on the distribution of loads through the composition of a succession of "Gaussian patches," a mathematical model which bears no relation to physical reality. On the other hand, it has been argued by Jones [8] that the SDG representation of atmospheric turbulence does take account of the phase correlations in measured turbulence. According to Jones this is accomplished by explicitly modeling the associated ramp-shaped gust components, and expressing the statistical description of the atmosphere in the form of probability distributions of patterns comprising both single and multiple ramp components.

Associated mathematical tools have been developed both for matching the SDG model to measured turbulence data, and for predicting the associated aircraft response statistics [5]. Furthermore, it has been argued that the SDG model of turbulence is unique in that it is the only approach for which mathematical tools have been developed that assign numerical parameters, specifying probabilities, so as to match data from measured severe gust encounters obtained during routine operational flying by civil airlines.

2.2   Methodology of the Search for a Worst-Case Gust

The approach used in the current study for discovering worst-case gusts is to combine the SDG model with a GA. This formulation has at least two very desirable attributes. First, the approach is applicable to both linear and nonlinear aircraft. Second, the method is easily adapted to account for a variety of definitions of "worst-case" gusts, e.g., the designer can easily incorporate a variety of criteria for determining exactly what a worst-case gust is.

The fundamental methodology relies on both a model of the SDG method and of a GA. It is implemented with the genetic-based search procedure that maximizes the peak response in a given aircraft load quantity, thereby determining the associated critical gust profile. The flow diagram shown as Figure 1 gives an overview of the SDG constrained optimization loop, which has been implemented for this study.

This diagram shows how the SDG search procedure has been formulated from four main stages that fit together within an overall framework. The stages include the generation of an excitation waveform from a gust generation function. The generated waveform is then passed through a "Modified von Karman" gust pre-filter for conversion into atmospheric turbulence. The generated atmospheric turbulence serves as a forcing function to a particular type of aircraft model. The dynamic response which occurs in the form of one of the four loads (wing root bending moment, wing root torque, engine lateral acceleration, or c.g. normal acceleration) is input into the optimization loop in order to search for the worst-case gust. The principal roles fulfilled by each of these modules are described in the following sections.

2.2.1 Waveform Construction

An excitation waveform is generated, as a function of time, from the set of discrete coefficients output from the optimization process. These coefficients are used in conjunction with a pulse function to define the excitation waveform that exists prior to the gust pre-filter. The pulse function used to model the discrete gust components is:

$$f(t) = \exp(-16\ t^2) + \exp[-4(-71/3 + 208\ t^2 - 640\ t + 2048\ t^6/3)]$$
$$\text{for } -0.5 \leq t \leq 0.5 \qquad (1)$$
$$f(t) = 0 \qquad \text{for } t < -0.5, t > 0.5$$

where

$t \quad = (tinp - tpos)\ /\ tlen;$
$tinp$ = time at which function is to be evaluated (sec.)
$tpos$ = position of the center of the pulse shape (sec.)
$tlen$ = length of pulse (sec.)

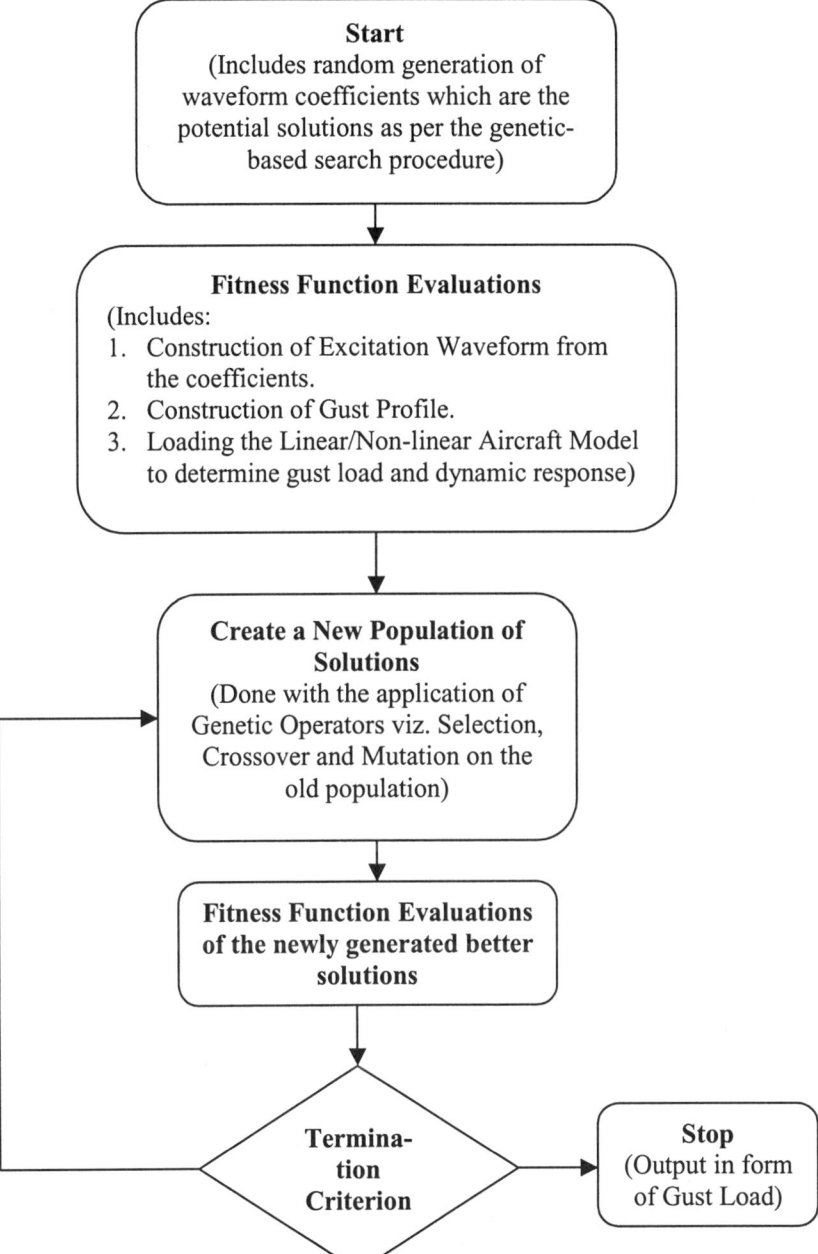

Figure 1. Schematic block diagram of the SDG constrained optimization loop.

The pulse function produces gust patterns that can comprise up to four pulse wavelets and these are free to overlap provided this does not increase the complexity of the gust pattern beyond that of the case being modeled. The "white space" waveform, $g(t)$, is reconstructed at discrete time intervals by substituting the estimated parameters from the search routine, into the following equation:

$$g(t) = \sum_{i=1}^{n} a_i f(\frac{t - b_i}{c_i}) \tag{2}$$

where

t = the time (sec.)

n = the number of pulses in the "white space" input pattern

a_i = the relative amplitude of the ith pulse in ft/s, where a_1 is fixed as ± 1.0

b_i = the time of the start of a pulse relative to the required simulation end time

c_i = the duration of the pulse (sec.)

The waveform, $g(t)$, is generated and the fitted pulses are free to overlap. Also the pulse amplitudes are parameterized as relative amplitudes where the first pulse is used as a reference. A typical example of a four pulse excitation wavefront is shown in Figure 2. Table 1 below shows the parameters associated with each pulse in the wavefront as discussed above.

Table 1. Parameters associated with each pulse in the waveform.

Parameter	First Pulse	Second Pulse	Third Pulse	Fourth Pulse
Amplitude (a_i)	$a_1 = 1.0000$	$a_2 = 0.7128$	$a_3 = -0.1626$	$a_4 = -0.7860$
Start Time (b_i)	$b_1 = 4.7278$	$b_2 = 7.1866$	$b_3 = 8.6974$	$b_4 = 5.3796$
Width (c_i)	$c_1 = 4.7278$	$c_2 = 1.3005$	$c_3 = 0.2111$	$c_4 = 0.8002$

2.2.2 Modified von Karman Gust Pre-Filter

This step provides a model of the atmosphere in the form which has been previously used for the von Karman PSD model of turbulence [13]. It converts the excitation waveform into atmospheric turbulence. The "gust generation" function incorporates the modified von Karman gust pre-filter defined by the following frequency response function:

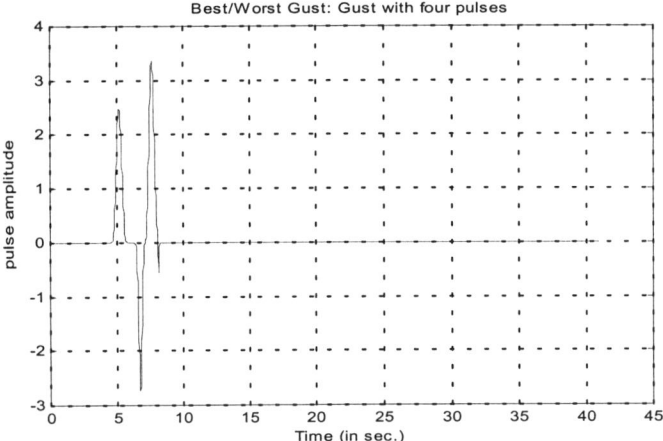

Figure 2. Excitation waveform.

$$H(i\omega) = \sqrt{\frac{L}{\pi V}} \frac{1 + \sqrt{\frac{8}{3}}(T_2 i\omega)}{(1 + T_2 i\omega)^{\beta}}$$

(3)

where

L = the scale length chosen to be (2500ft) 750m

V = the speed of the aircraft (TAS) (feet/sec)

β = the transfer function parameter (takes the value 10/6)

This frequency response function produces a one-sixth power law for the amplitudes of gusts whose gradient distance lies well below the turbulence scale length, L. Figure 3 shows the atmospheric turbulence when the excitation waveform shown in Figure 2 is passed through the gust generation function which incorporates the modified von Karman gust prefilter as discussed above.

2.2.3   Aircraft Model Simulation

The aircraft model simulates the gust response of the aircraft and control law model, which is described in a latter section in this chapter. The maximum peak value of the gust response in the selected load quantity (wing root bending moment, wing root torque, engine lateral acceleration, or c.g. normal acceleration) is returned to the optimization algorithm as a measure of the progress of the search. The four response quantities viz. wing root bending moment, wing root torque, engine lateral acceleration, and c.g. normal acceleration produced when the gust profile shown in Figure 3 is used as the forcing function to a linear aircraft model, are shown in Figure 4.

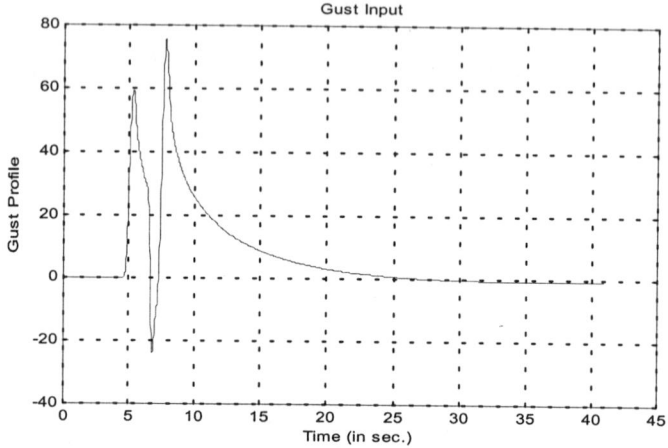

Figure 3. Gust profile.

2.3 Linear Aircraft Model

A linear aircraft model is employed to demonstrate the applicability of the proposed approach. Though it should be noted that the approach is valid for non-linear aircraft models as well.

A simplified linear model of a wide-bodied transport aircraft is used in the current study [12]. Sufficient characteristics of the full aircraft model are included to test the key elements of the SDG process and the GA search procedure, while sensibly limiting the order of the model and therefore the computing time necessary for a given simulation during this development study.

The flexible aircraft model chosen is a vertically symmetric model which has the rigid body heave and pitch freedoms, and the three flexible aircraft modes representing:

- first wing bending;
- engine lateral vibration;
- engine pitch/inner wing torsion.

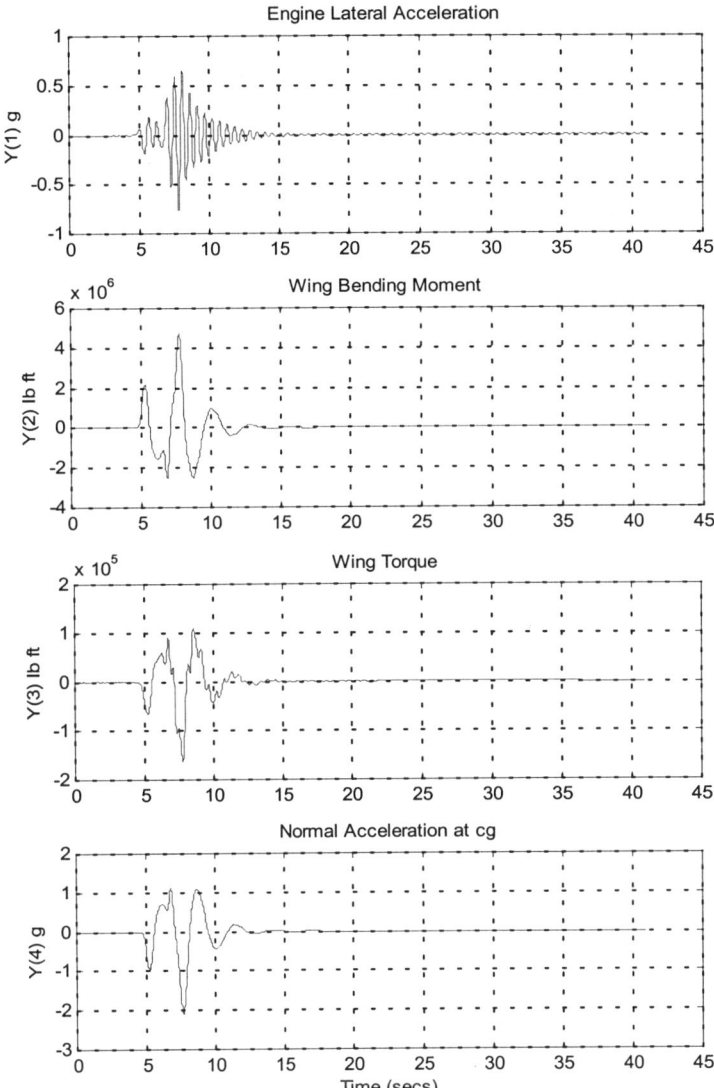

Figure 4. Linear aircraft gust response.

A structural damping set to 3% of critical damping has been assumed in all flexible modes.

The outputs from the aircraft model have also been reduced to include only a limited set of load quantities, again selected to provide a test for the key elements of the SDG method and the GA search procedure. The output quantities and their respective units are:

- engine lateral acceleration (g) [positive: left port];
- wing bending moment (lb. ft) [positive: right starboard wing tip down];
- wing torque (lb. ft) [positive: right starboard wing leading edge up];
- c.g. normal acceleration (g) [positive: upwards].

The wing bending moment was selected because it is a load quantity that is significantly affected by a Load Alleviation System based on a normal acceleration feedback. In addition, it includes the superposition of a higher frequency oscillation on a lower frequency mode. The engine lateral acceleration was chosen because it represents a lightly damped oscillation at a single frequency rather like a "tuning fork," and the wing torque was chosen because it is a load quantity that includes dynamic response from two closely coupled modes.

Three external inputs to the aircraft model are provided. The inputs are: vertical gust velocity, spoiler deflection and aileron deflection. However, as described by Pfeil [12] the dynamics of the control rotation modes have been eliminated from the model; only the static effects of the control rotations have been retained.

The aircraft aerodynamics are assumed to be instantaneous, and lift growth and gust penetration effects are not modeled. The steady state aerodynamics have therefore been evaluated at the single flight condition defined by:

Altitude   = 25000 ft
Mach   = 0.84
V_{EAS} = 571.9 ft/s (V_{TAS} = 854.3 ft/s)
All-up-weight = 290000 lb.

A more complete description of the aircraft equations of motion, their implementation and some example test case results could be found in Pfeil [12].

3   An Approach Using Genetic Algorithm

The optimization algorithm chosen for this study is the GA, which is an algorithm based on the principles of genetic search. Specifically, a floating point GA is employed to determine the worst-case, or critical, gust profile for specific loading conditions on specific aircraft models. The motivation for using a GA to solve this complex problem lies in the search characteristics of the GAs, which are probabilistic optimization approaches based on the mechanics of natural genetics [2]. GAs have been shown to be both robust and effective across a spectrum of nonlinear problems. They are able to rapidly locate near-optimal solutions in complex problem domains without requiring either derivative information or a continuous search space.

Also GAs provide an ideal platform for solving such a complex problem because they work with evaluation functions of arbitrary complexity. Objective, or cost, function complexity is tolerable by GAs because they treat the cost function as a "black box." All that is required of the black box is a measure of performance for the candidate solution. Hence, the contents of the evaluation function black box can contain the nonlinearities of the actual problem.

Genetic algorithms may be split into two different categories – binary GAs and floating point (or real value) GAs. Both types of GAs utilize the same basic genetic operators to drive their searches – selection, crossover, and mutation. Selection takes the current population of serach strings (that have already been evaluated and given a quantitative "fitness value"), chooses the strings with better fitness values, and places these strings in a "mating pool." Here, the strings are paired up, and a percentage of the pairs are "mated" with each other through crossover (splitting each string in half and combining one part of one string with one part on the other string). Thus, two new search strings are formed that part resemble the "parents" from which they were created. Once all of the new strings have been created, random search strings are chosen and "mutated" just slightly. Once these three steps are completed, the new GA population is evaluated by a "fitness function" and assigned a "fitness value." These search strings are then subjected to selection, crossover, and mutation once again. This combined process of exploiting knowledge about a search space (selection) and exploring a search space (crossover, mutation) is what drives the performance of a genetic algorithm. Over time, "bad" search strings disappear from a population, while "good" search strings live on and reproduce with other good strings to hopefully form better strings.

The process described above is essentially the same for both binary and floating point GAs. However, one major difference does exist between the two search methods. Binary GAs use a binary coding scheme to represent parameters in a search space. This is done to allow information existing in combinations of bits to be exchanged between strings by a GA during crossover. Floating point GAs, on the other hand, use the actual floating point values of the search space as their search parameters.

Floating point GAs have been used in earlier studies [15] and recently it has also been shown by Michalewicz [9] that more natural problem representation are more efficient and produce better results. Michalewicz has done extensive experimentation comparing float and binary GAs and has shown that the float GA is an order of magnitude more efficient in terms of CPU time. He has also shown that a real-valued representation moves the problem closer to the problem representation, which offers higher precision with more consistent results [9].

In float GAs as discussed above, binary strings are not used. Instead, the variables are directly used. Clearly, this results in slight modifications of the crossover and mutation operators. However, the basic application of each genetic operator stays the same.

The real-valued crossover operator termed *arithmetic crossover* developed by Michalewicz [9] and used in the implementation of float GA in this study is described as follows. *Arithmetic crossover* produces two complimentary linear combinations of the parents according to the following equations:

$$X' = rX + (1-r)Y$$
$$Y' = rX + (1-r)Y \qquad (4)$$

where X and Y denote individuals (parents) from the population and $r = U(0,1)$ is a random number generated from a uniform distribution.

The real-valued mutation operator is different from the binary mutation operator. There has been several real-valued mutation operators that have been developed and some of them are: uniform mutation, non-uniform mutation, multi-non-uniform mutation, boundary mutation and so on. The *uniform mutation* operator is defined and the reader is advised to refer Michalewicz [9] for complete definition of various other operators mentioned above. Let a_i and b_i be the lower and upper bound, respectively, for each variable i. Uniform mutation randomly selects one variable, j, and sets it equal to an uniform random number $U(a_i, b_i)$ according to the following equation:

$$x_i' = U(a_i, b_i), \quad \text{for } i = j$$
$$x_i' = x_i, \quad \text{otherwise} \tag{5}$$

Float point GA is used to search for the worst gust parameters, i.e. a gust described using four pulses. Eleven parameters or decision variables characterize the four-pulse excitation wavefront. The first pulse is characterized by two parameters: the start time of the pulse x_1 (width and start time is same for the first pulse); and its amplitude x_2. The subsequent remaining three pulses are defined by three parameters each viz. start time of the pulse, width of the pulse, and the amplitude, which act as design or decision variables and define different gust profiles. It is important to note that the GA searches for the pulses with the following parameter ranges shown below in Table 2.

Table 2. Upper and lower limits on the decision variables.

Parameter Range	First Pulse	Second Pulse	Third Pulse	Fourth Pulse
Start Time	$1.0 \leq x_1 \leq 5.0$	$5.0 \leq x_3 \leq 8.0$	$2.0 \leq x_6 \leq 8.0$	$2.0 \leq x_9 \leq 7.0$
Width	$1.0 \leq x_1 \leq 5.0$	$1.1 \leq x_4 \leq 2.0$	$0.1 \leq x_7 \leq 1.0$	$0.1 \leq x_{10} \leq 1.0$
Amplitude	$0.1 \leq x_2 \leq 1.0$	$0.15 \leq x_5 \leq 0.95$	$-0.95 \leq x_8 \leq -0.15$	$-0.95 \leq x_{11} \leq -0.15$

The excitation waveform is evaluated such that pulses are free to overlap, and the pulse amplitudes are parameterized as relative amplitudes where the first pulse is used as a reference. The parameters range as indicated above in Table 2 are decided so that the GA searches for the solution (gust waveform).

Float GA was run with the four fitness functions separately. The four fitness functions used to achieve the optimization goals are defined as follows:

- Fitness function 1 = $w1$ (wing bending moment), where $w1 = 10^{-6}$
- Fitness function 2 = (Engine Lateral Acceleration)
- Fitness function 3 = $w2$ (wing torque), where $w2 = 10^{-5}$
- Fitness function 4 = (Aircraft Normal Acceleration)

Table 3 shows the float GA parameters used for performing the simulation runs on each of the fitness functions. The crossover and mutation operator used for float GA is discussed earlier in this section.

The results for each of the test cases are explained in the following sections.

Table 3. Floating point genetic algorithm parameters.

Population size	100
Number of generations	200
Selection scheme	Tournament
Crossover scheme	Arithmetic crossover
Mutation scheme	Uniform mutation

4 Results of Approach on Linear Aircraft Model

Simulation runs were completed using a float GA for four specific objective functions stated in section three above. The results of these simulations are summarized in this section.

4.1 Wing Bending Moment

Figure 5 shows the convergence of float GA for this case. It is seen that the GA converges in 100 generations. The worst gust parameters obtained after 100 generations for maximizing wing bending moment are shown in Table 4.

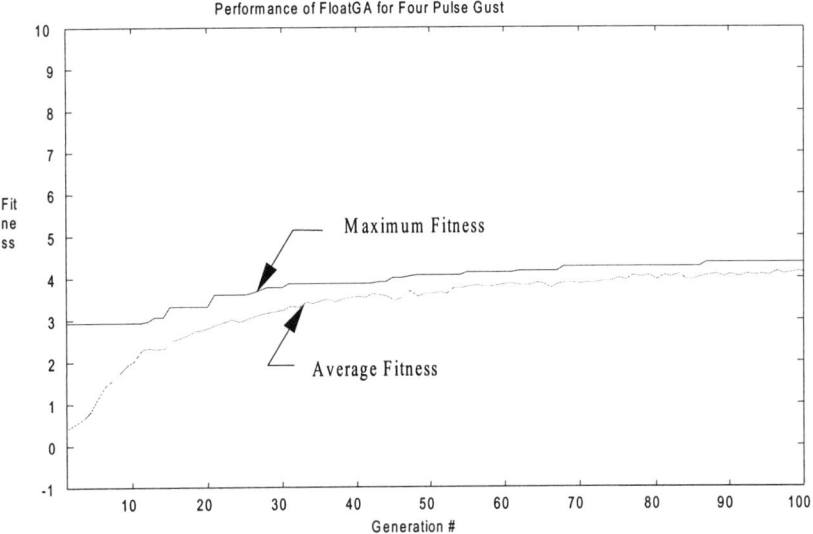

Figure 5. Float GA convergence for wing bending moment.

Table 4. Worst gust parameters for wing bending moment.

Parameter	First Pulse	Second Pulse	Third Pulse	Fourth Pulse
Start Time	$x_1 = 4.4063$	$x_3 = 6.6566$	$x_6 = 2.6487$	$x_9 = 3.0501$
Width	$x_1 = 4.4063$	$x_4 = 1.5097$	$x_7 = 0.3967$	$x_{10} = 0.1075$
Amplitude	$x_2 = 0.9999$	$x_5 = 0.9500$	$x_8 = -0.1500$	$x_{11} = -0.5032$

Figures 6 and Figure 7 show the best/worst excitation waveform and the corresponding gust profile respectively.

Figure 8 shows the linear aircraft response to the above worst gust. The wing bending moment found: 4.367e+006 lbft and occurs at time: 6.84 secs as is also seen from Figure 8.

4.2   Engine Lateral Acceleration

Figure 9 shows the convergence of float GA for lateral acceleration response quantity. In this case the run converges in 200 generations. The worst gust parameters obtained after 200 generations for maximizing engine lateral acceleration are shown in Table 5.

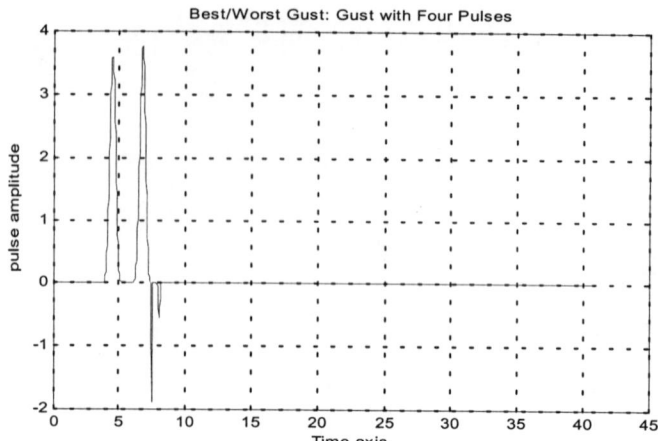

Figure 6. Excitation waveform.

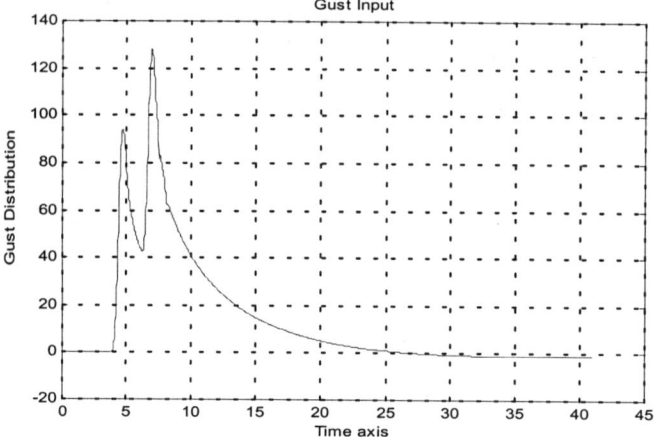

Figure 7. Gust profile.

Table 5. Worst gust parameters for engine lateral acceleration

Parameter	First Pulse	Second Pulse	Third Pulse	Fourth Pulse
Start Time	$x_1 = 4.5318$	$x_3 = 5.2776$	$x_6 = 4.0749$	$x_9 = 3.7537$
Width	$x_1 = 4.5318$	$x_4 = 1.1000$	$x_7 = 0.3487$	$x_{10} = 0.7041$
Amplitude	$x_2 = 0.9944$	$x_5 = 0.8136$	$x_8 = -0.9500$	$x_{11} = -0.1500$

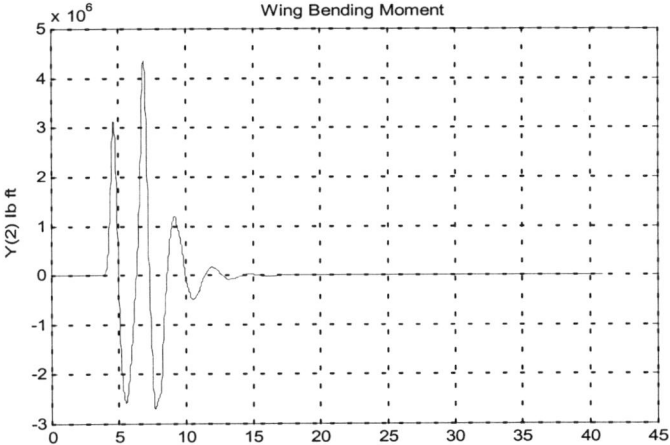

Figure 8. Linear aircraft gust response.

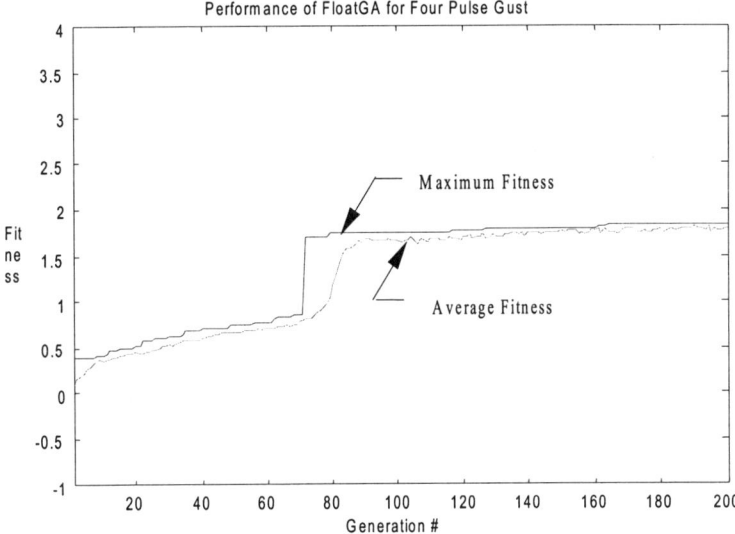

Figure 9. Float GA convergence for engine lateral acceleration.

Figures 10 and 11 show the best/worst excitation waveform and the corresponding gust profile respectively.

Figure 12 shows the linear aircraft response to the above worst gust. The engine lateral acceleration found: 1.83e+000g and occurs at time 8.12 secs. as is also seen from the Figure 12.

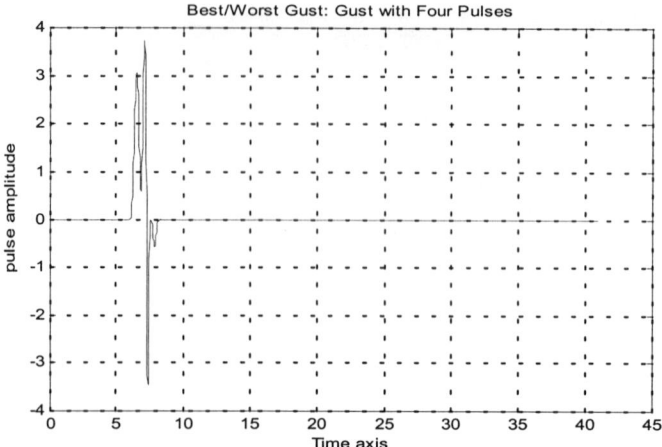

Figure 10. Excitation waveform.

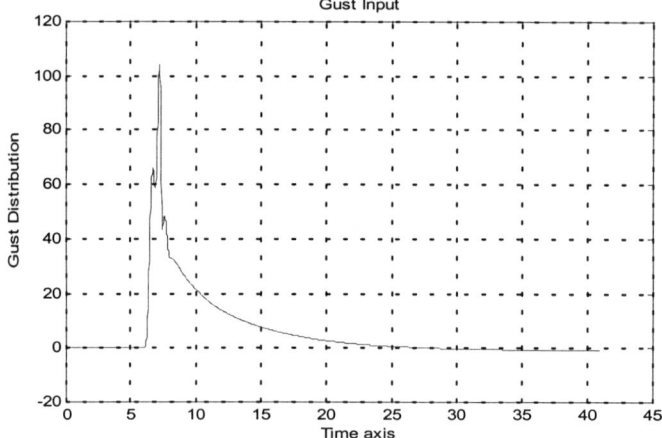

Figure 11. Gust profile.

4.3 Wing Torque

Figure 13 shows the convergence of float GA for wing torque response quantity. In this case the run converges in 100 generations. The worst gust parameters obtained after 100 generations for maximizing the wing torque are shown in Table 6.

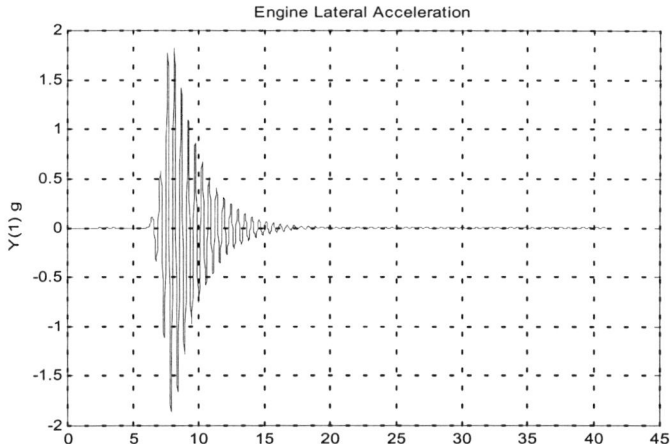

Figure 12. Linear aircraft gust response.

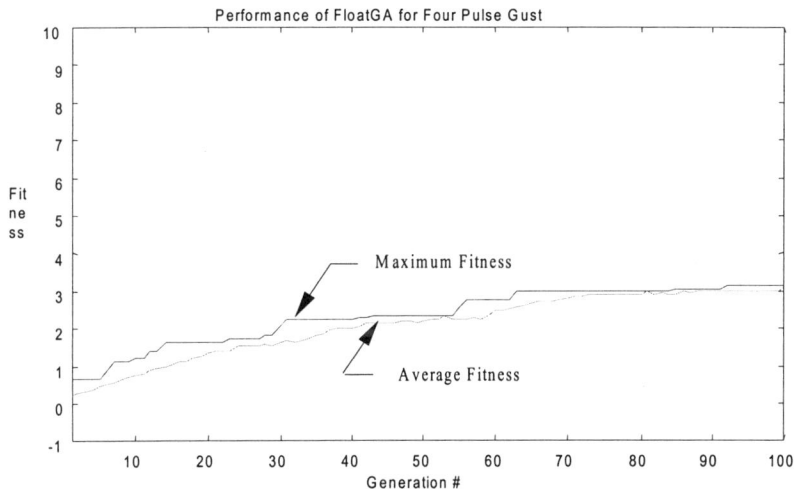

Figure 13. Float GA convergence for wing torque.

Table 6. Worst gust parameters for wing torque.

Parameter	First Pulse	Second Pulse	Third Pulse	Fourth Pulse
Start Time	$x_1 = 4.5992$	$x_3 = 5.0074$	$x_6 = 3.1891$	$x_9 = 3.1791$
Width	$x_1 = 4.5992$	$x_4 = 2.0000$	$x_7 = 0.8046$	$x_{10} = 0.3630$
Amplitude	$x_2 = 1.0000$	$x_5 = 0.9500$	$x_8 = -0.9499$	$x_{11} = -0.1500$

Figures 14 and 15 show the best/worst excitation waveform and the corresponding gust profile respectively. Figure 16 shows the linear aircraft response to the above worst gust. The wing torque found: 3.117e+005 lbft and occurs at time: 8.36 secs. as is seen from Figure 16.

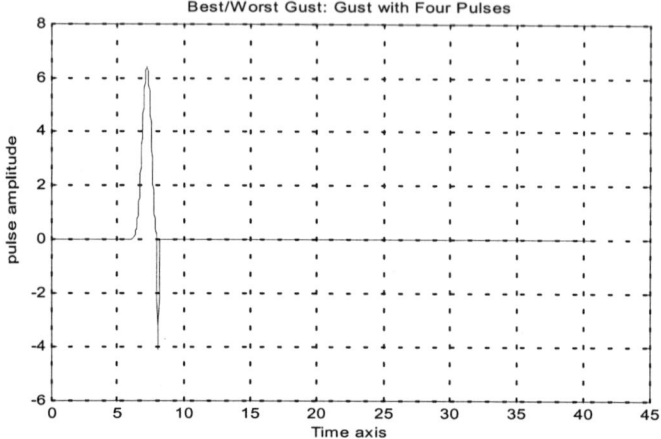

Figure 14. Excitation waveform.

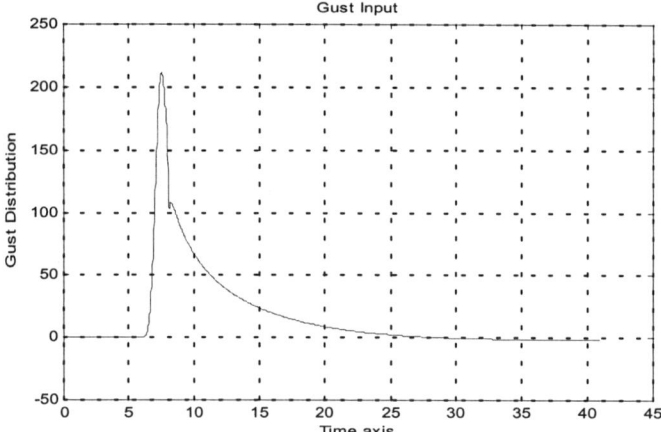

Figure 15. Gust profile.

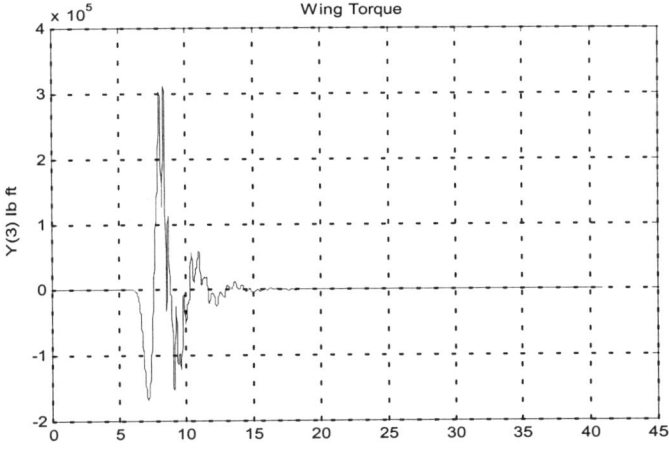

Figure 16. Linear aircraft gust response.

4.4   Aircraft Normal Acceleration

Figure 17 shows the convergence of float GA for normal acceleration response quantity. In this case also the run converges in 150 generations. The worst gust parameters obtained after 150 generations for maximizing the normal acceleration are shown in Table 7.

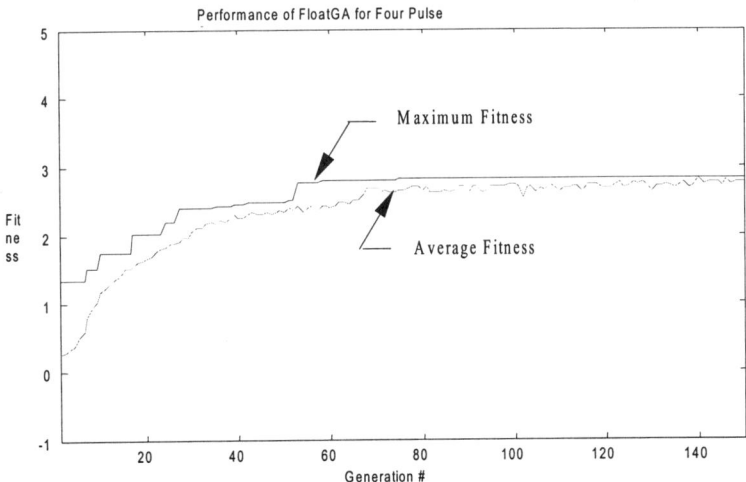

Figure 17. Float GA convergence for normal acceleration.

Table 7. Worst gust parameters for normal acceleration.

Parameter	First Pulse	Second Pulse	Third Pulse	Fourth Pulse
Start Time	$x_1 = 4.2846$	$x_3 = 6.5470$	$x_6 = 2.2713$	$x_9 = 2.2708$
Width	$x_1 = 4.2846$	$x_4 = 2.0000$	$x_7 = 0.8201$	$x_{10} = 0.8170$
Amplitude	$x_2 = 0.9999$	$x_5 = 0.9470$	$x_8 = -0.9500$	$x_{11} = -0.1785$

Figures 18 and 19 show the best/worst excitation waveform and the corresponding gust profile respectively.

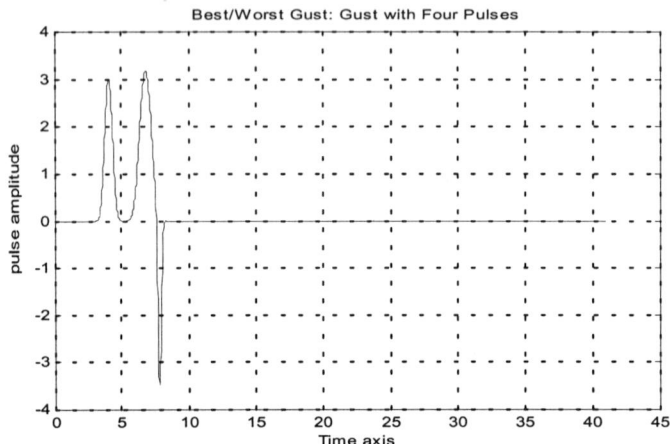

Figure 18. Excitation waveform.

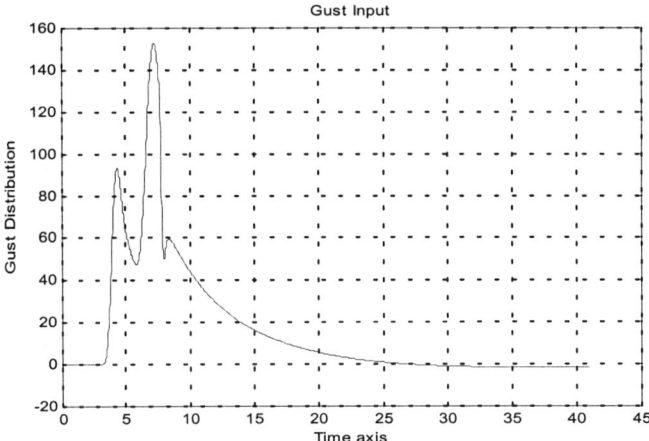

Figure 19. Gust profile.

Figure 20 shows the linear aircraft response to the above worst gust. The aircraft normal acceleration at c.g. found: 2.823e+000g and occurs at time: 7.86 secs. as is also seen in Figure 20.

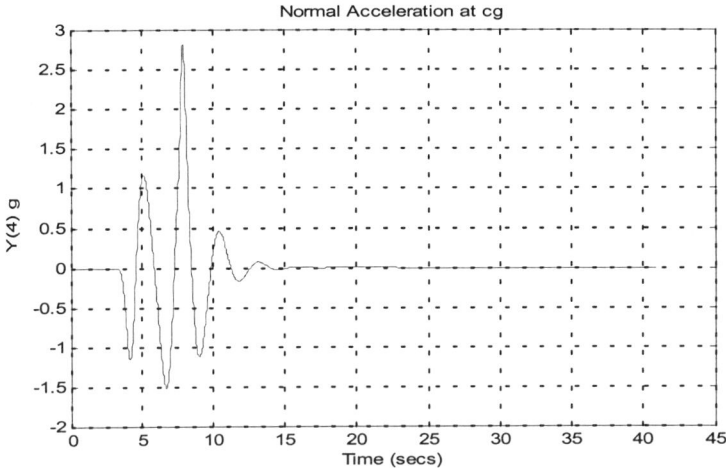

Figure 20. Linear aircraft gust response.

5   Summary and Conclusion

The objective of this study was to further develop a robust method of analysis to the problem of identifying worst-case gusts for the loading of a generic aircraft model based on the search characteristics of a non-traditional optimization algorithm viz. genetic algorithm. Table 8 summarizes the results obtained with float GA and also compares the results with those obtained using the wavelet analysis method [6], [7].

Table 8. Comparison of results.

Load Quantity	Maximized peak value using wavelet analysis method	Maximized peak value using float-GA
Engine Lateral Acceleration (g)	1.4774	**1.830**
Wing Bending Moment (lb ft)	3.1902E+06	**4.367E+06**
Wing Torque (lb ft)	2.4078E+05	**3.117E+05**
Aircraft Normal Acceleration (g)	1.3288	**2.823**

Acknowledgments

The authors would like to acknowledge the support of the Federal Aviation Administration (FAA) through University of Alabama Contract number: SDL-231-LE-22 and the Sterling Dynamics Ltd. for the research.

References

[1] Etkin, B. (1981), "Turbulent wind and its effect on flight," *Journal of Aircraft*, vol. 18, no. 5, pp. 327-345.

[2] Goldberg, D.E. (1989), *Genetic Algorithms in Search, Optimization, and Machine Learning*, Reading, MA: Addison Wesley.

[3] Jones, J.G. (1973), "Statistical Discrete Gust Theory for Aircraft Loads," Royal Aircraft Establishment, Farnborough, England, U.K., RAE TR-73167.

[4] Jones, J.G. (1984), "A Relationship between the Power-Spectral-Density and Statistical-Discrete-Gust Methods of Aircraft Response Analysis," Royal Aircraft Establishment, Farnborough, England, U.K., RAE TM Space 347.

[5] Jones, J.G. (1989), "Statistical discrete gust method for predicting aircraft loads and dynamic response," *Journal of Aircraft*, vol. 26, no. 4, pp 382-392.

[6] Jones, J.G. (1992), "Aircraft response to turbulence: recent research using a wavelet model," presented at *Gust Specialists Meeting*, Dallas, Texas, April 15.

[7] Jones, J.G., Foster, G.W., and Earwicker, P.G. (1993), "Wavelet analysis of gust structure in measured atmospheric turbulence data," *AIAA Journal of Aircraft*, vol. 30, pp. 94-99.

[8] Jones, J.G. (1994), "The case for a future airworthiness requirement for aircraft limit loads based on the statistical discrete gust

(sdg) method," presented at *Special Gust Specialists Meeting*, Hampton, Virginia, September.

[9] Michalewicz, Z. (1994), *Genetic Algorithms + Data Structures = Evolution Programs*, AI Series, Springer-Verlag, New York.

[10] Northy, D.O. (1943), "Analysis of the Factors which Determine Signal/Noise Discrimination in Radar," RCA Lab., Princeton, NJ, Rept. PTR-6C, June.

[11] Papoulis, A. (1970), "Maximum response with input energy constraints and the matched filter principle," *IEEE Transactions on Circuit Theory*, IEEE, New York, vol. CT-17, no. 2, May, pp. 175-182.

[12] Pfeil, A.L. (1996), "An Investigation of a Proposed Airworthiness Requirement Based on the SDG Method: Aircraft Mathematical Model," Stirling Dynamics Ltd. Report No. SDL-231-TR-1, Issue 1, June.

[13] Rosenberg, G., Cowling, D.A., and Hockenhull, M. (1993), "The deterministic spectral procedure for gust response analysis of nonlinear aircraft modes," (Stirling Dynamics Ltd. Report No. 233), *International Forum on Aeroelasticity and Dynamics*, Strasbourg, France, May.

[14] Wingrove, R. and Bach, R. (1994), "Severe turbulence and maneuvering from airline flight records," *Journal of Aircraft*, vol. 31, no. 4, pp. 753-760

[15] Wright, A. (1991), "Genetic algorithms for real parameter optimization," in Rawlins, G.J.E. (Ed.), *op. cit.*, pp. 205-220.

[16] Watson, G.H. and Jones, J.G. (1993), "Positive wavelet representation of fractations of fractals and chaos," in Crilly, A.J. et al. (Eds.), Springer Verlag.

[17] Zeiler, T.A. (1997), " Matched filter concept and maximum gust loads," *AIAA Journal of Aircraft*, vol. 34, no. 1, January-February.

Chapter 9

A Stochastic Dynamic Programming
Technique for Property Market Timing

T.C. Chin and G.T Mills

The problem we address in this chaper is to develop and test a market timing strategy for a property investor who has to decide the allocation of investment funds between the risk-free savings deposit and the comparatively risky property investment. We develop such a market timing strategy using a stochastic dynamic programming technique for optimal decision approach in the property market in Singapore. Using the information on the cash and property assets held by the investors as well as the predicted price movement, this strategy generates recommendation on investment action with the objective to attain superior investment returns. A recommended investment action is one of the four possible investment actions in a typical property investment scenario, namely: buy, hold, sell and wait. The investment performance of this market timing strategy is compared with that of the buy-and-hold strategy. Results from the simulation study for the 20-year period from 1977/Q1 to 1996/Q4 have been encouraging. The proposed market timing strategy is shown to capable of achieving superior investment returns in the Singapore property market.

1   Introduction

A traditional view on home-ownership, as noted more than five decades ago by Dean [1], states that the answer to the question on whether buying is more economical than renting depends very little on the comparison of ownership and renting at any particular point in time; it depends very much upon when the purchase is made as well as the trends of rents and purchase prices over the period after the purchase is made. This view recognises the importance of timing for buying property. This general view on the importance of timing being the central theme of this chapter conjectures the importance of timing in achieving superior return on investment in the private residential property market in Singapore. In this chapter, the importance of timing for property investment is not only limited to buying decision but also extended to selling decision. The market timing strategy, which shall be developed in this chapter, is the investment strategy that incorporates such timing decisions for buying and selling investment property assets.

Market events have generated an increasing interest in the property market in Singapore especially over the last two decades. Many investors and investment consultants are striving for the opportunity to exercise their investment talents in the property market in addition to the usual financial and futures markets in bonds, bills and equities. In this chapter, we demonstrate the potential of a stochastic dynamic programming approach to optimal decision making for market timing strategy in the property market in Singapore. This market timing strategy synthesises a set of investment recommendations that incorporate investor's financial capability with the forecasts of market price movement.

Market timing is defined as an investment strategy in which funds allocated to a designated set of assets are adjusted on an on going basis in response to changes in the investor's financial capability and the forecast of market price movement. In the present study, the designated set of assets is restricted to only two classes, namely, the savings deposit and the residential property in Singapore. Funds allocation adjustments occur quarterly. The size of the allocation adjustment is assumed to be 100% of available funds to a particular class of asset.

This chapter is organised as follows: Section 2 reviews the theoretical considerations. Section 3 specifies the market timing model. Section 4 incorporates the stochastic dynamic programming technique. Section 5 describes the data used in the simulation study. Section 6 discusses the investment performance and evaluation test. Finally, Section 7 concludes.

2   Review of Theoretical Considerations

William Sharpe [2] has indicated that at least 70% accuracy rate in market timing is required to make the practice worthwhile. Since achieving a 70% accuracy rate is extremely difficult, Sharpe's study suggests that funds managers and investors should minimize trading and emphasizes a buy-and-hold strategy. Shifts in the investment funds tend to add value only when there is a relatively high degree of confidence in the assessment of the investment environment.

However, subsequent studies in [3]-[5] report some success in asset allocation which is essentially a market timing strategy. Regressions of returns and dividend yields to track expected returns was used in reference [6] while a strategy of tactical asset allocation using logit analysis has been investigated in [7]. These studies show that their strategies can achieve greater terminal wealth with less variability of returns.

Our market timing strategy is different from those of the above studies in two aspects. Firstly, we employ a stochastic dynamic programming approach to optimal investment decision for the market timing strategy. Secondly, our strategy takes into account the investor's financial capability in terms of the cash available and the number of assets owned. The incorporation of investor's financial capability is similar to the idea used in a study of housing market strategy which includes the economic ability of individual home buyers [8].

There is more to managing a market timing investment scheme than just forecasting market movements. Our market timing strategy hinges on the theoretical framework of optimal decision. An optimal decision problem comprises a dynamic process to be controlled, a set of state variables describing the evolution of the dynamic process and a set of decision input to drive the dynamic process. Here, the dynamic process is represented by the investment process, the state variables are represented by the cash/property assets owned by the investor and the forecast of property price, and the decision input is the action taken on an investment recommendation in the property market. The optimal criterion is the maximisation of investment returns. The investment recommendation generated by the proposed market timing strategy is formulated as a non-linear function of the state variables with unknown coefficients. Specifically, stochastic dynamic programming procedure is used for the determination of these unknown coefficients. The market timing strategy therefore transforms the state variables into an optimal investment recommendation through the non-linear function.

In particular, the output of the market timing strategy is a set of recommended optimal investment action: BUY, HOLD, SELL and WAIT. The objective function of the optimal decision problem is the maximisation of terminal wealth at the end of the investment horizon. The constraint to this optimisation problem is represented by the sets of state transition equations describing the dynamics of the property investment process. In the context of stochastic dynamic programming formulation, we are interested in the one-step return function from the current decision stage and the optimal recurrent return function from the next decision stage to the end of the investment horizon. Investment action is taken at each decision stage. In this case, the investment action is taken at the start of each quarter throughout the investment horizon.

To begin with, we first assume perfect liquidity in the sense that buyers are readily available in the market. The "imperfect" liquidity constraint, if incorporated in our optimal decision approach, will make the problem more realistic. However, it must be noted that such a liquidity constraint is difficult to specify in the real-world situation and furthermore it is influenced by numerous external non-quantifiable factors.

Quarterly investment action has been chosen for our market timing strategy simply because of the availability of such property price index that is published quarterly. We can easily optimise the investment over longer period, e.g., a year, but the infrequent investment action is not a desired feature of optimal market timing model.

The market timing model developed in this chapter for the optimisation of investment action is different from other asset allocation decision, such as the modern portfolio theory approach. Here we maximise the expected return without explicit consideration of the risk involved in the investment decision. Furthermore, investor's financial strength has been specifically incorporated in our market timing strategy but not in the modern portfolio approach.

It can take many years before an investment manager can be certain that his or her strategic investment decisions have resulted in statistically abnormal returns [9]. Investment managers should, therefore, be interested in an approach that helps to improve the possibility of achieving superior abnormal returns. Strategic property investments are ex ante decisions. Whether these decisions turn out to be profitable or not can only be determined, after the event, by performance measurement. It should be evident that strategic decisions should be considered within the framework of achieving positive net present values (NPV). For a property investment to generate a positive NPV it must be bought when it is under-priced and sold when it is over-priced. This suggests an approach which tries to identify periods when the market as a whole is under of over priced and then focuses on individual property selection. This results in the strategy for buying and selling; and in this chapter, such an approach is the market timing strategy.

Evidence available in the literature shows that property market is not efficient. Therefore, the possibility of accomplishing superior return without exposing investment funds to commensurate higher risk is much better in the property market than in other investment markets (see referene [10]). Previous research indicates that relatively little attention has been given to, and few works have been done, in the area of analysing the role of a market timing strategy in property investment. In the property market, the amount of money involved is typically large. In view of these three reasons, it therefore makes good economic sense to research in this area of significant financial impact.

Study has shown that developing a successful strategy for investment in property is not easy [11]. Abnormal returns, net of transaction expenses, are difficult to achieve even though there is widespread belief among analysts and valuers that property markets are inefficient. This difficulty is compounded by the fact that reliable data on property performance is usually difficult to obtain. However, it is possible to make use of publicly available data in the way that may help investors making their investment decisions. If abnormal returns are difficult to achieve on a consistent basis, then the methods of analysis that give the investor some competitive advantage are worth pursuing.

3 Specification of Market Timing Model

The goal of a market timing model is to generate optimal investment recommendations for investors regarding the timing of the entry into or exit from the property market. This chapter evaluates the performance of the market timing model in the property market in Singapore for the 20-year period from the first quarter of 1977 to the last quarter of 1996. It is a common perception that a typical property market cycle in Singapore will last between 8 to 10 years (see, for examples, [12] and [13]). This 20-year period is considered adequate to cover two typical cycles of the property market.

Market timing must possess, as one of its features, the capability to predict the future behaviour of the market more accurately than the consensus implied by the market model. The recommendations of the market timing model are the decisions to BUY, SELL, HOLD or WAIT. These recommendations take into consideration of the comparison of the market's expectation and the model's expectation, i.e., the prediction of the price movement of the property price. If the model's expectation is higher than the market's expectation, then it recommends BUY or HOLD. On the other hand, if the model's expectation is lower than the market's expectation, then it recommends SELL or WAIT. In addition, it also takes into account the financial capability of the investor.

The market timing model is developed and constructed using a stochastic dynamic programming procedure (see reference [14] for an excellent theoretical discussion on stochastic dynamic programming). The intended task of this market timing model is to transform the observed input into the desired output. The input is the available current information comprising cash and the property assets owned by the investor as well as forecast of the property price movements. The desired output is a set of recommended optimal investment decisions, namely: BUY, HOLD, SELL and WAIT investment decisions.

Those organisations which offer strategic property investment advice based largely on forecasting market movements will find it difficult to show that their advice can result in consistent abnormal returns unless it is supported by a model which provides some estimate of expected returns. In other words, if everyone has similar forecasts concerning the way markets are expected to move in the future, the only way that abnormal returns can be identified is when the investor can perform better than average. Merely placing bets on properties which are expected to appreciate in price in the future is no guarantee of abnormal returns. A good investment strategy should, therefore, try to improve the probability of achieving abnormal returns in all phases of the market. The emphasis on trying to identify abnormal returns is important. Abnormal returns are equivalent to achieving positive net present values (NPV). This would happen if an asset can be sold for a price that is more than its value, or bought at a price less than its value. If investors are successful in achieving this objective then they should be able to generate positive abnormal returns and, therefore, positive NPV.

Property prices are expected to move in and out of being under- and over-priced over time. However, investors should not expect many properties to be grossly mispriced for long periods. This does not mean, however, that opportunities offering superior returns do not exist. There may be times when it is possible to locate a grossly mispriced property. What is difficult to know, however, is whether the property is really mispriced in economic sense, or whether the price, after allowing for expenses, merely compensates for risk. It must be pointed out that the focus of this chapter is not solely on under- or over-pricing of property, but rather on the timing of the acquisition and liquidation of property investments or equivalently to determine the optimal holding period if the property is owned or the optimal waiting period is the investor is ready to invest. The strategic decisions are to determine superior returns that is attainable as compared to the buy-and-hold strategy.

The various features unique to a typical investment include the cash inflows from the stream of rental incomes if the property is rented out and the sale proceed when the property investment is liquidated. On the other hand, the cash out-flows include the purchase price, the transaction expenses, and other incidental expenses related to mortgage interest and property maintenance expenses. Furthermore, a well-known feature of property investment is that it is comparatively for an investor to obtain from a bank or financial institution a high extent of leverage. Unlike many other regular debt financing where only interest payment is required before the loan matures, property mortgage loans are often structured such that the loan is amortised over the loan term with periodic repayments. It follows that the extent of financial leverage diminishes systematically over time while the cost of financing, i.e., mortgage interest rate remains unchanged. Hence, the risk of financial distress to the mortgagee also diminishes over time. Finally, property investment is considered reversible in the sense that the investment is not a suck cost because the investment can be liquidated when the property is sold.

From the financial analysis of investment property that produces income to the investor, four types of returns, namely: cash flow, tax shelter, equity build-up and capital appreciation can be identified [15]. Cash flow from investment properties varies from positive to negative depending on some fundamental determinants such as rental levels, financing levels and expense levels. Financing levels, which are influenced by the prevailing mortgage interest rate, typically constitute the largest drain on cash flow from rental receipts. Investors have historically been satisfied with negligible or negative cash flow from operation because they have little up-front equity in the investments and they have the ultimate aim of reaping major gains from the capital appreciation and tax shelter.

In order to construct a model for the property investment process dynamics in the contexts of stochastic dynamic optimisation, two investment asset classes are considered in this chapter. One asset class is the risk-free cash equivalents, represented by the savings deposits and the other class is the risky asset represented by the investment property. The property price movements are driven by some stochastic exogenous forces that influence the market. The interest rate of the savings deposit, the appreciation or depreciation rate of the property price and the rental income together constitute the important parameters of the market model. Wealth is defined in terms of stochastic variables.

Some property investment analysts argue that the termination decision results from a portfolio revision analysis, individual or institutional [16]. If the investment portfolio is monitored on a continuous basis, for example annually or quarterly, the dynamic asset allocation problem can be stated in three levels as follows:

   (i)   How much of which asset classes should be bought or sold?

(ii) Within the asset class, how much of which asset types or subclass should be bought or sold?

(iii) What specific assets should be bought or sold?

This problem can be highly complex for large portfolios. However, for most individual investor's portfolio, the problem reduces to manageable portfolio, such as cash equivalents, stocks, bonds, and property. In the case when only property is considered, the portfolios of most individual investors are usually small. If a portfolio has but a few specific property assets that are not well diversified, the risk of the portfolio will be greater than that of a property portfolio with more assets. Therefore, periodic revaluation becomes even more important.

Individual property investment should be re-evaluated regularly in most cases. This process can be relatively simple or highly complex. It must be noted that an investment's failure to perform as expected for part of the holding period does not necessarily mean that it should be terminated. The relevant period of any investment analysis is the future and not the past. Rarely does any investment under-perform or over-perform realistic expectation consistently for long periods. The essence for devising a market timing strategy is that expectations must be adjusted as additional and new information becomes available. During each period, the expected rate of return should be recalculated from anticipated future incomes with expectations adjusted as necessary, and the current after-tax equity. Equation (1) is the current after-tax equity of a property investment sold in period t may be defined as the sale proceeds that are net of selling expenses, the unpaid outstanding mortgage, and all taxes paid.

$$E_t = P_t - S_t - M_t - T_t \qquad (1)$$

where E_t = net after-tax equity in period t
P_t = probable selling price of property
S_t = selling expenses
M_t = outstanding mortgage
T_t = all taxes incurred

If the property is successfully sold, the net amount the seller will be retained for use in alternative investments [17]. The after-tax return on equity is generally recognised as the most significant measure of property returns because it explicitly considers the time value of money and the investor's ability to leverage the yield to a high degree, and then shelter the return from income taxes [18].

The basic after-tax present value model is analysed as follows. In order to specify the NPV model, the various parameters relevant to the model are now defined. Let the expected rate of capital appreciation be c, and let the initial net rental rate from the property be δ. At period t, the property is therefore expected to be worth $(1+c)^t$ and getting a rental of $\delta(1+c)^t$. Here, it is assumed that the rental grows at the same rate as capital appreciation. Mortgage loan is available at an interest rate of r, which is assumed fixed throughout the term of the mortgage loan, denoted T. Let D denotes the amount of mortgage loan obtained to finance the property purchase. Since this mortgaged loan is to be amortised over the period T, the constant periodic mortgage repayment, denoted by M, is given by $M = Dr(1+r)^T/[(1+r)^T -1]$. Further, let α (<1) be the transaction expenses incurred in the form of stamp duties, legal fees, maintenance and other incidental expenses. Let h be the holding period for the property so that at time $t = h$, the property is sold at the price of $(1+c)^h$. Note that $h = 0$ is equivalent to the decision not to invest in the property; and that $h = \infty$ represents the case where the property is purchased and held forever. Finally, if we define k as the required rate of return on the investment and define $x = min\{h, T\}$, we can specify the net present value of the terminal wealth resulting from the property investment, denoted by $V(h)$, for $h \geq 0$, in equation (2) as follows:

$$V(h) = \sum_{t=o}^{h} \frac{\delta(1+c)^t}{(1+k)^t} - \sum_{t=0}^{x} \frac{M}{(1+k)^t} + \frac{(1+c)^h}{(1+k)^h} -$$

$$\left[D(1+r)^x - \sum_{t=0}^{x} M(1+r)^t \right] \frac{1}{(1+k)^x} - (1+\alpha - D) \tag{2}$$

The various terms on the right-hand-side of the above are those defined above earlier. It is noted in equation (2) that future cash flow is discounted to the net present value at a constant rate of return on equity, k. In contrast to the deterministic IRR and NPV models, a risk analysis model is stochastic; that is, the value of many of the input variables are uncertain and must be statistically defined with associated probability distributions, rather than as single-point estimates (e.g., see [19]-[21]).

Property investment can be conceptualised as money flows over time. Five types of key decision makers in the property investment setting have been defined [15]. These decision makers are the developer, the joint venture partner, the construction lender, the mortgage lender and finally the equity investor. One common goal links all these decision makers is that each of them is attempting to maximise returns relative to associated risks. The primary focus of this chapter is on the equity investor. Like the securities investor and the equity investor considered in this study make decision primarily on whether or not to buy, and if so, when and how much? The property investment model described above cannot be used on its own for the formulation of a market timing strategy because of the presence of uncertain property prices and rents. Furthermore, it is not possible to incorporate the decisions to buy, sell, wait and hold in the typical property investment model. Therefore, a suitable market timing model needs to be appropriately specified formulated.

In this chapter, the concern is with the allocation of funds to two investment asset classes; i.e., private residential property representing the risky asset and savings deposit representing the risk-free asset. Decision theory and investment fit together in the following way. A particular investment market provides information to a decision maker who may use the information in choosing an action, for example, to get in or out of the market. The action may influence, at least partially, subsequent events. The decision maker examines the relationships among the state variables, the prediction of random state variables (such as price movements) and the selected action. A stochastic dynamic programming procedure is employed in this chapter to determine the relationships. The optimal investment decision, therefore, is one that maximised the terminal wealth, taking into accounts all incomes, expenses and costs. In the formulation of the market timing model for the property market, the incomes, expenses and costs are categorised into four groups, namely: transaction related costs, holding related costs, surplus related costs and shortage related costs.

The investment performance of the market timing model is analysed using the quarterly return on investment. Quarterly returns are calculated from the quarterly property price index published by the statutory board the *Urban Redevelopment Authority of Singapore*. Evaluation of timing performance involves examining whether the model is able to provide recommendations (either entry into or exit from the property market) that will be profitable to the investors.

To accomplish the task of market timing model development, the specific methodology is now elaborated. First of all, the outputs of the model, i.e., BUY, SELL, HOLD and WAIT are expressed mathematically as functions of the state variables representing the cash available (denoted by S_1), property assets owned (denoted by S_2) and the forecast of price movement (denoted by S_3). The above four investment decisions are categorised into the two types of decisions, namely: the *Asset Buying Decision* (either BUY or WAIT) and the *Asset Selling Decision* (either SELL or HOLD).

Consider the definition of *Asset Buying Decision* which is denoted by U_1 and expressed as a non-linear function of the state variables in equation (3) as follows:

$$U_1 = F_1 (X\ G_1) = 1 \text{ for BUY decision}$$
$$0 \text{ for WAIT decision} \qquad (3)$$

where F_1 is a signum function taking on value of either 1 or 0 depending on the argument;

X is the vector of individual and combination of state variables, i.e., $X = [S_1, S_2, S_3, S_1^2, S_2^2, S_3^2, S_1S_2, S_1S_3, S_2\ S_3]$;

G_1 is the vector of decision coefficients associated with the asset buying decision, i.e., $G_1 = [g_{11}, g_{12}, g_{13}, g_{14}, g_{15}, g_{16}, g_{17}, g_{18}, g_{19}]^T$.

Similarly, the definition of ASSET SELLING DECISION is denoted by U_2, and expressed as a non-linear function of the state variables in equation (4) as follows:

$$U_2 = F_2 (X\ G_2) = 1 \text{ for SELL decision}$$
$$0 \text{ for HOLD decision} \qquad (4)$$

where F_2 is a signum function taking on value of either 1 or 0
   depending on the argument;

   X is the vector of individual and combination of state variables,

   i.e., $X = [S_1, S_2, S_3, S_1^2, S_2^2, S_3^2, S_1 S_2, S_1 S_3, S_2 S_3]$;

   G_2 is the vector of decision coefficients associated with the asset
   selling decision,

   i.e., $G_2 = [g_{21}, g_{22}, g_{23}, g_{24}, g_{25}, g_{26}, g_{27}, g_{28}, g_{29}]^T$.

As can be seen from Equations (3) and (4) above, the specification
problem for the construction of the desired market timing model
reduces to the problem of finding the optimal values for the decision
coefficients vector G_1 and G_2. In this chapter, a stochastic dynamic
programming procedure is used to determine the corresponding values
of the elements of these two coefficients vectors.

In the formulation of a stochastic dynamic programming procedure, the
investment horizon is discretised into a sequence of N quarterly
investment decision stages. At each decision stage, the state variables is
spaced into a large number of design points. For each design point, an
optimal decision is determined using the stochastic dynamic
programming procedure. Let this optimal decision be denoted by U_1^*
for an optimal Asset Buying Decision and U_2^* for an optimal Asset
Selling Decision, respectively. Two sequences of such optimal
decisions, U_1^* and U_2^* for all the decision stages from stage 1 to stage
N can therefore be found.

From the knowledge of these optimal decision sequences, the estimated
decision coefficients for vectors G_1 and G_2 can readily be estimated
using a multivariate regression technique. Let the estimated decision
coefficients vector be denoted by $\underline{G1}$ and $\underline{G2}$; and these can be
estimated as follows:

$$\underline{G1} = (X^T X)^{-1} X^T U_1^* \tag{5}$$

$$\underline{G2} = (X^T X)^{-1} X^T U_2^* \tag{6}$$

As we can see, three distinctive phases can be identified for the
formulation process of the market timing model. The desired function
of the model is to generate the asset buying and asset selling decisions.
Firstly, these investment decisions are mathematically formulated as

non-linear functions of the state variables, individually and in combination. Secondly, the corresponding decision coefficients are estimated using a multivariate regression technique that regresses the optimal policies on the state variables. Thirdly, these optimal policies are to be determined by the stochastic dynamic programming procedure. The optimal objective for the maximisation of terminal wealth using the market timing strategy is therefore explicitly embedded in the formulation of the stochastic dynamic programming procedure.

4   Stochastic Dynamic Programming

The market timing strategy is designed to generate a recommended optimal investment action which is formulated as a function of the cash available, the asset held by the investor as well as the forecast of the property price movements. The coefficients of this function are determined such that an objective criterion, i.e., terminal wealth, is maximised. These coefficients are determined by regressing a set of optimal decision on the state variables. This set of optimal decision is the optimal policy derived from the stochastic dynamic programming procedure in which the various relevant property investment parameters and the definition of the objective function are required to be pre-specified. This objective criterion is expressed in terms of the income, expenses, and relevant investment parameters. Hence, the market timing problem has been cast into a maximisation problem where the objective is to maximise the return on investment, i.e., the terminal wealth, subject to the constraints of the investment dynamics. In this manner, we can conceptualise the proposed market timing procedure as a two-phase approach. The first phase is the off-line global optimisation planning using dynamic programming and the second phase is the on-line execution of optimal decision.

The off-line global optimisation planning phase consists of the following steps:

(1) Specify the objective function and the constraints represented by a set of state equations,

(2) Apply stochastic dynamic programming procedure to extract the optimal decision policy,

(3) Apply multivariate regression technique to the optimal decision policy to obtain the decision coefficients that relate the state variables to the optimal decision.

The on-line execution of optimal decision phase consists of the following steps:

(4) Observe the state variables representing the cash available, the asset held, and the property price prediction,

(5) Apply optimal decision according to step (3) as a function of the state variables observed in step (4).

The required investment dynamics are expressed as a set of state transition equations that balance the amount of money and the number of the properties owned by the investor from one stage to the next after an investment action has been taken. We shall now consider the definitions of the following state transition equations (7), (8) and (9), the return function equation (10) and the overall objective function equation (11).

$$S_{1next} = [(1+r)S_{1current}.(1-U_{1current})] + [(1-se)S_{2current}S_{3current}U_{2current}] \quad (7)$$

$$S_{2next} = S_{2current} + U_{1current} - S_{2current}\,U_{2current} \quad (8)$$

$$S_{3next} = S_{3current}\,(1 + \Delta S_{3next}) \quad (9)$$

$$\phi_k(S_k, U_k, I_k) = -beU_1 - seU_2S_2 + rS_1 + yS_2S_3$$
$$- mg(S_3 - S_1)U_1 - it.r.S_1 - pe.S_2.S_3 \quad (10)$$

$$J_k(S_k) = \text{Maximise } [\phi_k(S_k, U_k, I_k) + J_{k+1}(S_{k+1})] \text{ with respect to } U_k \quad (11)$$

where $S_{1current}$ = cash available to the investor at the current stage
     S_{1next} = cash available to the investor at the next stage
     $S_{2current}$ = number of properties owned at the current stage
     S_{2next} = number of properties owned at the next stage
     S_{3next} = predicted property price at the next stage
     $S_{3current}$ = actual observed property price at the current stage
     ΔS_{3next} = the predicted per unit price change
     $U_{1current}$ = investment action = 'buy' at the current stage
                [Note that $(1-U_{1current})$ means 'wait' or 'do not buy']
     $U_{2current}$ = investment action = 'sell' at the current stage
                [Note that $(1-U_{2current})$ means 'hold' or 'do not sell']

S_k = the vector of state variable at decision stage k,
U_k = the vector of decision variable at decision stage k,
I_k = the vector of property price movement at decision
                     stage k,
$\phi_k(S_k,U_k,I_k)$ = the one stage return function,
$J_{k+1}(S_{k+1})$ = the overall objective function.
be   = buying expenses
se   = selling expenses
pe   = property maintenance expenses
r   = risk-free interest rate of savings deposit
y   = rental yield
mg   = mortgage interest rate
it   = income tax rate

Cash available at the next stage, i.e., equation (7), is equal to the sum of the following two components:

(1) the cash available at the current stage incremented by the risk-free interest if the current stage investment action is to 'wait', i.e., do not buy; and

(2) the sale proceeds from a number of properties sold at the current market price at the current stage decremented by the transaction expenses.

The number of property owned by the investor at the next stage, i.e., equation (8), is the sum of the following three components:

(1) the number of property owned by the investor at the current stage;
(2) the investment action at the current stage is 'buy'; and
(3) the number of property sold at the current stage.

Equation (9) defines the prediction of property price movement. Equation (10) defines the one-stage return function or the immediate reward representing the sum of the following seven components:

(1) the buying expenses if the investment action taken is 'buy';
(2) the selling expenses if the investment action taken is 'sell';
(3) interest earned from the savings deposit of cash available;
(4) rental income from the property owned;
(5) mortgage repayment for mortgage taken up to finance the 'buy';
(6) personal income tax for the interest earned from the deposit; and
(7) property related maintenance expenses.

Equation (11) defines the overall objective function which comprises of two parts: the first part is the immediate one-stage return function, and the second part is the optimal future return function representing the return function from the next stage onwards until the end of the investment horizon.

Dynamic programming algorithm is used to recursively solve the above overall objective function equation (11) subject to the constraints of the state transition equations (7), (8) and (9) and the one stage income function equation (10). The recursive solution process is repeatedly executed, in a backward recursive sequence, starting from the last decision stage until the first decision stage, i.e., from stage N to stage 1. As a result of solving the recursive equation starting from stage N to stage 1, we obtain the sequence of optimal decisions in terms of the optimal policy $\{U_1^*, U_2^*, \ldots, U_N^*\}$. This optimal policy will yield an optimal terminal wealth at stage N, i.e., the end of the investment horizon. The market timing strategy decision coefficients are then estimated using the multivariate regression technique for the optimal decision policy on the state variables.

The resultant market timing strategy generates the following optimal decision functions:

$$Y_i = F_i (Z_i) \tag{12}$$

where Y_i is the recommended optimal investment action
Z_i is the set of input representing consisting of state variables
F_i is the function representing the optimal decision rules

F_i is the optimal decision function determined from the first phase of the market timing strategy modelling process by the stochastic dynamic programming procedure. The expression for the market timing input Z_i is as follows:

$$Z_i = [S_1 \; S_2 \; S_3 \; S_1^2 \; S_2^2 \; S_3^2 \; S_1S_2 \; S_1S_3 \; S_2S_3] \times G \tag{13}$$

where $G = [\; g_1 \; g_2 \; g_3 \; g_{11} \; g_{22} \; g_{33} \; g_{12} \; g_{13} \; g_{23}]^T$ is the vector whose elements are the decision coefficients determined from the multivariate regression technique.

5   Data Used in the Simulation Study

As explained in the previous section, the optimal decision approach to the construction of a market timing strategy comprises the following four steps:

(1)   Defining the goal of the market timing strategy.
(2)   Designing an algorithm for transforming the input into the output with the aim of achieving the goal defined in step (1).
(3)   Implementing the market timing algorithm.
(4)   Comparing the performance of the market timing strategy with a known strategy.

The evaluation of the investment performance of the market timing strategy requires a comparison of the relative performance of the active market timing strategy with that of the passive buy-and-hold strategy. We use the quarterly residential property price index published in the *Property Statistics Series for Price and Rental Indices* by the Urban Redevelopment Authority of Singapore. The price indices are the widely used indicators of price movement for the property market and are compiled from caveats lodged with the Registrar of Titles while rental information is collected through surveys. The publication is available five weeks after the end of each quarter.

Quarterly data series on the property investment and the interest rates of the savings deposit are gathered for the period from 1977/Q1 through 1996/Q4 (a total of 80 quarterly data points for each series). This twenty-year period, containing approximately two property market cycles, is deemed to be sufficiently long to justify the use of statistical analysis.

6   Performance and Evaluation Tests

A market timing strategy's approach of recommending investment action (when to be in the market and when to be out of the market) may not be optimal for either the bull-market or the bear-market if the costs of error in mis-recommendation for bullish and bearish periods are not equally weighted. For example, if the cost of error in mis-recommendation during the bullish periods (i.e., wrongfully recommending exit from the bullish market) is higher than the cost of mis-recommendation error during the bearish periods (i.e., wrongfully recommending entry into the bearish market), then only the exit recommendation is optimal while the entry recommendation is not optimal. Alternatively, if the cost of mis-recommendation error during the bearish periods (i.e., wrongfully recommending entry into the bearish market) is higher than the cost of mis-recommendation error during the bullish periods (i.e., wrongfully recommending exit from the bullish market), then only the recommendation to enter the market is optimal while the recommendation to exit the market is not optimal.

The investment performance of our market timing strategy can be evaluated by analyzing the error of mis-recommendation, i.e., wrongfully recommending entry into market for bearish periods and recommending exit from the market for bullish periods. Similarly, selling too early (or waited too long to buy) during bullish periods and buying too early (or holding too long before selling) during bearish periods are errors of mis-recommendation. These errors can be classified as Type I (false alarm) and Type II (failure to detect) errors as follows:

(1)   Type I Errors: (Exit from market wrongly; false alarm):
  • Sell too early during bullish periods
  • Wait too long to buy during bullish periods

(2)   Type II Errors: (Entry into market wrongly; failure to detect):
  • Buy too early during bearish periods
  • Holding too long before selling during bearish periods

The correct recommendations are buying (or holding) during the bullish periods and selling (or waiting) during the bearish periods.

In general, a strategy may not be optimal in recommending entry into the market when the mis-recommendation cost of Type I error (in exiting the market) is higher than the mis-recommendation cost of Type II error (in entering the market). Under such circumstances, predicting a rare occurrence of bear markets incorrectly is relatively costly; thus, attempting to predict bullish markets correctly at the expense of predicting bearish markets incorrectly may not be appropriate, see [7] for reference.

An investor using a market timing strategy would like to know the investment performance of such market timing strategy in comparison with the alternative passive investment approach such as the buy-and-hold strategy. A 72-quarter (i.e., 18-year) data sample is gathered for the assessment of the investment performance of the proposed market timing strategy in comparison with the theoretically-ideal perfect timing and the buy-and-hold strategies. Observation and estimation of each of the three state variables representing the input data are fed into the market timing decision function to generate the optimal investment recommendation.

We begin by building a base strategy using the stochastic dynamic programming procedure for an investment horizon of 72 quarters from 1979/Q1 through 1996/Q4. Given these base strategy coefficients, we use the strategy to generate an investment recommendation for the quarterly decision stages throughout the investment horizon.

Table 1 presents the overall comparison of investment performance by various timing strategies for the period from 1979/Q1 to 1996/Q4. Three investment strategies that are examined are the theoretically-ideal perfect timing strategy, our market timing strategy and the buy-and-hold strategy. For the 72-quarter investment horizon, the property price moves up in 49 quarters, moves down in 21 quarters and stays flat in 2 quarters. The property price is observed to move in eight up-trends and eight down-trends. The market attained a net present value of 4.66 times the initial capital invested; this figure translates to an equivalent annual returns of 8.93%. The buying and selling expenses are assumed to be, respectively, 5% and 2% of the transacted price. The risk-free rate of return is assumed to be stay constant at 4% per annum.

Examination of Table 1 reveals that our market timing strategy executed 7 buy decisions and 7 sell decisions throughout the investment horizon. In comparison with the perfect timing strategy which executed 23 buy decisions and 8 sell decisions, our market timing strategy has missed out 16 buy and 1 sell opportunities. The reason of missing out these buy opportunities is due to the fact that the investor does not have the required cash to buy even though the market offers the buy opportunities. Despite the fact of missing out the opportunities to buy and sell, the equivalent annual returns of our market timing strategy is relatively attractive at 16.42% in comparison with returns of 13.54% attained by the buy-and-hold strategy. It is also observed that the performance of our market timing strategy is closer to that of the perfect timing strategy than the buy-and-hold strategy. The above results show the superior performance of our market timing strategy.

Table 1. Overall comparison of investment performance for various strategies.

Investment Strategies	Perfect Timing	Market Timing	Buy-and-Hold
Number of BUY decisions:	23	07	01
Number of SELL decisions:	08	07	01
NPV of Terminal Wealth:	20.09	15.43	09.84
Equivalent Annual Returns (%):	18.14	16.42	13.54
Type I Error:			
Number of incorrect decisions (SELL or WAIT) in UP quarters:	0	16	22
Type II Error:			
Number of incorrect decisions (BUY or HOLD) in DOWN quarters:	0	01	07
Total Number of incorrect decisions:	0	17	29
% of incorrect decisions:	0%	20.83%	40.28%

Number of UP quarters:	49 (in 8 UP trends)
Number of DOWN quarters:	21 (in 8 DOWN trends)
Number of FLAT quarters:	02
Total Number of quarters:	72
NPV of Market Returns:	4.66 times the initial capital invested
Equivalent Annual Returns:	8.93%
Risk-free Returns per year:	4%
Buying Expenses:	5% of transacted price
Selling Expenses:	2% of transacted price
Cash Downpayment	0%

6.1 Performance of Market Timing Strategy

To test the investment performance of the proposed market timing strategy with respect to the variations of investment horizon, transaction expenses and the cash required for down payment, we evaluated three investment strategies based on their end-of-period terminal wealth. Strategy A is the theoretically-ideal perfect market timing strategy. Strategy B is our market timing strategy. Strategy C is the buy-and-hold strategy. The "Market Index" which represents the property price index is included as a reference for the purpose of comparison. The difference between the buy-and-hold strategy and the market index is that expenses and incomes are incorporated in the buy-an-hold strategy whereas there is neither expenses nor incomes for the market index.

6.2   Comparison for Various Investment Horizons

In order to evaluate the investment performance for different lengths of investment horizon, simulation studies were conducted for three categories of investment horizon, namely: short, medium and long investment horizons. For the purpose of this Chapter, short horizon is represented by 1-year or 2-year period; medium horizon is represented by 3-year or 5-year period; and long horizon is represented by 10-year or 13-year period. Table 2 presents the summary of results of this comparison of investment performance. Buying and selling expenses are fixed, respectively, at 9% and 1% of the transacted price. The risk-free return from the savings deposit is 4% per annum.

From the inspection of Table 2, it can be seen that our market timing strategy achieves a high rate of returns relative to that achieved by the perfect timing strategy for all categories of investment horizon. The results also reveal that the market timing strategy is performing better for long investment horizon.

It is necessary to highlight the occurrence of poorer performance of the perfect timing strategy for the 3-year horizon as compared to our market timing strategy as seen in Table 2. Such apparently poor performance is due to the high transaction expenses for the frequent shifting of funds back and forth between the asset classes. The perfect timing strategy has been designed with the assumption of perfect information of market movement and that a buy (or sell) decision is executed if the market is moving up (or down) in the subsequent quarter regardless of the impact of the transaction expenses on the overall investment return.

Table 2. Comparison investment performance for different investment horizons.

	Market Timing	Perfect Timing
(A) Short Investment Horizons:		
1-Year:		
NPV of Terminal Wealth:	1.0212	1.0250
Mean Annual Returns:	0.0212	0.0250
SD of NPV:	0.0019	0.0109
Efficiency Ratios (NPV/SD):	551	94
2-Year:		
NPV of Terminal Wealth:	1.3277	1.3320
Mean of Annual Returns:	0.1523	0.1541
SD of NPV:	0.0077	0.0092
Efficiency Ratios (NPV/SD):	172	144
(B) Medium Investment Horizons:		
3-Year:		
NPV of Terminal Wealth:	1.5959	1.4842
Mean of Annual Returns:	0.1686	0.1407
SD of NPV:	0.0628	0.0593
Efficiency Ratios (NPV/SD):	25	25
5-Year:		
NPV of Terminal Wealth:	2.0569	2.1699
Mean of Annual Returns:	0.1552	0.1676
SD of NPV:	0.2321	0.1043
Efficiency Ratios (NPV/SD):	8	20
(C) Long Investment Horizons:		
10-Year:		
NPV of Terminal Wealth:	3.3069	4.0500
Mean of Annual Returns:	0.1270	0.1501
SD of NPV:	0.6499	0.9118
Efficiency Ratios (NPV/SD)	5	4
13-Year:		
NPV of Terminal Wealth:	4.0559	4.6486
Mean of Annual Returns:	0.1137	0.1249
SD of NPV:	0.2262	0.3182
Efficiency Ratios (NPV/SD):	7	14

Risk-free Returns per year:	4%
Buying Expenses:	9% of transacted price
Selling Expenses:	1% of transacted price
Cash Downpayment	0%

6.3 Comparison of Efficiency Ratios

Efficiency ratio is defined in this study as the ratio of the mean value of the NPV to the respective standard deviation. NPV is the net present value of the terminal wealth in terms of the multiples of the initial capital invested. The efficiency ratio, therefore, can be viewed as the uncertainty-adjusted mean value of NPV since standard deviation is a measure of uncertainty. Higher values of efficiency ratios imply higher confidence in the computed value of NPV values.

Comparison of efficiency ratios for various investment strategies including the market index is also presented in Table 2. The results in Table 2 indicate that the proposed market timing strategy has the efficiency ratios comparable with those of the perfect timing strategy except for the very short (1-year) and very long (13-year) investment horizons. Thus, the results suggest that our market timing strategy is not only effective but also efficient. Hence, the results provide further evidence in support of the proposition of the superior investment performance of the proposed market timing strategy.

6.4 Comparison for Various Transaction Expenses

Table 3 presents the results of comparison of investment performance for different transaction expenses. Three sets of transaction expenses are specified for this comparative study, namely: (a) both the buying and selling expenses are set at 0%; (b) buying and selling expenses are set at, respectively, 9% and 1% of the transacted price; and (c) buying and selling expenses are set at, respectively, 12% and 5% of the transacted price. The investment performance is examined for short, medium and long investment horizons.

Table 3 reveals that higher transaction expenses lead to lower rate of returns, as expected, across all investment horizons. It is further observed that the investment performance of our market timing strategy with respect to the investment horizon is dependent on the transaction expenses. For zero transaction cost, it is noted that our market timing strategy performs better for medium investment horizon. As transaction expenses increase, the investment performance improves for longer horizon.

Table 3. Comparison of investment performance for different transaction expenses.

	Market Timing			Buy-and-Hold		
Buying Expenses	0%	9%	12%	0%	9%	12%
Selling Expenses	0%	1%	5%	0%	1%	5%
(A) Short Horizons (2-Year):						
NPV of Terminal Wealth:	1.55	1.33	1.22	1.36	1.26	1.19
Mean of Annual Returns:	0.24	0.15	0.11	0.17	0.12	0.09
SD of NPV:	0.006	0.009	0.003	0.006	0.006	0.006
(B) Medium Horizons (5-Year):						
NPV of Terminal Wealth:	2.60	2.17	1.92	1.95	1.85	1.77
Mean of Annual Returns:	0.21	0.16	0.13	0.14	0.13	0.12
SD of NPV:	0.05	0.10	0.12	0.13	0.13	0.12
(C) Long Horizons (10-Year):						
NPV of Terminal Wealth:	4.84	4.05	3.83	2.88	2.79	2.70
Mean of Annual Returns:	0.17	0.15	0.14	0.11	0.10	1.10
SD of NPV:	0.84	0.91	0.87	0.43	0.43	0.41

Risk-free Returns per year:	4%
Cash Downpayment	0%

6.5 Comparison for Various Cash Downpayments

In many market timing studies, the financial strength of the investor is excluded in the formulation of market timing strategy. An investor who is not financially strong may not be in the position to reap the potentially profitable gains offered by a bullish market in terms of excellent buy opportunities. Our market timing strategy explicitly incorporates the financial strength of the investor in terms of the credit rating in securing a property mortgage loan to finance the purchase of a property. This credit rating is reflected as market value of property less cash downpayment. It is equivalent to the amount of loan borrowed to finance the purchase of the property. Property mortgage loans are known to be an efficient investment leverage (or gearing) instrument. If it is used correctly, then leverage can significantly enhance the investment returns. On the contrary, if it is applied at the wrong time, leverage can lead to enormous financial losses. An investor with high credit rating can secure a larger percentage of property mortgage loan and hence requires lower or no cash downpayment requirement. The investor who requires lower or no cash downpayment stands in good stead to reap the benefits of higher returns on investment, provided the right investment strategy is employed.

Table 4 presents the results of the comparison of investment performance for different percentage of cash downpayments required for the property purchases. This comparative study essentially examines the effects of investor's financial credit rating on the investment performance of the various timing strategies. Here, the transaction expenses are set to zero since the main purpose is to compare the impact of cash downpayments.

Table 4. Comparison investment performance for difference cash downpayments.

Cash Down-payment	NPV of Terminal Wealth		Mean Annual Returns		SD of NPV	
	Perfect Timing	Market Timing	Perfect Timing	Market Timing	Perfect Timing	Market Timing
0%	5.0931	4.8458	0.1768	0.1709	0.7920	0.8419
10%	4.8185	4.4955	0.1703	0.1622	0.7667	0.9048
20%	4.6394	4.2747	0.1659	0.1564	0.6731	0.9180
30%	4.5145	4.1854	0.1627	0.1539	0.5405	0.8596
40%	4.3359	4.0667	0.1580	0.1506	0.5335	0.8653
50%	4.2078	3.3590	0.1545	0.1288	0.5565	0.3794
60%	4.1363	3.1720	0.1526	0.1224	0.5204	0.4130
70%	3.9880	3.0077	0.1484	0.1164	0.2516	0.4243
80%	3.8085	2.9108	0.1431	0.1128	0.2838	0.4417
90%	3.7085	2.8112	0.1400	0.1089	0.3018	0.4596
100%	3.6060	2.8119	0.1368	0.1089	0.3202	0.4595

Buying Expenses:	0%
Selling Expenses:	0%
Risk-free Returns per year:	4%
Investment Horizon:	10 years

From Table 4, it can be seen that as the required cash downpayment percentage increases, the returns on investment decreases correspondingly. In other words, the returns on investment is inversely proportional to the required cash downpayment from the investor. This result implies that if the investor has high financial credit rating and hence is able to apply leverage by borrowing higher amount of money to finance the purchase, then he is capable of achieving higher returns on investment by the proposed market timing strategy. It appears that when the required cash downpayment is lower than 80%, the returns on investment from the market timing strategy is better than that from the buy-and-hold strategy. This observation infers that the market timing proposed in this study is effective in exploiting leverage advantage to enhance the returns on investment.

7   Conclusions

Recent studies, such as [22]-[24], use market data to predict the property price movements, thereby suggesting the feasibility of integrating such information into the market timing strategy. While several other studies, e.g., [5] and [7] show some success in market timing, these studies, however, do not take into consideration the financial strength of the investors in their market timing strategies. A successful market timing strategy is very much dependent on the financial strength of the investors who aim to reap maximum benefit from the market. The market timing strategy developed in this chapter incorporates the information on the cash available to the investor, the number and values of the property assets owned by the investors and the forecast of the property price movements. Other information required by the strategy includes the relevant property investment parameters such mortgage interest rates, tax rates, transaction expenses, holding expenses, rental incomes and risk-free rates.

The results presented in the previous section offers a number of implications for property investment. Firstly, the results suggest that the proposed market timing strategy is able to add value to the returns on investment for the property investor. Risks are controlled and returns are enhanced. Lost opportunities for gains in the bullish markets are more than compensated by avoiding losses during the market downturns. The benefits from the proposed market timing strategy has demonstrated to be feasible, at least in the type of market environment in the Singapore property market during the last two decades. Secondly, the analysis of results highlights the important impact of the transaction expenses. The results imply that investors who incur significant transaction costs should refrain from making frequent shifts of funds back and forth from one investment class to another based on the modest changes in the market movement. Thirdly, the financial strength of the investor must be taken into consideration for a market timing strategy to be realistic.

The market timing strategy proposed in this study is both easy to use and yet inexpensive to implement. Input to the strategy can be readily obtained from public sources and investor' own database. The proposed market timing strategy adds a quantitative dimension to complement the subjective judgement.

The study has been simplified for two classes of assets, namely the savings deposit and the property assets. This concept can be easily extended to include other asset classes. We restrict the number of independent variables and our procedure allows shifts only at the beginning of each quarter. Furthermore, in the present study, it is assumed that ready buyers are always available in the market. However, this may not be the case in the actual market. There may not be any buyer for a particular property and hence a possible extension of the present study is to incorporate such a liquidity scenario into the strategy in the forms of the probability of a successful property sale. By assuming perfect liquidity, we are able to evaluate the stochastic dynamic programming approach to formulate market timing strategy. The introduction of liquidity constraint will definitely deteriorate the investment performance of our market timing strategy. The investigation of such impact of liquidity constraint on the investment performance would be a topic for further research.

References

[1] Dean, J.P. (1945), *Home Ownership: Is It Sound?* Harper and Brothers, New York.

[2] Sharpe, W. (1975), "Likely gains from market timing," *Financial Analysts Journal*, vol. 31, pp. 60-69.

[3] Arnott, R.D. and von Germetten, J.N. (1983), "Systematic asset allocation," *Financial Analysts Journal*, vol. 39, pp. 31-38.

[4] Sorensen, E. and Arnott, R.D. (1988), "The risk premium and stock market performance," *Journal of Portfolio Management*, vol. 15, pp. 50-55.

[5] Vandell, R.F. and Stevens, J.L. (1989), "Evidence of superior performance from timing," *Journal of Portfolio Management*, vol. 16, pp. 38-42.

[6] Fama, E. and French, K. (1988), "Dividend yields and expected stock returns," *Journal of Financial Economics*, vol. 20, pp. 3-25.

[7] Nam, J. and Branch, B. (1994), "Tactical asset allocation: can it work?" *Journal of Financial Research*, vol. 17, No 4, pp. 465-479.

[8] Smith, L.B., Rosen, K.T. and Tallis, G. (1988), "Recent developments in economic strategys of housing markets," *Journal of Economic Literature*, vol. XXVI, pp. 29-64.

[9] Brown, G R (1991), *Property Investment and the Capital Markets*, E & F N SPON, London.

[10] Bauer, R.J. (1994), *Genetic Algorithms and Investment Strategies*, John Wiley and Sons, New York.

[11] Brown, G.R. (1996), "Buy-sell strategies in the Hong Kong commercial property market," *Journal of Property Finance*, vol. 7, no. 4, pp. 40-42.

[12] Low, N.K. (1989), *Singapore Economic Trends and Property Prices*, BT Brokerage Research, Singapore.

[13] Wen, I. (1995), *What are the Factors that Determine the Property Price*, Shin Min Daily News, Singapore, 30 Oct.

[14] Puterman, M.L. (1994), *Markov Decision Process: Discrete Stochastic Dynamic Programming*, John Wiley, New York.

[15] Pyhrr, S.A., Cooper, J.R., Wofford, L.E., Kapplin, S.D., and Lapides, P.D. (1989), *Real Estate Investment: Strategy, Analysis, and Decision*, 2nd ed., John Wiley & Sons, New York.

[16] Curcio, R.J. and Gaines, J.P. (1977), "Real estate portfolio revision," *Journal of American Real Estate and Urban Economics Association*, Winter, pp. 399-410.

[17] Miller, N.G. and Sklarz, M.A. (1987), "Pricing strategies and residential property selling prices," *Journal of Real Estate Research*, Fall, pp. 12-26.

[18] Wendt, P.F. and Cerf, A.R. (1979), *Real Estate Investment Analysis and Taxation*, McGrawhill, New York, p. 53.

[19] Pyhrr, S.A. (1973), "A computer simulation model to measure the risk in property investment," *American Real Estate and Urban Economics Association Journal*, June, p. 57.

[20] Kapplin, S.D. (1976), "Financial theory and the valuation of real estate under condition of risk," *The Real Estate Appraisal*, September-October, pp. 28-37.

[21] Peiser, R.B. (1984), "Risk analysis in land development," *American Real Estate and Urban Economics Association Journal*, p. 1.

[22] Geltner, D. and Mei, J. (1995), "The present value strategy with time-varying discount rates: implications for commercial property valuation and investment decisions," *Journal of Real Estate Finance and Economics*, vol. 11, no. 2, pp. 119-135.

[23] Mei, J. and Liu, C.H. (1994), "Predictability of real estate returns and market timing," *International Journal of Real Estate Finance and Economics*, vol. 8, no. 2, pp. 115-135.

[24] Liu, C.H. and Mei, J. (1992), "The predictability of returns on equity REITs and their co-movement with other assets," *Journal of Real Estate Finance and Economics*, vol. 5, no. 4, pp. 401-408.

Chapter 10

A Hybrid Approach to
Breast Cancer Diagnosis

M. Sordo, H. Buxton, D. Watson

In vivo [31]P Magnetic Resonance Spectroscopy (MRS) is a non-invasive technique for the observation of phosphorus-containing metabolites and intracellular pH. MRS plays an important role in the investigation of cell biochemistry and offers a reliable means for detection of metabolic changes in breast tissue. However, the scarcity of ^{31}P MRS data and the complexity of interpretation of relevant physiological information impose extra demands that preclude the applicability of most statistical and machine learning techniques developed so far. To overcome such constraints, we propose Knowledge-Based Artificial Neural Networks (KBANNs) [1], a hybrid methodology that combines knowledge from a domain in the form of simple rules with connectionist learning. This combination allows the use of small sets of data (typical of medical diagnosis tasks) to train the network. The initial structure is set from the dependencies of a set of known domain rules and it is only necessary to refine these rules by training. In this chapter, we present KBANNs with a topology derived from knowledge elicited from the domain of metabolic features of normal and malignant mammary tissues. KBANN performance is assessed over the classification of 26 *in vivo* [31]P spectra of normal and cancerous breast tissues. Results confirm the suitability of KBANNs as a computational aid capable of classifying complex and limited data in a medical domain. The present study is part of an ongoing investigation into normal and abnormal breast physiology, which may help in the non-invasive early detection of breast cancer [2], [3].

1   Introduction

Classification of complex data has been addressed by various statistical and machine learning techniques. Although these methodologies have been successfully applied in a variety of domains, there are some classification tasks, particularly in medicine, which require a more powerful yet flexible and robust technique to cope with the extra demands of limited datasets and complexity of interpretation. ^{31}P Magnetic Resonance Spectroscopy is a non-invasive technique for the investigation of cell biochemistry and tissue energetics. ^{31}P MRS can be used to observe a few but critical metabolites, which are closely related to energy demand/yield processes involved in cell replication. Variation in these metabolite levels can provide clinicians with important biochemical information about the tissues under study. However, identification and interpretation of such relevant physiological information is non-trivial and requires a high level of expertise. To fully explore the potential of ^{31}P MRS in clinical applications it is desirable to provide clinicians with computational aids to help them in the identification and interpretation of variations in tissue metabolism.

Artificial Neural Networks (ANNs) have proved to be a useful tool in pattern recognition and classification tasks in diverse areas including clinical medicine [4]. Despite their wide applicability, the large amount of data required for training makes them an unsuitable classification technique when the available data is scarce. To overcome this drawback, new techniques have been developed (see [5] for a review). Among them, Knowledge-Based Artificial Neural Networks (KBANNs), a hybrid methodology that allows the combination of empirical and symbolic learning techniques, is a promising approach for classification of complex and scarce data.

The observation of metabolic changes of both normal and malignant breast tissues by *in vivo* ^{31}P MRS provides a real-life framework to assess the advantages of KBANNs over a pure connectionist approach in the classification of such complex and scarce data. Successful classification of ^{31}P MRS data could lead to a better understanding of subtle biochemical mechanisms in both normal and abnormal tissues and could be of considerable value for researchers in breast physiology and medical practitioners in general [6].

Section 2 presents a description of the KBANN methodology. Section 3 describes the disrupted mechanisms of cancerous cells with emphasis on cancerous cells in the mammary gland. Also, metabolic features of such tumours are described at some length.

Section 4 focuses on knowledge elicitation and refinement processes. Domain knowledge required by the symbolic component of KBANNs was extracted from a review of important contributions and transformed into a knowledge base of hierarchically structured inference rules. The elicitation process was guided by a variation of an interviewing technique [7]. Knowledge base refinement [8] and teachback interviewing [9] techniques were applied for further refinement of the knowledge base. (A more detailed description of the complete elicitation process can be found in [5] and [10].). The outcome of the elicitation process is a knowledge base that reflects the expertise of our particular medical domain, that is, metabolic features of breast cancer.

Section 5 describes the *in vivo* ^{31}P MRS data set, whereas Section 6 presents the KBANN topology derived from the domain model. Section 7 evaluates the performance of both KBANNs and "knowledge-free" networks over the classification of *in vivo* ^{31}P MRS data of normal and malignant mammary tissues. It also presents an analysis of the final configuration of weights to determine whether initial knowledge was refined through learning. Finally, Section 8 summarises and discusses results.

2   KBANNs

The core motivation for research in hybrid systems lies in the synergistic combination of methodologies that could lead to more robust, flexible approaches which are capable of making the most of any available source of knowledge to constrain the search space and, at the same time, to guide the search itself to improve generalization as much as possible.

This motivation led G. Towell [1] to implement Knowledge-Based Artificial Neural Networks (KBANNs) as a hybrid methodology consisting of two approaches: symbolic and connectionist. The symbolic module is knowledge intensive. It contains the description of a domain theory in the form of a hierarchically structured set of rules. The connectionist module uses a localist representation with one node for each concept in the domain model via a direct mapping of the knowledge structure into a neural network structure.

There are two main elements involved in the KBANN methodology. First, the symbolic approach, represented by a domain theory describes the important features of an object required to recognise it as a member of a particular class. The domain theory could be described as [11]:

> "… a collection of rules that describes task-specific inferences that can be drawn from the given facts. For classification problems, a domain theory can be used to prove whether or not an object is a member of a particular class."

Second, the empirical learning approach represented by a connectionist methodology, is example-driven. It requires a set of examples to learn to distinguish between objects belonging to different categories. Both approaches have limitations, but when they merge into a hybrid system, drawbacks of one method are complemented by strengths of the other. Figure 1 shows how symbolic and connectionist modules integrate the KBANN methodology.

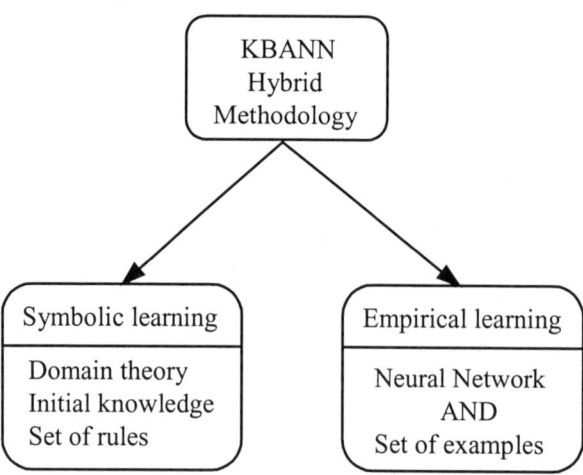

Figure 1. The two components of KBANN methodology.

2.1   KBANN Methodology

The description of the methodology given here is similar to that presented by Towell [1]. To a first approximation, the KBANN learning algorithm can be summarised as follows (see Table 1):

Table 1. KBANN learning algorithm.

Given:	• An approximately correct domain theory with a set of rules and features. • Set of examples.
Do:	• Map knowledge structure into a neural network structure. • Train knowledge-based network with the set of examples. • After training, the network is ready to classify unseen examples.

The whole process consists of two algorithms (see Figure 2). The "Rules-to-Network" algorithm maps the structure of an approximately correct domain theory, with all the rules and their dependencies, into a neural network structure. The defined network is then trained using the backpropagation learning algorithm.

2.1.1   "Rules-to-Network"

The translation process between a knowledge base containing information about a domain theory and the initial structure of a neural network is the main task of the "Rules-to-Network" algorithm. Rules in the knowledge base are acyclic and hierarchically structured. The hierarchical structure *indicates contextual dependencies or other useful conjunctions within example descriptions* [1]. Rules must appear as Horn clauses in a LISP-like notation such as:

rule(rule$_i$, ((conclusion) ← ((antecedent$_1$) &...& (antecedent$_n$))))

There are clear correspondences between elements of the knowledge base and the neural network that must be preserved. For example, supporting facts or features in the knowledge base correspond to the input units of a neural network. Intermediate conclusions in the knowledge base correspond to hidden units, whereas final conclusions in the knowledge base correspond to the output units of a network. Dependencies among the rules correspond to highly-weighted connections among units in the network.

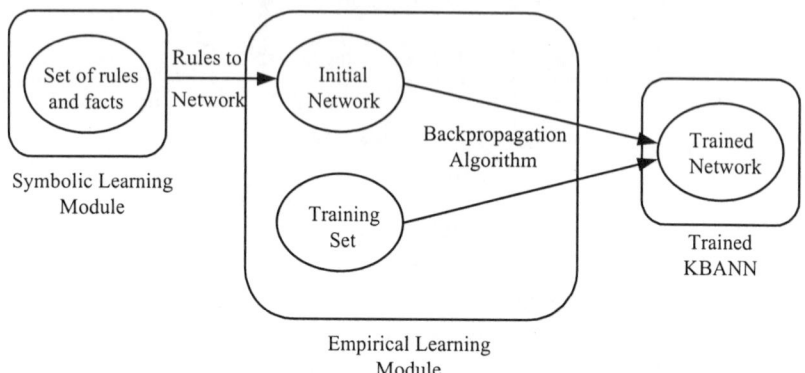

Figure 2. KBANN algorithms.

This translation procedure identifies important features that appear as antecedents in the rules, as well as intermediate and final conclusions. It also identifies important features not defined in the rules, but believed to be important for a correct classification and maps them as input units in the network. Table 2 summarises the translation process.

Table 2. "Rules-to-Network" algorithm.

1. Rewrite rules to eliminate disjuncts.
2. Map rule structure into a neural network.
3. Add important features not defined by the rules.
4. Label units in the KBANN according to their level.
5. Link nodes according to dependencies between rules.
6. Add links not defined by the rules to nodes in adjacent layers.
7. Set initial biases.
8. Perturb weights and biases.

The current implementation of the KBANN methodology supports binary, nominal, hierarchical, linear and ordered values (as in Towell's implementation) with the addition of real-valued features. Handling real-valued features preserves continuity along the implicit hierarchical relationships amongst real-valued features. It also eliminates unnecessary interpretation and preprocessing that could distort or lose vital information. Furthermore, since KBANNs are trained with a supervised learning algorithm, both input and output are presented to the network, so it is possible to establish the appropriate mapping between the presented input and the desired output. The format used consists of three elements. The first is the feature label; the second, the feature type; and the third is a list of possible values that a feature can take:

$$(\text{feature-label, feature-type (value}_1, \text{value}_2,...,\text{value}_n))$$

2.1.2   Empirical Module

Once the "Rules-to-Network" algorithm maps the structure of a domain theory into a neural network structure, the empirical learning algorithm (Backpropagation [12]) refines the initial domain theory by learning from the available examples.

2.1.3   Overview of KBANN Features

Previous sections described the elements involved in the KBANN methodology. However, they said nothing about the important features, advantages and disadvantages of this hybrid methodology. The final part of this section is devoted to these important issues.

As seen before, KBANNs consist of two approaches. The symbolic approach, represented by a set of hierarchically structured rules, contains knowledge from a domain. Rules must be propositional, non-recursive and variable free. These requirements do not constrain the KBANNs, but on the contrary, they provide a tractable, yet knowledge-rich, localist-representation network structure, which clearly reflects the dependencies among concepts established by the rules. The initial knowledge prevents the network from learning from scratch and guides it through the space of possible solutions. Prior knowledge helps to overcome the problem of accuracy, scarcity and complexity of available data, whereas the empirical approach, represented by a neural network learning algorithm, refines the initial domain theory.

The superior performance of KBANNs over ANNs (networks without any embedded knowledge in their structure) has already been demonstrated in previous experimental binary-valued tests by Towell and Shavlik [13]-[17]. KBANN performance was particularly superior to ANNs when both approaches were trained with small data sets. It is neither the symbolic nor the empirical approaches alone that are responsible for KBANN performance, but the synergistic combination and interaction of both merged into one hybrid system that provides KBANN strengths. It is, as stated by Towell [1]:

> "The combination of structure and weight both focuses the network on significant input features and provides the set of derived features that are likely important. Structure alone provides only the potential for derived features and weight alone provides only the significant inputs. It is the combination of the structure and weights that supplies the derived features that give KBANNs its power."

3 Metabolic Features of Cancerous Breast Tissues

The division and replacement of cells in normal tissues is controlled by feedback mechanisms that stimulate or inhibit the growth of normal cells. However, in the case of tumours, this mechanism is incapable of controlling the production of new cells and the division is done without any regard to the need for replacement, disrupting the structure of normal tissue. Alterations in cell membrane composition can occur with malignancy and can influence the levels of phospholipid precursors.

The main advantage of MRS is its ability to inspect tumours non-invasively *in vivo*. Since MRS can obtain information about both normal and malignant tissues, it is possible to observe alterations in metabolism due to disrupted biochemical processes in cells. Changes observed in phospholipid metabolite concentrations, associated with differences in cell proliferation in malignant tissues, have served as the basis for the identification of relevant features present in malignant but not in normal tissues.

Smith et al. [18] have shown that tumours with high proliferation rates have higher concentrations of phosphocholine (PC) than slower growing ones. Similar results have been reported by other authors [19]-[23]. Kalra et al. [22], Leach et al. [24] and Ting et al. [25] found that an increase in phosphomonoesters (PMEs) and phosphodiesters (PDEs) is associated with cell activity present in rapid cell replication related to tumour growth and embryonic tissues.

Thus, under certain circumstances, the PME peak can provide relevant information about cell activity that could help in the interpretation of a spectrum from a tissue of suspected malignancy. Merchant et al. reported that the relative concentration of PE was significantly enhanced in malignant tissue compared to benign. They also reported that the most frequently observed metabolic characteristic of cancers is an increased PME and since PC and PE (the major components of the PME peak) are intermediaries in the pathway of membrane synthesis, this elevation reflects alterations in membrane metabolism [26]. High concentrations of PDEs are indicators of necrotic fraction in tumours as a consequence of phospholipid degradation [21], [27], [28]. A prominent PME resonance level has also been observed in breast tumours [29]. Hence, PME could be considered as a diagnostic discriminant between benign and malignant tissues [30]-[32].

In oncology, ^{31}P MRS can observe energy metabolism of cells by monitoring high-energy phosphates such as adenosine tri-phosphate (ATP), which provides energy for cellular processes; inorganic phosphate (Pi), the by-product of energy expenditure; and phosphocreatine (PCr), an energy reservoir. It also can determine tissue pH by measuring Pi chemical shift [24]. Levels of energy-rich compounds such as PCr and ATP tend to decline with tumour growth, whereas Pi tends to increase. In spite of the variety of methodologies used in the studies reviewed above, there are some common characteristics that reflect altered metabolic states in malignant tissues. High levels of PMEs, PDEs and Pi, being PMEs the most important marker, are considered indicators of tumour progression and response to treatment by most workers.

ATP is required in cellular processes, whereas PCr acts as an energy reservoir. These two metabolites and Pi are closely associated with cell activity [30]. Thus, low levels of ATP, low PCr and high Pi are indicators of increased cell activity related to tumours. These observations can provide some guidelines for the classification of mammary tumours. Further investigation of these techniques would be of considerable value for the understanding of cancer cell function and the role of phospholipid metabolism in tumour growth. Common characteristics reflecting altered metabolic states in malignant tissues are summarized in Table 3.

Table 3. Summary of important metabolic features of breast tumours.

Marker	Feature
PMEs	• Important marker to distinguish between normal and cancerous tissues. • Elevated levels associated with malignant tumours. • Levels related to cell proliferation rate. • Sensitive indicator of response of breast carcinomas to treatment.
PE and PC	• PE and PC are major contributors to PME peak. • Elevation in PE and PC levels reflect alterations in cell metabolism. • Malignant tumours have low PE/PC ratio. • Decrease in PE and PC levels indicate good response to treatment. • High PE and PC levels are related to tumour growth.
PDEs	• Elevated levels are characteristic of malignant tumours. • Reduction in PDE levels reflect tumour necrosis.
GPE and GPC	• GPE and GPC are major contributors to PDE peak. • GPC is a marker of hormonal receptor status of breast carcinomas. • High GPE and GPC concentrations in tumours indicate phospholipid degradation.
Organic Phosphate	• High Po and low pH are features associated with ischaemia.
Inorganic Phosphate	• Pi chemical shift indicates pH. • Pi level increases with tumour growth. • Pi level is higher in tumours than in normal tissues.
NTP	• Levels tend to decline during tumour growth. • Higher NTP concentrations in carcinomas than in benign tumours. • Lower levels of NTP in postmenopausal breast tumours compared to healthy breast.
ATP	• High levels of ATP in tumours. • ATP provides energy for cellular processes. • ATP levels are higher in tumours than in normal tissue.
PCr	• Low PCr levels in tumours. • PCr is an energy reservoir. • PCr levels decline during tumour growth. • Higher PCr levels in benign compared to malignant tumours. • High PCr levels result from successful endocrine therapy.
PH	• Tumours have abnormal pH levels.

4   Knowledge Elicitation and Refinement

Knowledge acquisition (KA) is a process that involves eliciting, analysing, interpreting, formalizing and transforming into a suitable machine representation, knowledge that experts use for solving problems [33].

During the knowledge acquisition process, a set of statements about metabolic features of breast cancer was defined. The extracted statements were grouped into different categories based on the metabolic feature they reflected. Similar statements were merged into a single statement. The simplified statements were transformed into a set of hierarchically structured *If...then* inference rules required by the symbolic module of KBANNs. These rules reflected the causal relationships of metabolic features of breast cancer. The final set of rules was the result of a process that involved elicitation of knowledge from several authors, and refinement of the knowledge base.

The knowledge acquisition process was guided by a variation of an interviewing technique [7] since, in this case, experts provided their knowledge through published work. Refinement of the domain model in Table 4 was done in collaboration with the experts[1]. A combination of knowledge base refinement [8] and teachback interviewing [9] techniques was applied for the refinement of the domain model. Adjectives, such as *high* and *low*, referring to metabolite levels, energy levels or cell metabolic activity were eliminated, because they force the representation of knowledge into a more rigid and restricted format. This requires previous conceptualisation from the user, reducing the generalization capabilities inherent in neural networks. Hence, the decision of whether a metabolite level or cell activity was *high* or *low* was left entirely to the network, based solely on the value associated with that entry. For a complete explanation of the elicitation and refinement process see [5] and [10]. Objects and relations from simplified excerpts define the corresponding domain model presented in Table 4.

---

[1] CRC Clinical Magnetic Resonance Research Group at the Royal Marsden Hospital, Sutton, UK.

Table 4. Domain model of breast cancer metabolic features.

Features	PDE, PME, Pi, PCr, ATP
Intermediate goals	Energy levels
Final goals	Tumour
Relations	1. *If* (high) PDE *then* tumour
	2. *If* (low) PCr *then* tumour
	3. *If* (high) ATP *then* tumour
	4. *If* (high) PME *then* tumour
	5. *If* (high) energy *then* tumour
	6. *If* (high) PME *then* abnormal membrane metabolism
	7. *If* (high) energy *then* abnormal metabolism
	8. *If* (high) ATP *and* (high) Pi *and* (low) PCr *then* (high) energy
	9. *If* γ-ATP *and* α-ATP *and* β-ATP *then* total ATP

Nine relations extracted from the domain model presented in Table 4 form the initial knowledge base presented in Table 5. These relations focus on the metabolic alterations in malignant mammary tissues reflected in a ^{31}P spectrum (references attached at the end of each rule indicate the sources of knowledge). From these 9 rules in Table 5, rules 5-9 were considered redundant and were deleted. Rule 6 was equivalent to rule 4, whereas rules 5, 7-9 did not provide any relevant information.

Table 5. Initial knowledge base of breast cancer metabolic features.

R1: *If* (high) PDE *then* tumour [6], [25], [31], [34].
R2: *If* (low) PCr *then* tumour [28], [30], [31], [34].
R3: *If* (high) ATP *then* tumour [28].
R4: *If* (high) PME *then* tumour [6], [21], [22], [25], [28]-[32][34]-[37].
R5: *If* (high) energy *then* tumour.
R6: *If* (high) PME *then* abnormal membrane metabolism [6], [25], [28], [30]-[32], [35], [36].
R7: *If* (high) energy *then* abnormal metabolism.
R8: *If* (high) ATP *and* (high) Pi *and* (low) PCr *then* (high) energy [6], [28], [30], [31], [34], [38], [39].
R9: *If* γ-ATP *and* α-ATP *and* β-ATP *then* total ATP.

While discussing the relevance of the antecedents in rule 3, one of the experts explained that from the three components of ATP, β-ATP is the metabolite that best reflects energy levels and gives the true amount of ATP. Thus, it was agreed to rewrite rule 3 as:

R3: *If* β-ATP *then* tumour.

Further analysis indicated that rules 4, 3, 1 and 2 (in this order) are the most important. Thus, after refinement, the final knowledge base for metabolic features of breast tumours is presented in Table 6. It contains rules and features believed to be important for a correct classification of breast tumour spectra. These rules and features appear in the appropriate notation described in Section 2.1.1.

Table 6. Final knowledge base of breast cancer metabolic features.

goal(tumour).

rule(1, ((tumour, ?x) ← (PDE level, value))).
rule(2, ((tumour, ?x) ← (PCr level, value))).
rule(3, ((tumour, ?x) ← (β-ATP level, value))).
rule(4, ((tumour, ?x) ← (PME level, value))).

feature(1, (PME, real, (value))).
feature(2, (PDE, real, (value))).
feature(3, (Pi, real, (value))).
feature(4, (PCr, real, (value))).
feature(5, (α-ATP, real, (value))).
feature(6, (β-ATP, real, (value))).
feature(7, (γ-ATP, real, (value))).

5 ^{31}P MRS Data

A data set with 26 *in vivo* ^{31}P MR spectra from normal and cancerous mammary tissues was provided by the CRC Clinical Magnetic Resonance Research Group at the Royal Marsden Hospital, Sutton, UK. It consists of 16 control cases obtained from four female premenopausal volunteers (ages 21-45), all with regular menstrual cycles and none using the contraceptive pill. Four ^{31}P spectra from each volunteer, one from each phase of the menstrual cycle, were acquired using a small surface coil positioned over the left breast, while the volunteer lay supine (a full description can be found in [40]). The remaining 10 cases were acquired from five female patients (one premenopausal, four postmenopausal) with breast carcinoma. All patients were under treatment with chemotherapy, radiotherapy or tamoxifen. Two spectra per patient were provided: one measurement prior to treatment, and the second from a later stage of treatment to assess tumour response. Tumours were located by palpation. A 5cm surface coil was placed over the tumour while patients lay supine. A series of measurements were performed and the resulting set of FIDs were fitted to a $\sin^2$ function. Spectra were Fourier transformed and processed using Lorentzian model functions to model PDE, PCr, PME, Pi, γ-, α- and β-ATP peaks. Tumour volume was assessed by either clinical or MRI section analysis (a full description of acquisition methods can be found in [24]). Figure 3 shows a breast tumour spectrum after processing. The seven metabolites extracted from each spectrum are indicated. They correspond to the normalized area under the peak of PME, Pi, PDE, PCr, γ-ATP, α-ATP and β-ATP metabolites.

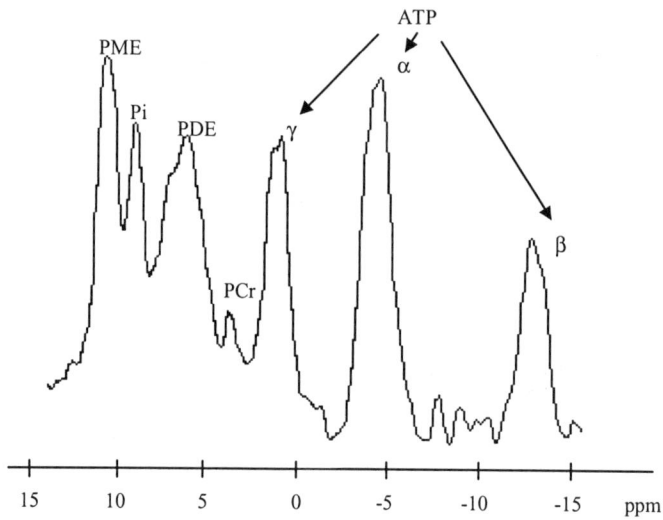

Figure 3. ^{31}P spectrum from a breast tumour.

6   KBANN Topology

The four relations and seven features in the domain model were converted into Horn clauses in a LISP-like notation (Table 6 shows the knowledge base reflecting these dependencies in the domain model). The "Rules-to-Network'" (RTN) algorithm (described in Section 2.1.1) uses this knowledge base to define the KBANN topology. There were no disjunctive rules with more than one antecedent, thus no rewriting was necessary. The RTN maps the dependencies in the domain model into the initial structure of a KBANN, adds the important features not defined in the relations, and labels all units according to their level (Figure 4). Low-weighted links are added to connect units in adjacent layers. Initial weights and biases are set and perturbed. The resulting topology for a KBANN defined from the breast cancer domain model in Table 6 is depicted in Figure 5. This network was then trained using the backpropagation learning algorithm.

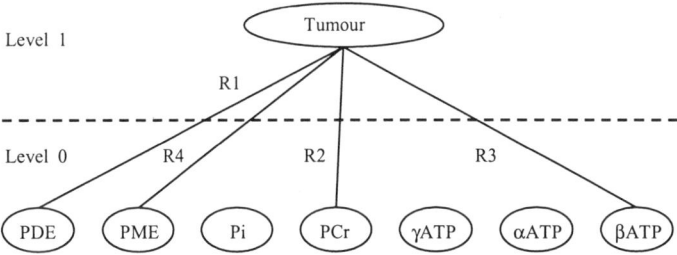

Figure 4.   Definition of the initial KBANN structure for the breast cancer problem derived from the rules and additional features in the knowledge base.

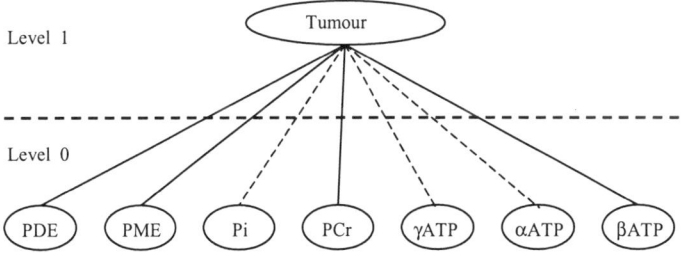

Figure 5.   KBANN structure for the breast cancer problem with the addition of low-weighted links (indicated by dashed lines) to adjacent units after node labelling.

7 Results

This section presents results of the empirical tests with "knowledge-free" ANNs (networks with no embedded knowledge in their structure) and KBANNs on the classification of 26 *in vivo* ^{31}P MR spectra of normal and cancerous breast tissues.

7.1 Knowledge-Free Networks

840 knowledge-free feedforward networks were trained with the complete set of 26 *in vivo* ^{31}P MRS using the backpropagation learning algorithm. All these networks had 7 inputs corresponding to the normalized peak area of PME, PDE, PCr, Pi, α-ATP, β-ATP and γ-ATP metabolites, and one output to indicate whether the input vector corresponds to a normal or a cancer spectrum. The number of hidden units varied from 1 to 20. For each topology, 42 networks were trained with different learning rate, momentum and seed values. The seed number was used to generate the initial random noise added to weights and biases. Table 7 presents the observed average/pattern error of 0.2276 and standard deviation of 0.0298 for all these networks. The average correct classification was 62.42%, that is 16.23 cases from the set of 26.

Table 7. Knowledge-free and knowledge-based networks performance for the metabolic changes in cancerous mammary tissues classification problem.

		Trained Networks			
Type of Network	Features	Average pat/error	Standard deviation	Correct Classification Cases (avg)	(%)
Knowledge-free	No rules	0.2276	0.0298	16.23	62.42
Knowledge-based	4 rules	0.0500	0.0179	22.71	87.36

7.2 KBANNs

42 KBANNs with the topology derived from the set of four relations, seven features and one goal (in Table 6 and depicted in Figure 5) were trained with different learning rate, momentum and random initial seed values. Table 7 and Figure 6 present a comparison in performance for both KBANNs and ANNs. Figure 6a indicates the error values at the beginning of the learning process. Zone A corresponds to the initial

error for KBANNs, whereas Zone B corresponds to the ANNs initial error. Figure 6b depicts how even from the beginning, KBANNs are more accurate than ANNs.

The 42 KBANNs correctly classified an average of 22.71 cases from the set of 26. This is equivalent to 87% correct classification rate, with an average pattern error of 0.0500 and standard deviation of 0.0179 (see Table 7). From the aforementioned 42 KBANNs, the best five were retrained and tested using the leave one out method. All them had average pattern/error values well below the average 0.0500 (Table 8).

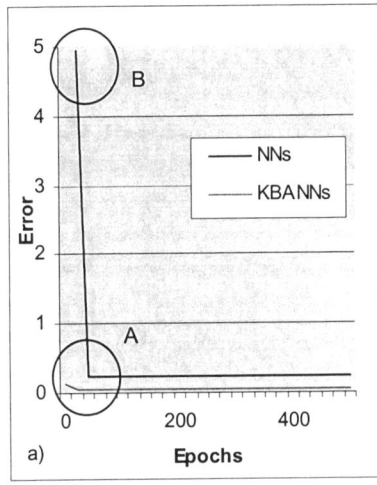 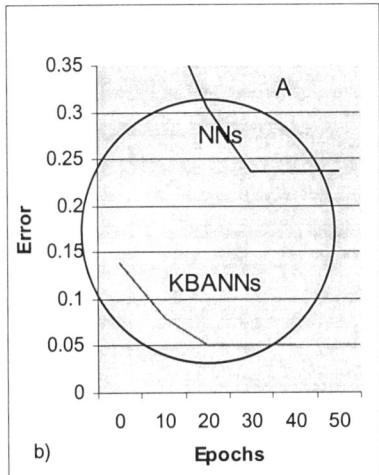

Figure 6. KBANN and ANN performance in the classification of *in vivo* [31]P spectra of normal and cancerous breast tissues. The two circles (A and B) in Figure 6a indicate the accuracy of KBANNs and ANNs before learning from the available examples. Figure 6b illustrates how KBANNs with a simpler structure, such as the one derived from the set of four rules and seven features can perform much better than ANNs with more complex topologies, but with no initial knowledge from the domain.

These networks were labelled cKBANN-1, cKBANN-2, cKBANN-3, cKBANN-4 and cKBANN-5. For each KBANN, a 26 fold cross-validation run test was set to evaluate their performance over new cases. Hence, each KBANN was trained with 25 cases and was tested over one unseen case. cKBANN-1 was trained with learning rate = 0.1, momentum = 0.8 and seed = 9 (for the added random noise to weights and biases). It showed average pattern errors of 0.0232 (std = 0.0045) and 0.1218 (std = 0.1976) during training and testing respectively. It correctly classified 18 of 26 cases (69%). cKBANN-2 was trained with learning rate = 0.3, momentum = 0.8 and seed = 1. It correctly classified 18 of the 26 cases, with an average pattern error of 0.0273 (std = 0.0150) and 0.1202 (std = 0.1967) for training and testing respectively. cKBANN-3 was trained with learning rate = 0.3, momentum= 0.8 and seed = 3. Average pattern errors of 0.0240 (std = 0.0110) and 0.1177 (std = 0.1979) were observed during training and testing respectively. It correctly classified 18 (69%) cases. cKBANN-4 presented an average pattern error of 0.0271 (std = 0.0173) during training and 0.1203 (std = 0.1968) during testing. It was trained with learning rate = 0.3, momentum = 0.8 and seed = 7. It correctly classified 18 (69%) of the 26 cases in the data set. Finally, cKBANN-5, trained with learning rate = 0.3, momentum = 0.8 and seed = 9, showed average pattern errors of 0.0347 (std = 0.0152) and 0.1248 (std = 0.2003) during training and testing respectively. It showed the best classification rate of 73% and correctly classified 19 of the 26 cases.

Table 8. KBANN performance during cross-validation for the breast cancer classification problem.

Network	Training		Testing		
	avg. pat/error	std	avg. pat/error	std	Correct (%)
cKBANN-1	0.0232	0.0045	0.1218	0.1976	18/26 (69.23%)
cKBANN-2	0.0273	0.0150	0.1202	0.1967	18/26 (69.23%)
cKBANN-3	0.0240	0.0110	0.1177	0.1979	18/26 (69.23%)
cKBANN-4	0.0271	0.0173	0.1203	0.1968	18/26 (69.23%)
cKBANN-5	0.0347	0.0152	0.1248	0.2003	19/26 (73.08%)

Further analysis of KBANN performance was carried to identify the nature and quantity of the errors. For this purpose, a confusion matrix for each KBANN was defined. A confusion matrix contains as many rows and columns as there are classes. Numbers in the diagonal indicate the cases that were correctly classified. Quantities in off-diagonal positions represent misclassifications. The values off-diagonal define two types of error. One error, called *type I error* (light rectangle in Figure 7) occurs when it is concluded that a condition is present when, in fact it is not. The second, called *type II error* (dark rectangle in Figure 7) occurs when it is concluded that a condition is not present when, in fact it is. This terminology is introduced for the purpose of defining the possible types of errors that can occur during classification. Hence, classification of a normal spectrum as cancerous is a type I error, whereas a cancer spectrum classified as normal is a type II error (see Figure 7).

It is possible to derive four concepts from the above definitions. Type I error can be interpreted as a "false positive response". In other words, a false positive response occurs when a condition is diagnosed as present when in fact it is absent. Following the same line of reasoning, a "true positive response" will occur when a present condition is identified as such (a cancer spectrum is classified as cancer). Type II error can be interpreted as a "false negative response". It occurs when a cancer spectrum is classified as normal. A "true negative response" occurs when a condition is not present and it is indeed absent. This is the case of a normal spectrum classified as normal. Figure 7 gives a graphical example of these concepts.

Desired/Actual	Normal	Cancer
Normal	true negative	false positive
Cancer	false negative	true positive

☐ type I error   ▨ type II error

Figure 7. Possible classification outcome for the breast cancer problem.

Five confusion matrices are depicted in Figure 8, one for each cKBANN. Recalling from the previous paragraph, numbers in the diagonal indicate the cases that were correctly classified. Quantities in off-diagonal positions represent misclassifications. Confusion matrices for cKBANN-1, cKBANN-2, cKBANN-3 and cKBANN-4, depicted in Figures 8a, 8b, 8c and 8d respectively, show that these four networks correctly classified 18 of the set of 26 cases. 12 of the 16 normal cases were correctly classified as normal, whereas four of them were misclassified as cancer (type I error). Six of the 10 cancer spectra were correctly classified, but the remaining four cases were classified as normal (type II error). cKBANN-5 correctly classified 13 of the 16 normal cases as normal, and six of the 10 cancerous cases as cancer (Figure 8e). All KBANNs classified 60% of the cancer cases as malignancies (6 of 10 cases). Four of the five KBANNs correctly classified 75% of normal (12 of 16) cases. The remaining KBANN correctly classified 13 of 16 normal cases (81%). The total number of cases correctly classified was 69% for the first four and 73% for the remaining KBANN.

D/A	Normal	Cancer
Normal	12	4
Cancer	4	6

a) cKBANN-1

D/A	Normal	Cancer
Normal	12	4
Cancer	4	6

b) cKBANN-2

D/A	Normal	Cancer
Normal	12	4
Cancer	4	6

c) cKBANN-3

D/A	Normal	Cancer
Normal	12	4
Cancer	4	6

d) cKBANN-4

D/A	Normal	Cancer
Normal	13	3
Cancer	4	6

e) cKBANN-5

Figure 8. Confusion matrices for KBANNs and the cancer classification problem. Values in the diagonal indicate cases correctly classified. Values off-diagonal correspond to misclassifications.

7.3 Discussion

The two circles (A and B) in Figure 6a indicate the accuracy of ANNs and KBANNs before learning from the available examples. Figure 6b illustrates how KBANNs with a simpler structure, such as the one derived from the set of four rules and seven features can perform much better than ANNs with more complex topologies, but with no initial knowledge from the domain. Accuracy around 70% (Table 8) supports previous assertions as to the applicability of KBANNs for classification problems in complex medical domains [2], [5].

In total, nine spectra were misclassified by KBANNs in cross-validation (see Table 9). Four of them are spectra taken from cancerous tissues, and five from healthy volunteers. The nature of these misclassifications is discussed in terms of the two types of error described above.

Table 9. Misclassified cases by KBANNs during cross-validation. • indicates misclassification of a cancer (C) or normal (N) case.

	Cases								
	6	7	9	10	11	14	19	21	26
Network/Type	C	N	C	N	N	C	N	N	C
KBANN-1	•	•	•	•		•	•	•	•
KBANN-2	•	•	•	•	•	•		•	•
KBANN-3	•	•	•	•	•	•		•	•
KBANN-4	•	•	•	•	•	•		•	•
KBANN-5	•		•	•		•	•	•	•

Type I error contains the false positive misclassifications corresponding to the five normal cases identified as cancerous (cases 7, 10, 11, 19 and 21). Cases 10 and 21 were misclassified by all five cKBANNs (see Table 9). Cases 10, 19 and 21 were acquired from volunteers during the early follicular phase of the menstrual cycle. Cases 7 and 11 were acquired from two volunteers during the late follicular phase of the menstrual cycle, when cell proliferation increases [40]. Cases 7 and 11 have low PCr levels, which according to rule 2, is a characteristic of tumours (see Figure 9b). Case 10 has high PDE (see Figure 9a), whereas normal case 19 has high PME levels (see Figure 9c). Again, both are features related to malignancies. Finally, case 21 has high β-ATP levels, which, according to rule 3, is characteristic of tumours (see Figure 9g).

Type II error indicates the false negative misclassifications of four cancer cases identified as normal (cases 6, 9, 14 and 26). Case 6 was acquired before treatment, whereas cases 9, 14 and 26 were acquired in a later stage of treatment. They all were misclassified by the five KBANNs (see Table 9), because none of them fully satisfied rule 3 (*If* (high) β-ATP *then* tumour). According to this rule, these four spectra should have high β-ATP levels to be considered tumours, and as observed in Figure 9g they do not comply with this condition. Failure to satisfy rule 3 may be due to the fact that β-ATP reflects tissue energetics and cell proliferation, but does not reflect the alterations in phospholipid pathways characteristic of tumours.

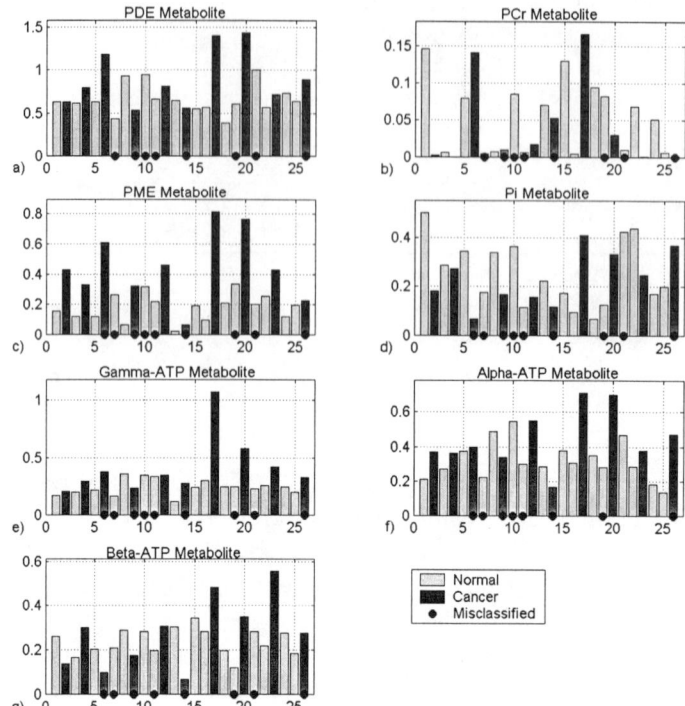

Figure 9. Normalized peak areas of seven metabolites extracted from the 26 *in vivo* ^{31}P spectra of normal and cancerous breast tissues. a) PDE; b) PCr; c) PME; d) Pi; e) γ-ATP; f) α–ATP; g) β–ATP.

It has been stated by Ting et al. that PME levels are higher in tumours than in normal tissues [25]. However, Negendank observed that a decrease in phosphocholine (PC) and phosphoethanolamine (PE) (recalling from Section 3, PC and PE are the major components of the PME peak) levels indicate good response to treatment [31]. Thus, a decline in the PME level observed in cases 9, 14 and 26 could be interpreted as a positive response to treatment (see Figure 9c).

Though it is well known that there are differences in the biochemistry of normal and cancerous cells, the underlying mechanisms that regulate normal and malignant tissue metabolism are not yet fully understood. Changes in metabolite levels (see Figure 9) respond not just to a single, but to a combination of factors. Variations in metabolite concentrations can be triggered by hormone events, as in pre-menopausal normal tissues, or by a disrupted cell physiology as in cancerous tissues. There are significant variations between normal and cancerous cells, particularly in the PME region. However, no distinct differences have been observed in the energetics (ATP levels) of normal and cancerous cells. These observations suggest that changes in the phosphorylation pathways are associated with malignancy, whereas the energy metabolism is more related to the proliferation rate of mammary epithelial cells [25].

7.4   Analysis of Final Connection Weights

This section analyses the final configuration of weights to determine whether KBANNs were capable of refining the initial knowledge embedded in their structure. Figure 10 depicts the Hinton diagram for the initial and final configuration of weights of KBANNs defined from the set of four relations and seven features in Table 6. The y-axis shows only the nodes with incoming connections from other nodes.

The x-axis displays all the nodes in the network structure. The magnitude of each weight or bias is represented by the square's size, whereas the colour represents the sign. Positive weights are represented by dark squares and negative weights by light squares.

The criteria for validation compare the initial and final weights against the bias. It also considers the logical connections in each relation. *High* and *low* labels appearing in the original set of rules are reintroduced here to help in the interpretation of the weights. These criteria are based on the logical connectives involved in each rule and the final incoming weights and bias of the consequent node. Accordingly, the consequent of a conjunctive rule is activated if the sum of all the incoming positive weights (antecedents) exceeds the bias of the consequent node. Negative incoming weights should have values near zero. The consequent of a disjunctive rule activates if one of the incoming positive weights is similar or greater than the bias regardless of the other incoming weights. Strong weights have values $|w| \geq |Bias|$. Weak weights have values $|w| \approx 0$. Weights $w \geq 0$ are positive, and $w < 0$ are negative.

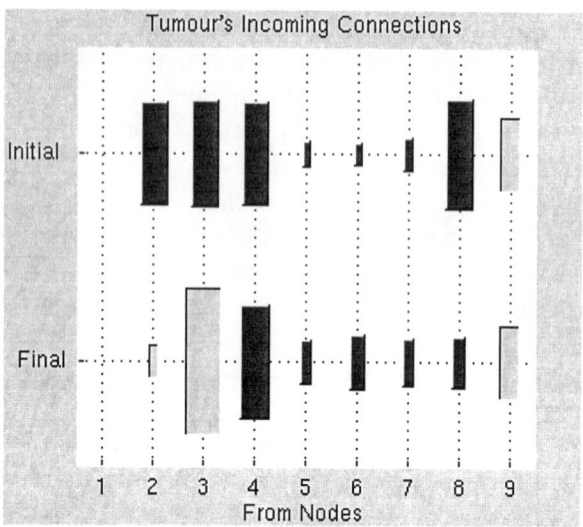

Figure 10. Hinton diagram for KBANN initial and final weights for classification of *in vivo* ^{31}P spectra of normal and cancerous breast tissues. Nodes appear in the following order: 1) tumour; 2) PDE metabolite; 3) PCr metabolite; 4) PME metabolite; 5) Pi metabolite; 6) γ-ATP metabolite; 7) α-ATP metabolite; 8) β-ATP metabolite and 9) bias. Only the tumour node has incoming links from the seven aforementioned metabolites. Weights and biases have been normalized against their bias. Positive weights are represented by dark squares and negative weights by light squares.

The initial and final configuration of weights and biases of tumour incoming connections is depicted in a Hinton diagram in Figure 10. Weights and biases have been normalized to facilitate interpretation. From the seven incoming connections to the tumour node, four connections are supported by the set of four disjunctive rules in Table 6.

Disjunctive rule 1 in Table 6 defines an initial strong positive incoming weight from the PDE metabolite. After learning, this weight changed into a weak negative weight. Although there is a change in the sign of this weight, its magnitude is not large enough to either excite or inhibit activation. Thus, a weak negative weight indicates that there is no strong evidence from data to support or reject rule 1. It also suggests that PDE levels may not be a good indicator of malignancy.

Disjunctive rule 2 in Table 6 defines an initial strong positive incoming weight from the PCr metabolite. A strong negative final weight precludes the activation of the tumour node. This weight suggests that indeed PCr levels in tumours are low. Learning from data endorsed the initial knowledge from rule 2. This assertion is also supported by [28] and [31].

Disjunctive rule 3 in Table 6 defines an initial strong positive incoming weight from β-ATP. After learning, this weight decreased in magnitude. Though it remained positive, it no longer exceeds the bias and cannot activate the tumour node. Therefore, there is no strong evidence to support or reject rule 3. This suggests that β-ATP may not be a good marker of malignancy. This assertion is supported by [25].

Disjunctive rule 4 in Table 6 defines an initial strong positive incoming weight from the PME metabolite. The final weight slightly increased in magnitude. This suggests that PME is indeed a good marker of malignancy. This assertion is supported by [25] and [35].

The three incoming connections from Pi, γ- and α-ATP were not supported by any initial knowledge from the domain. The initial weak positive connections showed no important changes after learning. This may indicate that these three metabolites are not good markers of malignancy. α-, γ- and β-ATP levels reflect tissue energetics, and cell proliferation. They do not reflect alterations in the phospholipid metabolism associated with malignancy [25].

An interpretation of new dependencies extracted after learning is presented as follows:

- From the seven metabolites, PME is the most reliable marker of malignancy. This is supported by a strong positive weight from the PME connection. This is also supported by [25], [35], and [37].

- A strong negative weight may indicate that low PCr levels are a good indicator of malignancy. Low PCr levels also reflect depletion of energy reservoir sometimes associated with increased proliferation rate of cancerous cells. This assertion is supported by [26], [28], and [31].

- A weak negative weight may indicate that PDE is not a good indicator of malignancy.

- Weak positive weights from γ-, α- and β-ATP suggest that these metabolites are good markers of tissue energetics and cell proliferation, but they are not good markers of malignancy. These assertions are supported by [25].

8   Conclusions

Findings presented in previous sections confirm KBANNs as capable of learning from small and complex data sets and to refine initial knowledge through empirical learning. Classification rates of 62% for ANNs and 87% for KBANNs, showed how KBANNs outperformed ANNs in the classification of 26 *in vivo* ^{31}P spectra of normal and cancerous breast tissues. In a separate set of cross-validation tests KBANNs maintained a good performance of 69-73%, with 60% true positive and 75-81% true negative classification rates. This superior performance relies upon the KBANN combination of symbolic and empirical learning. Figure 6 shows how knowledge from the domain embedded in the KBANN structure provides a more advantageous position at the beginning of the learning process.

Analysis of weights confirms the proficiency of KBANNs to refine the initial knowledge from the domain. The initial structure of KBANNs derived from the dependencies in the domain model was simple. All the dependencies directly linked the features (input nodes) with the goal (output node), with no intermediate goals. The lack of intermediate nodes in the KBANN topology restrained the generalization capabilities of the empirical module of KBANNs. Nevertheless, KBANNs were quite effective at refining these dependencies and extracting new knowledge from the presented samples. Rules 2 and 4 were supported by data, but there was not enough evidence to support or reject rules 1 and 3. Refinement of the initial knowledge indicates that PDE, Pi, γ-, α- and β-ATP may not be good indicators of malignancy. A summary of findings is presented in Table 10.

The classification of *in vivo* ^{31}P spectra of normal and cancerous breast tissues clearly shows the impossibility of fully formalizing a causal model from such a complex real-life domain as the one presented here. However, KBANN hybrid methodology provides the means for refinement of "roughly" correct domain theories through empirical learning. Results presented confirm previous assertions as to the applicability of KBANNs for classification problems in complex medical domains, and even under constraints that preclude their generalization capabilities.

Table 10. Summary of extracted knowledge of metabolic features of breast cancer. The second row indicates knowledge confirmed by data. The third shows knowledge rejected by data. The fourth row contains rules that could neither be endorsed nor rejected by data. The fifth row presents new knowledge gathered during the learning process.

Goal	· Tumour
Confirmed	· Rule 2: Low PCr · Rule 4: High PME
Rejected	----
No Evidence	· Rule 1: High PDE · Rule 3: High β-ATP
New	· Low PCr levels indicate depletion of energy reservoir (supported by [26], [28], and [31]). · PME is a good marker of malignancy (supported by [25], [35], and [37]). · PDE may not be a good marker of malignancy. · Pi may not be a good marker of malignancy. · γ- α- and β-ATP may not be good markers of malignancy (supported by [25]).

KBANNs provide a means for analysing, interpreting and understanding the extracted metabolic dependencies from ^{31}P MRS. This knowledge can provide an extra dimension of information to fill the gaps in the understanding of metabolic alterations related to malignancies. These biochemical clues, combined with other techniques for the visualisation of the anatomical structure of the breast by means of MRI and X-Rays can be of important clinical value for diagnosis and can provide scientific information about the nature of this disease.

Acknowledgements

We would like to thank the CRC Clinical Magnetic Resonance Research Group at the Royal Marsden Hospital, Sutton, UK for their advice on magnetic resonance and metabolic changes in breast cancer tissues, and for providing the data for these experiments. We also thank Gabriela Ochoa for her comments on the final version of this document.

References

[1] Towell, G.G. (1991), *Symbolic Knowledge and Neural Networks: Insertion, Refinement and Extraction*, Ph.D. thesis, University of Wisconsin, Madison.

[2] Sordo, M., Buxton, H., Watson, D., Collins, D., Ronen, S., Leach, M., and Payne, G. (1998), "KBANNs for classification of normal ^{31}P MRS based on hormone-dependent changes during the menstrual cycle," *Fourth International Conference on Neural Networks and their Applications*, Marseilles, France.

[3] Sordo, M., Buxton, H., and Watson, D. (1999), "KBANNs and the classification of ^{31}P MRS of malignant mammary tissues," *Ninth International Conference on Artificial Neural Networks* Edinburgh, U.K.

[4] Baxt, W.G. (1995), "Application of artificial neural networks to clinical medicine," *Lancet*, vol. 346, pp. 1135-1138.

[5] Sordo, M. (1999), *A Neurosymbolic Approach to the Classification of Scarce and Complex Data*, D.Phil. thesis, University of Sussex, Brighton, U.K.

[6] Leach, M. (1994), "Magnetic resonance spectroscopy applied to clinical oncology," *Technology and Health Care*, vol. 2, pp. 235-246.

[7] Kuipers, B. and Kassier, J. (1987), "Knowledge acquisition by analysis of verbatim protocols," chapter in *Knowledge Acquisition*

for Expert Systems. A Practical Handbook, pp. 45-70, Plenum Press.

[8] Fox, J., Myers, C., Greaves, M., and Pegram, S. (1987), "A systematic study of knowledge base refinement in the diagnosis of leukemia," chapter in *Knowledge Acquisition for Expert Systems. A Practical Handbook*, pp. 73-89, Plenum Press.

[9] Johnson, L., and Johnson, N. (1987), "Knowledge elicitation involving teachback interviewing," chapter in *Knowledge Acquisition for Expert Systems. A Practical Handbook*, pp. 73-89, Plenum Press.

[10] Sordo, M., Buxton, H., and Watson, D. (1998), "A knowledge base for classification of normal breast ^{31}P MRS," in Ifeachor, E., Sperduti, A., and Starita, A. (Eds.), *3rd International Conference on Neural Networks and Expert Systems in Medicine and Healthcare*, Pisa, Italy.

[11] Mitchell, T.M., Keller, R.M., and Kedar-Cabelli, S.T. (1986), "Explanation-based generalization: a unifying view," *Machine Learning*, vol. 1, pp. 47-80.

[12] Rumelhart, D., McClelland, J., et al. (1986), *Parallel Distributed Processing. Explorations in the Microstructure of Cognition, Vol. 1: Foundations*, M.I.T.

[13] Towell, G., Shavlik, J., and Noordewier, M. (1990), "Refinement of approximate domain theories by knowledge-based neural networks," *VIII National Conference on Artificial Intelligence*, vol. 2, pp. 861-866.

[14] Towell, G. and Lehrer, R. (1993), "A knowledge-based model of geometry learning," in Hanson, S., Cowan, J., and Giles, C. (Eds.), *Advances in Neural Information Processing Systems*, vol. 5, pp. 887-894, Morgan Kauffmann, San Mateo, CA.

[15] Towell, G., and Shavlik, J.W. (1989), "Combining Explanation-Based and Neural Learning: an Algorithm and Empirical Results,"

Technical Report 859, University of Wisconsin, Computer Science.

[16] Towell, G. and Shavlik, J.W. (1994), "Knowledge-based artificial neural networks," *Artificial Intelligence*, vol. 70, nos. 1-2, pp. 119-165.

[17] Towell, G.G. and Shavlik, J.W. (1992), "Using symbolic learning to improve knowledge-based neural networks," *X National Conference on Artificial Intelligence*, pp. 177-182.

[18] Smith, T., Glaholm, J., Leach, M., Collins, D., et al. (1991), "A comparison of in vivo and in vitro ^{31}P NMR spectra from human breast tumours: variations in phospholipid metabolism," *British Journal of Cancer*, vol. 63, no. 4, pp. 514-516.

[19] Bishop, R. and Bell, R. (1988), "Assembly of phospholipids into cellular membranes: biosynthesis, transmembrane movement and intracellular translocation," *Ann. Rev. Cell Biol.*, vol. 4, pp. 579-610.

[20] Pelech, S. and Vance, D. (1989), "Signal transduction via phophatidylcholine cycles," *TIBS*, vol. 14, pp. 28-30.

[21] Ruiz-Cabello, J. and Cohen, J.S. (1992), "Phospholipid meta-bolites as indicators of cancer cell function," *NMR in Biomedicine*, vol. 5, pp. 226-233.

[22] Kalra, R., Wade, K., Hands, L., et al. (1993), "Phosphomonoester is associated with proliferation in human breast cancer: a ^{31}P MRS study," *British Journal of Cancer*, vol. 67, pp. 1145-1153.

[23] Su, B., Kappler, F., Szwergold, B., and Brown, T. (1993), "Identification of a putative tumor marker in breast and colon cancer," *Cancer Research*, vol. 53, pp. 1751-1754.

[24] Leach, M., Verrill, M., Glaholm, J., Smith, T., Collins, D., et al. (1998), "Measurements of human breast cancer using magnetic resonance spectroscopy: a review of clinical measurements and a

report of localised ^{31}P measurements of response to treatment," *NMR in Biomedicine*, vol. 11, no. 7, pp. 314-340.

[25] Ting, Y., Sherr, D., and Degani, H. (1996), "Variations in energy phospholipid metabolism in normal and cancer human mammary epithelial cells," *Anticancer Research*, vol. 16, pp. 1381-1388.

[26] Merchant, T., Meneses, P., Gierke, L., Otter, W.D., and Glonek, T. (1991), "^{31}P magnetic resonance phospholipid profiles of neo-plastic human breast tissues," *British Journal of Cancer*, vol. 63, pp. 693-698.

[27] Cohen, J., Lyon, R., Chen, C., Faustino, P., Batist, G., et al. (1986), "Differences in phosphate metabolite levels in drug-sensitive and -resistant human breast cancer cell lines determined by ^{31}P magnetic resonance spectroscopy," *Cancer Research*, vol. 46, pp. 4087-4090.

[28] Daly, P. and Cohen, J. (1989), "Magnetic resonance spectroscopy of tumors and potential in vivo clinical applications: a review," *Cancer Research*, vol. 49, pp. 770-779.

[29] Oberhaensli, R., Hilton-Jones, D., Bore, P., et al. (1986), "Bio - chemical investigation of human tumours in vivo with phosphorus-31 magnetic resonance spectroscopy," *Lancet*, vol. 2, pp. 8-11.

[30] Steen, R. (1989), "Response of solid tumors to chemotherapy monitored by in vivo ^{31}P nuclear magnetic resonance spectroscopy: a review," *Cancer Research*, vol. 49, pp. 4075-4085.

[31] Negendank, W. (1992), "Studies of human tumors by MRS: a review," *NMR in Biomedicine*, vol. 5, pp. 303-324.

[32] Gadian, D. (1995), *NMR and Its Applications to Living Systems*, (2nd edition), Oxford University Press.

[33] Kidd, A.L. (1987), "Knowledge acquisition – an introductory framework,"chapter in *Knowledge Acquisition for Expert Systems. A Practical Handbook*, pp. 1-16, Plenum Press.

[34] Redmond, O., Bell, E., Stack, J., Dervan, P., Carney, D., Hurson, B., and Ennis, J. (1992), "Tissue characterization and assessment of preoperative chemotherapeutic response in musculoskeletal tumors by in vivo ^{31}P magnetic resonance spectroscopy," *Magnetic Resonance in Medicine*, vol. 27, pp. 226-237.

[35] Degani, H., Horowitz, A., and Itzchak, Y. (1986), "Breast tumors: evaluation with P-31 MR spectroscopy," *Radiology*, vol. 161, pp. 53-55.

[36] Glaholm, J., Leach, M., Collins, D., Mansi, J., et al. (1989), "In-vivo ^{31}P magnetic resonance spectroscopy for monitoring treatment response in breast cancer," *Lancet*, vol. 8650, no. 1, pp. 1326-1327.

[37] Twelves, C., Porter, D., Lowry, M., Dobbs, N., et al. (1994), "Phosphorus-31 metabolism of postmenopausal breast cancer studied in vivo by magnetic resonance spectroscopy," *British Journal of Cancer*, vol. 69, pp. 1151-1156.

[38] Ng, T., Evanochko, W., Hiramoto, R., Ghanta, V., et al. (1982), "^{31}P NMR spectroscopy of in vivo tumors," *Journal of Magnetic Resonance*, vol. 49, pp. 271-286.

[39] Ross, B., Marshall, V., Smith, M., Bartlett, S., et al. (1984), "Monitoring response to chemotherapy of intact human tumours by ^{31}P nuclear magnetic resonance," *Lancet*, no. 1, pp. 641-646.

[40] Payne, G., Dowsett, M., and Leach, M. (1994), "Hormone-dependent metabolic changes in the normal breast monitored non-invasively by ^{31}P magnetic resonance (MR) spectroscopy," *The Breast*, vol. 3, pp. 20-23.

Chapter 11

Artificial Neural Networks as a Computer Aid for Lung Disease Detection and Classification in Ventilation-Perfusion Lung Scans

G.D. Tourassi, E.D. Frederick, and R.E. Coleman

Artificial intelligence (AI) has been established as a promising technology for computer-assisted medical decision making. Artificial neural networks (ANNs) are by far the most popular AI approach to the diagnostic interpretation of medical images. Several studies have shown that ANNs can be trained to perform diagnostic tasks, offering physicians a fast, consistent, and unbiased second opinion. This paper presents an application of ANNs in the field of nuclear medicine. Specifically, an ANN approach is developed for the diagnostic interpretation of ventilation-perfusion lung scans for patients with clinical suspicion of acute pulmonary embolism.

1   Introduction

The application of artificial intelligence (AI) techniques in medical imaging has become a thriving area of research. Given the clinical significance of patient data management and the increasingly digital nature of medical images, the utilization of AI in medical imaging has grown dramatically. One of the most popular areas of application is the diagnostic interpretation of medical images. Physicians typically perform the diagnostic task by successfully integrating the process of image perception with clinical reasoning. Nonetheless, there are frequently documented errors and variations in the human interpretation of medical images [1]-[4]. Studies suggest that some interpretation errors can be eliminated using a second reader approach [5]. Since AI has been shown to provide effective decision models, many researchers have recognized its potential as a diagnostic tool for clinical use.

The present study is an example of AI utilization in the field of nuclear medicine. Specifically, we present the application of artificial neural networks (ANNs) in the diagnostic interpretation of ventilation-perfusion (V-P) lung scans. V-P lung scans are the initial study performed to patients clinically suspected with acute pulmonary embolism (PE). The clinical significance of PE is well documented in the medical literature. PE is the third most common cause of death in the United States with more than 600,000 cases of PE per year [6]. Furthermore, the mortality rate due to PE has remained unchanged for decades with approximately 50,000 to 200,000 deaths per year [7], [8]. Perfusion lung scans are used as a screening tool because PE typically causes areas of decreased or absent perfusion. However, other lung diseases such as obstructive pulmonary disease can cause abnormal perfusion findings compromising the specificity of the scans. Consequently, ventilation scans and chest radiographs are added to improve diagnostic accuracy. Generally, V-P lung scans with mismatched areas (corresponding lung areas of abnormal perfusion and normal ventilation) are considered high risk for PE.

The interpretation of V-P scans is a clinical challenge [9]. Almost 70% of clinically suspected patients have low or intermediate probability V-P scans [10]. These scans are considered non-diagnostic since PE prevalence has been shown to vary substantially in this subgroup of patients (anywhere between 14% to 35%) [11], [12]. Since there are serious complications associated with anticoagulation treatment given to patients suspected with PE [13], further evaluation of patients with nondiagnostic scans is imperative before treatment is started. On the other hand, pulmonary angiography (the gold standard for PE diagnosis) is an invasive procedure with morbidity and mortality associated, which makes many physicians uncomfortable recommending it as a screening tool.

Given the clinical significance of the diagnostic problem, the application of AI techniques for the diagnosis of PE from V-P lung scans has attracted the researchers' interest over the past decade. Many efforts have focused on synthesizing physician-extracted V-P scan findings using AI decision algorithms [14]-[18]. Some studies proposed initial preprocessing of the scans for automated extraction of the V-P input findings [19], [20]. Depending on case difficulty, the studies showed that the diagnostic performance of the proposed computer aids was comparable or better to that of physicians.

In our study, we are proposing a computer-assisted diagnostic (CAD) tool for the diagnostic interpretation of V-P lung scans using multifractal texture analysis and ANNs. Contrary to the previously published studies, emphasis is placed on performing differential diagnosis, and not just diagnosis of PE. As such, the computer aid is designed to operate as a detection and classification tool for a variety of lung diseases that can compromise the diagnostic accuracy of the lung scans. Initially, the proposed CAD approach targets perfusion lung scans alone, aiming to improve their diagnostic specificity. Next, information extracted from the ventilation scans is added to test if it can further boost the CAD diagnostic performance.

The chapter is organized as follows. Section 2 provides a brief introduction to feed-forward artificial neural networks as applied in medical decision making. Section 3 decsribes in detail the patient database, the feature extraction technique, and the ANN architectures employed for the implementation of the proposed CAD tool. In addition, Section 3 addresses some data sampling and performance evaluation issues that are critical in CAD applications. Section 4 presents CAD performance evaluation results for a variety of diagnostic problems in lung scintigraphy. A brief summary and conclusions are presented in the final section.

2   Artificial Neural Networks

There are numerous applications of computer-assisted diagnosis in radiology over the past decade. The applications typically involve detection and localization of disease in medical images such as detection of breast cancer in mammograms, identification of lung cancer in chest radiographs, detection of diffuse liver disease in ultrasound images, or lesion detection and localization in Single Photon Emission Computed Tomographic images. Although most CAD applications utilize some form of artificial intelligence (i.e., artificial neural networks, fuzzy logic, genetic algorithms, or case-based reasoning), feed-forward ANNs are by far the most popular AI technology in the medical field.

The building block of an ANN is the artificial neuron [21]. A typical ANN consists of many, highly interconnected neurons and has a parallel architecture that imitates the structure of the human brain [22]. The connections among the neurons are called weights and store the knowledge acquired by the network during training. Several studies have shown that ANNs can duplicate important aspects of the human intelligence such as the ability to learn and generalize [23]. Actually, ANNs can be trained to decided the presence or absence of a disease the same way a physician learns. The network is presented with a number of patients, their clinical findings, and their final diagnoses. As learning occurs, the network continually modifies its weights to optimize its prediction of disease for every patient (the same way a physician becomes more experienced with time). Although data-driven just as humans, ANNs are task-specific and thus less flexible than the physicians who can transfer acquired knowledge from one diagnostic task to another. However, ANNs have all the advantages of computers providing physicians with decision tools which can be fast, precise, unbiased, consistent, and fatigue-tolerant.

In the past, ANNs have been proven extremely helpful in computationally complex decision problems. Furthermore, ANNs have been established as a promising alternative to statistical techniques [24]. Thus far, the most popular ANN architecture for medical applications has been the feed-forward backpropagation neural network [25]. This network has been successful with a wide variety of pattern recognition problems. Consequently, it is always the first choice for medical diagnostic tasks. The basic characteristics of a backpropagation neural network are:

(*i*) *multilayered, feed-forward architecture*: That is, the network accepts the input information (i.e., clinical findings) which it propagates forward throughout the hidden layers to produce its output (i.e., diagnosis).

(*ii*) *supervised learning*: The network is provided with sets of input-output cases. During the learning process the network tries to adjust its weights so that for every given input case it computes the desired output. This type of learning is called supervised because for every input case the network tries to reproduce the *known* output.

(*iii*) *generalized delta rule*: The backpropagation algorithm is used to adjust the ANN weights. Specifically, backpropagation tries to minimize the mean squared error (MSE) between the desired and the actual network output following an iterative gradient search technique [25]. Cost functions other than the MSE may be used for training the network.

Application of an ANN for a medical diagnostic task is composed of two phases. During the training phase, the ANN is presented with a dataset and modifies its weights until it converges to a set of weights that can sufficiently reproduce the correct answers. During the testing phase, the ANN weights remain fixed and the trained network is validated on a dataset different from the one it was trained. Generally, if a neural network has the proper architecture and sufficient training data it will be able to give correct answers for input data it has never seen before.

The next section demonstrates how an ANN approach has been developed for the diagnostic interpretation of ventilation-perfusion lung scans.

3   Materials and Methods

3.1   Overview of the AI Diagnostic Scheme

The proposed diagnostic scheme is designed to operate on regions of interest (ROIs) selected from patients' lung scans. For a specific lung location, three ROIs are extracted from: (i) the posterior view of the perfusion scan, (ii) the initial breath view of the ventilation scan, and (iii) the equilibrium phase of the ventilation scan. The scheme is based on a two-step approach. Initially, textural information is extracted from the selected ROI. Subsequently, the extracted information is used by a decision algorithm (i.e., an artificial neural network) designed to perform the diagnostic task. This two-step approach emulates a human observer. First, image texture analysis parallels the role of the human visual perception and the process of feature extraction is achieved. Then, the extracted features are merged into a diagnosis using a decision algorithm. The algorithm is intended to duplicate the process of clinical reasoning. There is a wide selection of decision algorithms available, such as rule-based models, traditional statistical analysis, or the more popular (and often more successful) neural networks. Figure 1 is a schematic overview of the proposed diagnostic algorithm.

3.2   Patient Data

The study is based on a small database of 45 patients who were admitted at Duke University Medical Center with the suspicion of acute pulmonary embolism between April 1, 1992 and July 1, 1993. All patients underwent ventilation-perfusion lung scanning and chest radiography. The patients' imaging studies were read by experienced physicians and risk stratified for PE according to the well-documented PIOPED criteria [12]. Forty-one out of the 45 scans were considered non-diagnostic for PE (7 low probability scans and 34 indeterminate scans). Due to indeterminate findings or discordant results all 45 patients were later referred to pulmonary angiography for final diagnosis. All angiograms were definitive. Thirteen out of the 45 patients were angiographically confirmed with PE.

The V-P images included in the study were acquired according to the PIOPED protocol [11]. For every patient a standard eight-view perfusion study was performed using approximately 4 millicuries of Tc-99m MAA. The ventilation scans were acquired with 15-30 millicuries of Xenon-133 using a posterior 100,000 count single breath image followed by a posterior equilibrium image and six washout phase images. Both ventilation and perfusion scans were obtained using a parallel hole, low energy, all-purpose collimator on gamma camera with a 38cm field of view. All views were acquired with a 128×128 pixel map and a 0-255 gray level range. The perfusion and ventilation scans were printed on 8×10 inch films.

The stored patient films were digitized for the needs of this study at 0.17mm resolution with 8 bit gray scale precision using an AGFA Arcus II transparency scanner. The study focused on the posterior view of the perfusion scans and on the initial breath view and equilibrium phase of the ventilation scans. An experienced nuclear medicine physician reviewed the patients' V-P lung scans and chest radiographs and manually extracted 24×24 pixel ROIs from the posterior view of the perfusion scans. Subsequently, the physician selected the corresponding 24×24 pixel ROIs in the ventilation scans. The size of the ROIs was determined by the physician (R.E.C) so that the ROIs could fit within the natural outline of the normal lung without overlapping. Figure 2 shows a representative V-P lung scan (posterior view of the perfusion scan, initial breath view and equilibrium phase view of the ventilation scan).

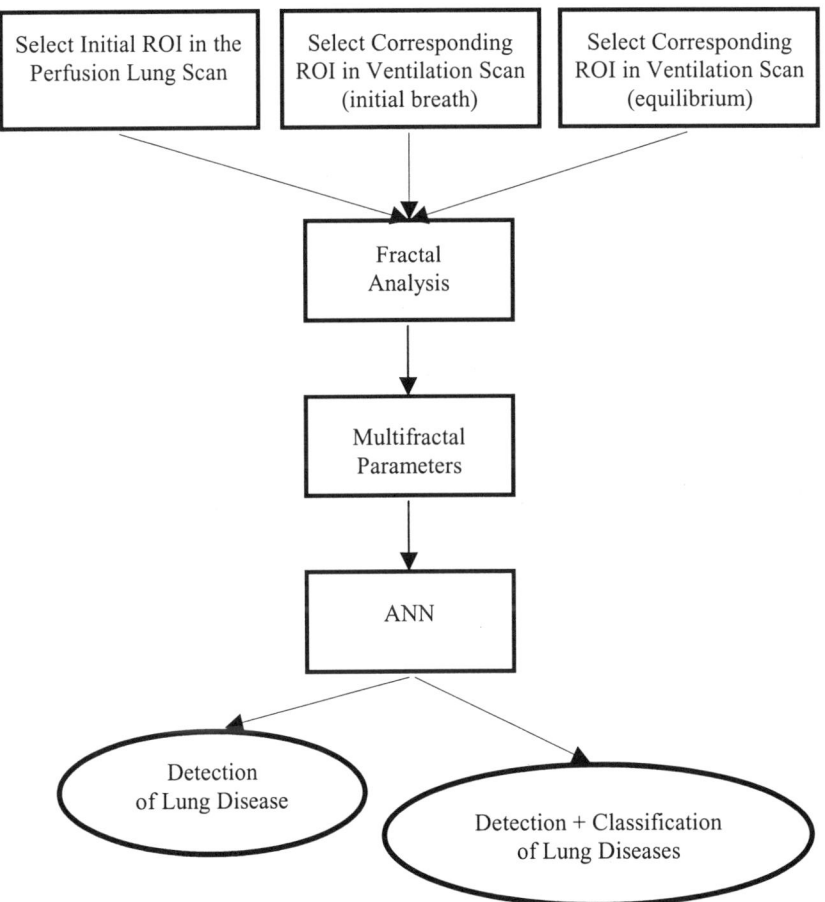

Figure 1. Overview of the diagnostic scheme.

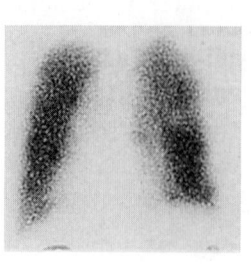 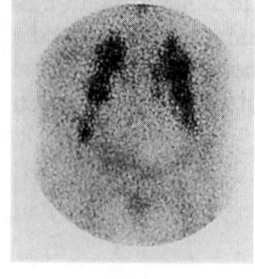 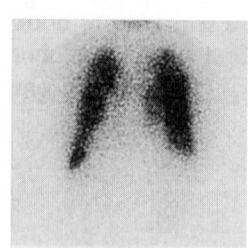

Perfusion Scan   Ventilation Scan   Ventilation Scan
                            Initial Breath   Equilibrium Phase

Figure 2. Selected views of a representative V-P lung scan.

The ROIs were selected according to following protocol. Each lung was divided into three zones (upper, middle, lower) and an ROI was selected for each lung zone for a subtotal of six ROIs per patient. Consequently, there were 270 ROIs in total. Each ROI was characterized as either normal or abnormal. Six lung diseases were considered in the diagnostic scheme: PE, obstructive pulmonary disease (OPD), parenchymal opacity (OPAC), pleural effusion (EFFU), atelectasis (ATEL), and other lung disease (OTHER). The lung disease characterization was based on the patient's complete V-P lung scans and chest radiographs. The pulmonary angiograms were then reviewed and the ROIs corresponding to the angiographic abnormalities were determined and labeled as PE. Some abnormalities were considered PE by the imaging studies but not by pulmonary angiogram. Those abnormalities were reviewed again by the physician and reassigned to another lung disease category based on the patients' complete imaging studies. There were 94 normal and 176 abnormal ROIs in total. Table 1 shows the detailed distribution of the abnormally perfused ROIs.

Table 1. Number and type of selected ROIs.

Type of Lung Disease	Number
Normal	94
Atelectasis (ATEL)	8
Pleural Effusion (EFFU)	10
Parenchymal Opacity (OPAC)	25
Obstructive Pulmonary Disease (OPD)	96
Other Lung Disease (OTHER)	15
Pulmonary Embolism (PE)	22

3.3   Multifractal Texture Analysis

Fractal analysis is a popular form of texture analysis with numerous applications in medical imaging. The analysis is based on the principle that an image property is a function of the scale parameter according to which it is measured. Their logarithmic transforms are linearly related. The slope of the fitted line determines the fractal dimension (FD) which is the most frequently used fractal measure in medical imaging [26].

There are various techniques available to estimate the FD of an image. The techniques typically differ in the definition of the image property and the scale parameter that are used. For this study, we used the circularly averaged power spectrum method to perform fractal analysis [27], [28]. According to this method, the two-dimensional power spectrum of each ROI is transformed into one dimension by averaging the spectrum as a function of the radial distance from zero frequency. The Fourier power spectrum is then plotted on a log-log scale as a function of the frequency:

$$P(f) \propto \left(\frac{1}{f}\right)^{\beta}.$$

The slope β of the linear regression of the log-log plot is linearly related to the fractal dimension (FD) of the image [27]:

$$FD = \frac{8 - \beta}{2}.$$

Figure 3 shows the power spectrum of a representative perfusion ROI plotted on a log-log scale as a function of the frequency. Two distinct linear portions were observed in the log-log plot. Thus, two FDs were calculated each one corresponding to a different portion. In addition, the coordinates for the point of intercept of the two linear segments were also extracted. The FD parameters were estimated using software written in the MATLAB environment (The MathWorks, Inc., Natick, MA).

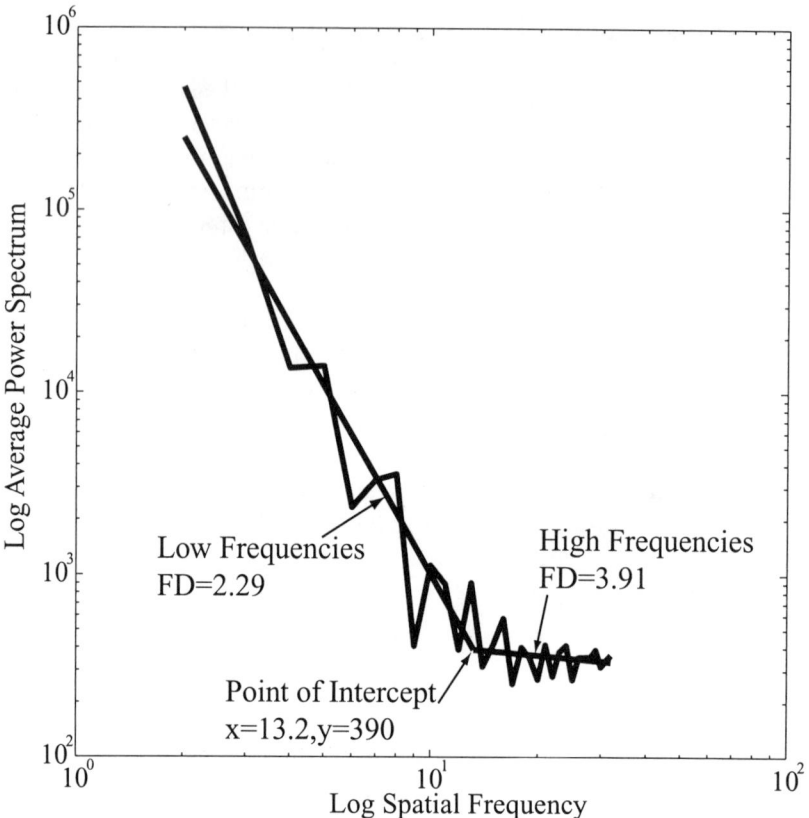

Figure 3. Representative power spectrum of a perfusion ROI plotted on a log-log scale as a function of the frequency.

3.4   Artificial Neural Network Predictions

The extracted fractal parameters were then merged into a final diagnosis using a connectionist ANN approach. Several different ANNs were implemented based on which lung scans were included in the analysis [perfusion scans only or perfusion and ventilation scans] and the diagnostic problem [detection or classification]. Initially, an ANN was developed to detect the presence or absence of lung disease in an ROI. Subsequently, another ANN was developed to decide not only the presence but also the type of lung disease present. Generally, all networks had a three-layer, feed-forward architecture and they were trained using the backpropagation algorithm with the sigmoid activation function [25]. The input layer had as many nodes as the number of the multifractal parameters combined for the task at hand. The number of output nodes varied between one and seven depending on if the ANN was designed to perform detection or classification. The number of hidden nodes was determined empirically. Preliminary studies varying the number of hidden nodes between two and sixteen showed that ANNs with eight hidden nodes had the best generalization performance. For all experiments, the weights were randomly initialized between -0.1 and 0.1. During the training sessions, the ANN weights were adjusted after each case presentation. Finally, the training and momentum coefficients were determined empirically and set at 0.4 and 0.2 respectively.

3.5   Performance Evaluation

Results are reported in the form of Receiver Operating Characteristic (ROC) curves for the detection problem [29]. Conventionally, ROC curves plot the true positive fraction (or sensitivity) vs. the false positive fraction (or 1-specificity) for a wide and continuous range of decision thresholds. They are considered the most meaningful form of performance evaluation in medical imaging because (i) they take into account the decision criterion level and (ii) they provide a measure of performance not affected by prior probabilities (i.e., how rare a disease is). Furthermore, ROC curves can be used to determine optimum criterion levels that maximize accuracy, average benefit or other measures of clinical efficacy.

The most frequently selected index of diagnostic performance in ROC analysis is the area (A_z) under the ROC curve. The area index varies between 0.5 (chance behavior) and 1 (perfect performance). Therefore, better performance is reflected in a higher ROC area index. We used the ROCKIT software package developed by Metz *et al.* at the University of Chicago (see http://www-radiology.uchicago.edu/krl/toppage11.htm) to fit ROC curves to the output responses of all decision models developed in this study. Confusion matrices were employed to report classification results. For comparison, linear decision models were developed for the same problems using linear discriminant analysis (LDA) and the SAS/STAT software (SAS Institute Inc., Cary, NC).

A critical issue in the clinical evaluation of ANNs with limited datasets is data sampling [30]. We employed the leave-one-out crossvalidation sampling scheme to optimize the utilization of the available data [31]. According to this method, all cases but one are used to train the network for a fixed number of iterations. Then, the trained network is tested on the one case left out. The same process is repeated until every case in the database has served as a test case. The leave-one-out cross-validation approach is considered an efficient data sampling scheme because it utilizes almost all available cases for the training phase of the ANN without compromising the statistical significance of the testing phase.

4   Results

4.1   Analysis of Perfusion Lung Scans

Initially, the study focused on the perfusion scans only. Various feed-forward ANNs were developed depending on the diagnostic problem at hand. Results are presented for the following problems: (i) normal vs. abnormal ROI, (ii) PE present vs. PE absent, and (iii) classification of lung disease.

4.1.1   Detection of Lung Disease

An ANN was developed to detect the presence or absence of lung disease in ROIs extracted from perfusion lung scans. This detection task is closer to clinical practice where physicians initially screen the perfusion scan to detect any potential abnormalities.

The ANN had an input layer with four nodes each one corresponding to a multifractal parameter extracted from the perfusion ROI. The target node was set to 0 (lung disease absent) or 1 (lung disease present). Figure 4 shows the ROC curve achieved by the ANN according to the leave-one-out sampling scheme. We have also included for comparison the ROC curve achieved by linear discriminant analysis (LDA) for the same dataset and sampling scheme. Furthermore, to demonstrate the utility of multifractal texture analysis, we have included the ROC curve achieved by the single fractal dimension (FD) estimated using the full frequency range for each ROI. To plot this curve, the estimated FD is used directly as the decision variable for further ROC analysis.

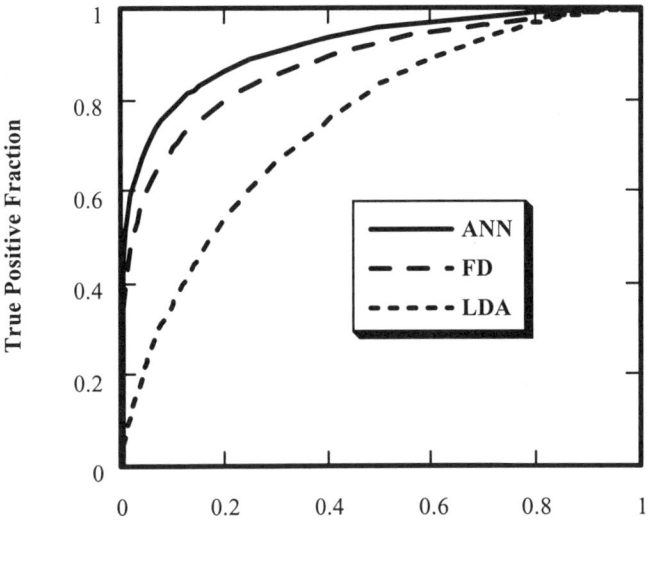

False Positive Fraction

Figure 4. ROC evaluation of all decision models for the detection of abnormal ROIs in perfusion lung scans.

The corresponding ROC area indices were the following:

$$A_z(\text{ANN}) = 0.92 \pm 0.02$$
$$A_z(\text{FD}) \quad = 0.87 \pm 0.03$$
$$A_z(\text{LDA}) = 0.75 \pm 0.03$$

The ANN had superior diagnostic performance than either the single fractal dimension or LDA. The differences were statistically significant at the 95% confidence level. For a decision threshold set at 0.5, the network achieved 89% sensitivity and 75% specificity.

4.1.2   Detection of Pulmonary Embolism

Subsequently, an ANN was developed for a more ambitious decision task. It was trained to detect the presence or absence of pulmonary embolism (PE) in a given ROI. This task is much more challenging than before because physicians typically decide the presence or absence of PE only when given additional information from ventilation scans and chest radiographs. The previous ANN was modified so that the target values were set to 1 for the 22 ROIs with PE present and 0 for the remaining ROIs. Therefore, the data set had 8.1% prevalence of disease (22/270 ROIs confirmed with PE). Figure 5 summarizes the diagnostic performance of all decision algorithms.

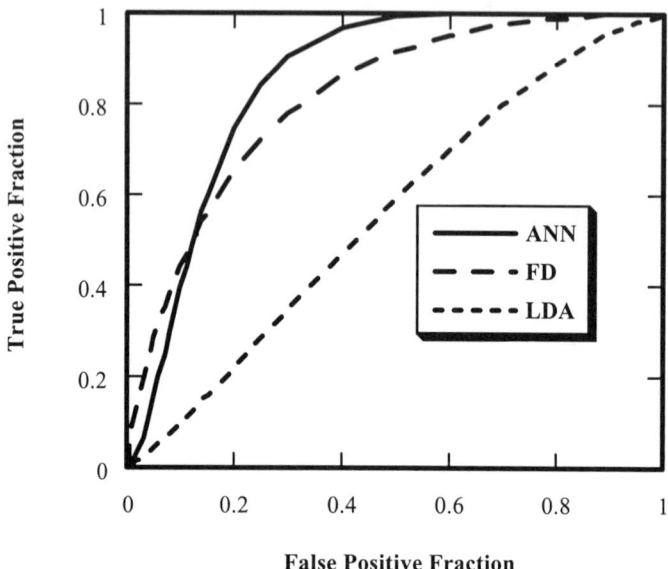

Figure 5. ROC evaluation of all decision models for the detection of PE in ROIs extracted from perfusion lung scans.

The corresponding ROC area indices were the following:

$$A_z(ANN) = 0.85 \pm 0.03$$
$$A_z(FD) \;\; = 0.81 \pm 0.04$$
$$A_z(LDA) = 0.56 \pm 0.06$$

As expected, the ANN performance declined due to the increased difficulty of the detection task. However, the ANN was still superior than either the single fractal dimension or LDA.

Compared to clinicians, the ANN showed competitive diagnostic performance. Specifically, the nuclear medicine physician who selected the ROIs for the needs of this study had 55% sensitivity (12 out of the 22 ROIs with PE were correctly identified) and 98.3% specificity (244 out of the 248 ROIs were correctly identified as normal or abnormal but not related to PE). In comparison, Figure 5 shows that for the same ROC operating point (sensitivity = 55%) the ANN has 84% specificity.

Even though this performance is noticeably lower than that of the experienced nuclear medicine physician, it should be noted that the physician's diagnosis is based on the full (multiple views) perfusion scan, the ventilation scan, and the chest radiograph and not only on the limited 24×24 pixel ROI extracted from the perfusion scan.

4.1.3   Classification of Lung Diseases

The final ANN was designed to perform differential diagnosis. Thus, it was developed to perform both detection and classification of lung diseases. The ANN had similar architecture as the ANNs described in the previous sections except that the output layer consisted of seven nodes. The following confusion table summarizes the classification results.

The ANN showed overall 45.9% (124/270) classification accuracy. Seventy-three out of 94 normal ROIs were correctly identified (specificity = 77.7%). Similarly, 154 out of 176 abnormal ROIs were characterized as abnormal (sensitivity = 87.5%). This performance is similar to the one achieved by the ANN developed in Section 3.1.1 (Figure 4). Similar comparison can be made between the classification network and the detection network developed in Section 3.1.2. Specifically, the classification ANN correctly recognized seven out of the 22 PE ROIs (sensitivity 31.8%). The corresponding specificity was 88.3% (219/248) which is similar to the 91% specificity achieved by the detection network for the same sensitivity (Figure 5).

Table 2. ANN classification results for lung disease characterization in perfusion scans.

TRUE CLASS	NORMAL	ATEL	EFFU	OPAC	OPD	OTHER	PE
NORMAL	73	0	0	5	9	7	0
ATEL	1	2	0	3	1	0	1
EFFU	0	1	3	0	0	0	6
OPAC	4	6	3	2	5	4	1
OPD	15	13	6	8	32	5	17
OTHER	2	1	0	2	1	5	4
PE	0	4	8	0	2	1	7

It is very encouraging that the ANN correctly identified all PE ROIs as abnormal. However, it appears that the network has difficulty discriminating between ROIs related to pleural effusion and pulmonary embolism.

4.2   Analysis of Perfusion-Ventilation Lung Scans

Thus far, a neural network approach appears to be a promising way to detect and even characterize lung diseases in perfusion lung scans. The ANN showed specificity competitive to that of an experienced nuclear medicine physician based on limited fractal information from selected ROIs. The next step is to test if this performance can be further improved by adding information from corresponding ROIs extracted from the patients' ventilation lung scans. Specifically, two more ROIs were extracted: (i) one ROI from the initial breath view, and (ii) one ROI from the equilibrium phase of the ventilation scan. Both ROIs corresponded to the same locations of each perfusion ROI. The ROIs were analyzed using the multifractal approach presented in Section 2.3. Four multifractal parameters were extracted per ventilation ROI for a total of twelve fractal features per lung location. Two ANNs were developed to merge the twelve parameters into a final diagnosis. Since the ventilation scans can assist primarily on the diagnosis of PE, results were presented for the PE detection problem and subsequently for the lung disease classification problem as in the previous two subsections. Results are based again on the leave-one-out sampling scheme.

4.2.1   Detection of Pulmonary Embolism

Figure 6 summarizes the results for the ANN utilizing textural information from both perfusion and ventilation scans. The corresponding ROC area index is slightly lower [A_z(ANN-VPscan) = 0.83 ± 0.03] than the one achieved by the ANN utilizing only the perfusion lung scan [A_z(ANN-P scan) = 0.85 ± 0.03].

It is interesting to observe that the ROC curves cross. Although the addition of the ventilation scan appears to degrade the overall ANN performance (as measured by the Az index), there is a certain range of specificities (>80%) where incorporating information from the ventilation scan improves the ANN sensitivity. Consequently, if it is clinically more important to utilize the computer aid as a highly specific tool, adding the ventilation scan improves significantly the sensitivity of the diagnosis.

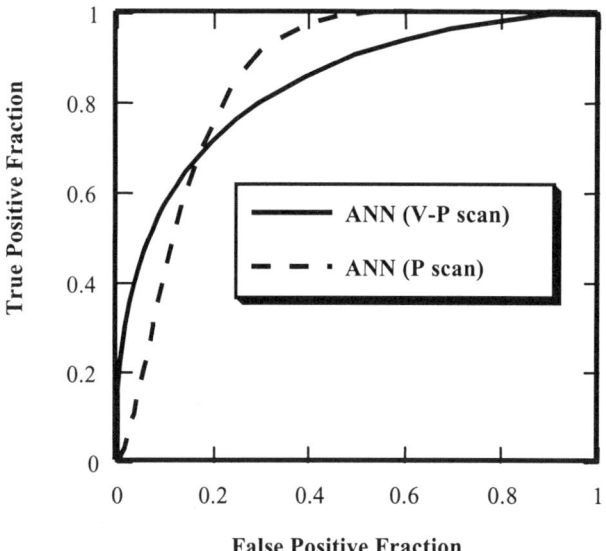

Figure 6. ROC evaluation of ANN models for the detection of PE.

4.2.2   Classification of Lung Diseases

Table 3 summarizes the classification results for the ANN network based on textural information extracted from V-P lung scans.

Table 3. ANN classification results for lung disease characterization in V-P lung scans.

TRUE CLASS	NORMAL	ATEL	EFFU	OPAC	OPD	OTHER	PE
NORMAL	81	1	0	4	6	2	0
ATEL	1	0	1	2	3	0	1
EFFU	0	2	3	1	0	0	4
OPAC	6	2	3	1	11	2	0
OPD	15	8	5	7	28	13	20
OTHER	4	0	1	0	4	3	3
PE	0	2	3	0	5	2	10

The ANN showed overall 46.7% (126/270) classification accuracy. The performance is very similar to the one achieved without the ventilation information. The addition of the ventilation scan appeared to improve the detection rate of normal ROIs (86.2%) reducing however the detection rate of abnormal ROIs (85.2%). The network classified better the PE ROIs (sensitivity 45.5% compared to 31.8%) preserving the specificity level (88.7% compared to 87.5%). Therefore, the addition of the ventilation scan appears to improve the detection rate of pulmonary embolism. Comparing Tables 2 and 3, it is apparent that the improvement is due to better discrimination between PE and pleural effusion.

5   Discussion

V-P scans are utilized as a non-invasive, risk stratification tool for patients clinically suspected with PE. However, not only is the diagnostic interpretation of the V-P scans a clinical challenge but studies have also shown that significant variation occurs among clinicians even when using identical sets of decision rules [32]. The purpose of this study was to investigate if ANNs can assist physicians in the diagnostic interpretation of V-P lung scans. An improvement in the diagnostic yield of V-P scans could have major clinical impact by potentially (i) reducing the prevalence of non-diagnostic scans, and (ii) reducing observer variability in lung scan interpretations.

Our study presented a computer-assisted diagnostic tool based on fractal texture analysis and ANNs. Although the application of ANNs in the diagnosis of PE has been a relatively popular topic, the majority of the studies used physician-extracted input variables. This study aimed to exploit the textural signature of the V-P lung scans as a preprocessing step for the CAD scheme. Specifically, textural features were extracted from the various scans to serve as inputs to an ANN trained to perform the final diagnosis. Automated extraction of the ANN input findings generally results in a more robust, unbiased, and reproducible CAD tool.

There are several unique aspects in the proposed CAD approach. First, the study was designed to evaluate incrementally the diagnostic contribution of each lung scan for a variety of clinically relevant decision tasks. Second, the diagnostic task was more broadly defined to include not only detection of PE but also detection and characterization of six possible lung diseases which can coexist and therefore can be an obstacle for the physicians when considering PE. Finally, the study was based on a clinically challenging patient sample. A previous study showed that not only the physicians' diagnostic performance was generally low ($A_z = 0.69 \pm 0.06$) in this patient population but there was also significant between-observer variability among physicians [16].

Considering the difficulty of the diagnostic task, the proposed CAD tool achieved diagnostic performance comparable to that of a very experienced nuclear medicine physician. The performance was impressive given the limited amount of lung scan information analyzed at a time. Physicians make their diagnosis by visually correlating multiple views of the perfusion scan, ventilation scan, and the chest radiograph. The CAD tool was designed to operate only on small size ROIs. Adding multiple views of the V-P lung scans could potentially improve the CAD performance. Given the promising results, a large scale study is required to better define the clinical role of the proposed computer aid.

In conclusion, CAD analysis has faced numerous criticisms in the past. Regardless, CAD research has made considerable progress by producing practical applications, which have shown to enhance the diagnostic performance of radiologists if used as a second opinion [33]-[35]. As the trend in radiology is increasingly digital, CAD interpretation of medical images will continue advancing by addressing well known clinical needs such as difficult to interpret images and cost-effective utilization of available resources.

Acknowledgments

This work was supported by grant RG 98-0324 from the Whitaker Foundation and in part by grant R29-HL-52826 from the National Heart, Lung, and Blood Institute.

References

[1] Brady, A.P., Stevenson, G.W., and Stevenson, I. (1994), "Colorectal cancer overlooked at barium enema examination and colonoscopy: continuing perceptual problem," *Radiology*, vol. 192, pp. 373-378.

[2] Anderson, R.E., Hill, R.B., and Key, C.R. (1989), "The sensitivity and specificity of clinical diagnostics during five decades: toward an understanding of necessary fallibility," *JAMA*, vol. 261, pp. 1610-1617.

[3] Renfrew, D.L., Franken Jr., E.A., Berbaum, K.S., Weigelt, F.H., and Abu-Yousef, M.M. (1992), "Error in radiology: classification and lessons in 182 cases presented at a problem case conference," *Radiology*, vol. 183, pp. 145-150.

[4] Robinson, P.J. (1997), "Radiology's Achilles' heel: error and variation in the interpretation of the Roentgen image," *BJR*, vol. 70, pp. 1085-1098.

[5] Anttinen, I., Pamilo, M., Soiva, M., and Roiha, M. (1993), "Double reading of mammography screening films – one radiologist or two?," *Clin Radiol*, vol. 48, no. 6, pp. 414-421.

[6] Gillum, R.F. (1987), "Pulmonary embolism and thrombophlebitis in the United States, 1970-1985," *Am. Heart J.*, vol. 114, pp. 1262-1264.

[7] Soskolne, C.L., Wong, A.W., and Lilienfeld, D.E. (1990), "Trends in pulmonary embolism death rates for Canada and the United States, 1962-87," *Can. Med. Assoc. J.*, vol. 142, no. 4, pp. 321-324.

[8] Lilienfeld, D.E., Chan, E., Ehland, J., Godbold, J.H., Landrigan, P.J., and Marsh, G. (1992), " Mortality from pulmonary embolism in the United States: 1962 to 1984," *Chest*, vol. 98, no. 5, pp. 1067-1072.

[9] Tapson, V.F. (1997), "Pulmonary embolism: the diagnostic repertoire," *Chest*, vol. 112, no. 3, pp. 578-580.

[10] The PIOPED Investigators (1990), "Value of the ventilation/ perfusion scan in acute pulmonary embolism: results of the prospective investigation," *JAMA*, vol. 263, pp. 2753-2759.

[11] Gottschalk, A., Juni, J.E., Sostman, H.D., Coleman, R.E., Thrall, J., McKusick, K.A., Froelich, J.W., and Alavi, A. (1993), "Ventilation-perfusion scintigraphy in the PIOPED study: Part I. Data collection and tabulation," *J. Nucl. Med.*, vol. 34, no. 7, pp. 1109-1118.

[12] Gottschalk, A., Sostman, H.D., Coleman, R.E., Juni, J.E., Thrall, J., McKusick, K.A., Froelich, J.W., and Alavi, A. (1993), "Ventilation-perfusion scintigraphy in the PIOPED study: Part II. Evaluation of the scintigraphic criteria and interpretation," *J. Nucl. Med.*, vol. 34, no. 7, pp. 1119-1126.

[13] Palareti, G., Leali, N., Coccheri, S., *et al.* (1996), "Bleeding complications of oral anticoagulant treatment: an inception-cohort, prospective collaborative study (ISCOAT). Italian Study on Complications of Oral Anticoagulant Therapy," *Lancet*, vol. 348, pp. 423-428.

[14] Scott, J.A., and Palmer, E.L (1993), "Neural network analysis of ventilation-perfusion lung scans," *Radiology*, vol. 186, pp. 661-664.

[15] Tourassi, G.D., Floyd Jr., C.E., Sostman, H.D., and Coleman, R.E. (1993), "Acute pulmonary embolism: artificial neural network approach for diagnosis," *Radiology*, vol. 189, pp. 555-558.

[16] Tourassi, G.D., Floyd Jr., C.E., Sostman, H.D., and Coleman, R.E. (1995), "Performance evaluation of an artificial neural network for the diagnosis of acute pulmonary embolism: effect of case and observer selection," *Radiology*, vol. 194, pp. 889-893.

[17] Fisher, R.E., Scott, J.A., and Palmer, E.L. (1996), "Neural networks in ventilation-perfusion imaging," *Radiology*, vol. 198, pp. 699-706.

[18] Scott, J.A., Fisher, R.E., and Palmer, E.L. (1996), "Neural networks in ventilation-perfusion imaging. Part II. Effects of interpretive variability," *Radiology*, vol. 198, pp. 707-713.

[19] Gabor, F.V., Datz, F.L., and Christian, P.E. (1994), "Image analysis and categorization of ventilation-perfusion scans for the diagnosis of pulmonary embolism using an expert system," *J Nucl Med*, vol. 35, pp. 797-802.

[20] Scott, J.A. (1999), "Using artificial neural network analysis of global ventilation-perfusion lung scan morphometry as a diagnostic tool," *AJR*, vol. 173, pp. 943-948.

[21] McCulloch, W.S., Pitts, W.H. (1943), "A logical calculus for the ideas immanent in nervous activity," *Bulletin of Mathematical Biophysics*, vol. 5, pp. 115-133.

[22] Rosenblatt, F. (1959), Principles of Neurodynamics, Spartan Books, New York, NY.

[23] Rosenblatt, F. (1958), "The perceptron: a probabilistic model for information storage and organization in the brain," *Psychological Review*, vol. 65, pp. 386-408.

[24] White, H. (1989), "Learning in artificial neural networks: a statistical perspective," *Neural Computation*, vol. 1, pp. 425-484.

[25] Rumelhart, D.E., Hinton, G.E., and Williams, R.J. (1986), "Learning internal representations by error propagation," in Rumelhart, D.E., McClelland, J.L. (Ed.), *Parallel Distributed Processing: Explorations in the Microstructures of Cognition*, Cambridge, MA: The MIT Press, vol. I, pp. 318-362.

[26] Fortin, C., Kumaresan, R., and Ohley, W. (1992), "Fractal dimension in the analysis of medical images," *IEEE Eng Med Biol*, pp. 65-71.

[27] Anguiano, E., Pancorbo, M., and Aguilar, M. (1993), "Fractal characterization by frequency analysis. I: Surfaces," *J Microscopy*, vol. 172, pp. 223-232.

[28] Aguilar, M., Anguiano, E., and Pancorbo, M. (1993), "Fractal characterization by frequency analysis. II: A new method," *J Microscopy*, vol. 172, pp. 233-238.

[29] Swets, J.A. (1988) "Measuring the accuracy of diagnostic systems," *Science*, vol. 240, pp. 1285-1293.

[30] Tourassi, G..D., and Floyd Jr., C.E. (1997) "Effect of data sampling on the performance evaluation of artificial neural networks for medical diagnosis," *Med Dec Making*, vol. 17, no. 2, pp. 186-192.

[31] Efron, B. (1983) "Estimating the error rate of a prediction rule: improvement on cross-validation," *J American Stat Assoc*, vol. 78, pp. 316-331.

[32] Scott, J.A., and Palmer, E.L. (1993) "Do diagnostic algorithms always produce a uniform lung interpretation?" *J Nucl Med*, vol. 34, pp. 661-665.

[33] Kegelmeyer, W.P., Pruneda, J.M., Bourland, P.D., Hillis, A.H., Riggs, M.W., and Nipper, M.P. (1994) "Computer-aided mammographic screening for spiculated lesions," *Radiol*, vol. 191, no. 2, pp. 315-317.

[34] Jiang, Y.L., Nishikawa, R.M., Schmidt, R.A., Metz, C.E., Giger, M.L., and Doi, K. (1999), "Improving breast cancer diagnosis with computer-aided diagnosis," *Acad Radiol*, vol. 6, no. 1, pp. 22-33.

[35] Chan, H.P., Sahiner, B., Helvie, M.A., Petrick, N., Roubidoux, M.A.. Wilson, T.E., Adler, D.D., Paramagu,l C., Newman, J.S., and Sanjay-Gopal, S. (1999) "Improvement of radiologists' characterization of mammographic masses by using computer-aided diagnosis: an ROC study," *Radiol*, vol. 212, no. 3, pp. 817-827.

Chapter 12

Neural Network for Classification of Focal Liver Lesions in Ultrasound Images

H. Yoshida

A novel method of multiscale texture analysis based on neural networks (NNs) has been developed and applied to the automated classification of benign and malignant focal liver lesions in ultrasound images. Our method is unique in the sense that it integrates a process of selection of multiscale texture features and a process of classification by a NN for effective classification. We developed an automated method that selects a set of multiscale texture features in the wavelet domain which maximize the performance of the NN for a given classification task. For the automated classification of benign and malignant focal liver lesions, regions of interest (ROIs) extracted from within the lesions were decomposed into subimages by wavelet packets. Multiscale texture features that quantify the homogeneity of the echogenicity were calculated from these subimages and were combined by a NN. A subset of the multiscale features was selected that yielded the highest performance in the classification of lesions. In an analysis of a set of ROIs extracted from hemangiomas (benign lesions), and from hepatocellular carcinomas and metastases (malignant lesions), the multiscale features yielded a high performance in distinguishing the benign from the malignant lesions. Therefore, our new multiscale texture analysis method based on NNs has the promise of increasing the accuracy of diagnosis of focal liver lesions in ultrasound images.

1 Introduction

Visually, echo texture on ultrasound images is represented by spatially distributed elements of localized gray-level variation that is limited by the local dynamic range of the equipment. The texture pattern is therefore highly scale-dependent, because the size of this local variation may change according to the echo texture. Furthermore, there may be various sizes of such elements in a texture. Therefore, echo texture patterns can be described efficiently by multiscale (or multiresolution) analysis methods such as the wavelet transform, which has been applied successfully to texture classification [1]-[3].

In this chapter, we present a novel multiscale texture feature analysis method, developed by the authors based on wavelet packets [6], [7] and a neural network (NN) [8], which can maximally distinguish benign from malignant focal liver lesions on ultrasound images. Focal liver disease is very common, with 25% to 50% of all patients who died with cancer having liver metastases at autopsy [4], and an estimated 1% to 4% of the general population alive today having hemangiomas, which are benign liver tumors [5]. The advantages of ultrasound in imaging of focal liver disease are well known; they include safety, relative accuracy, low cost, and availability. However, once a liver lesion is detected, whether during screening for malignancy or incidentally, its sonographic appearance may not provide the radiologist with adequate information for distinguishing a benign lesion, such as a hemangioma, confidently from a malignant lesion, such as a metastasis or hepatocellular carcinoma (HCC).

When there is uncertainty as to whether a lesion is benign or malignant, additional, more expensive imaging (CT, MRI, scintigraphy) or a biopsy, which carries risks of bleeding and infection, is often performed. Although the differentiation between benign and malignant lesions on ultrasound images is generally difficult, experienced radiologists can determine whether or not a focal lesion is normal by observing the echo textural pattern of the lesions in these images. For the sonographic differentiation between benign and malignant focal liver lesions to be a viable alternative to CT, MRI, scintigraphy, and/or biopsy, we need to extract subtle sonographic information that may be difficult to evaluate visually by radiologists in a consistently and objective manner. It is therefore expected that sophisticated computerized analysis of the echo texture pattern can yield an objective characterization of lesions, and that it may improve radiologists' accuracy in the diagnosis of focal liver lesions.

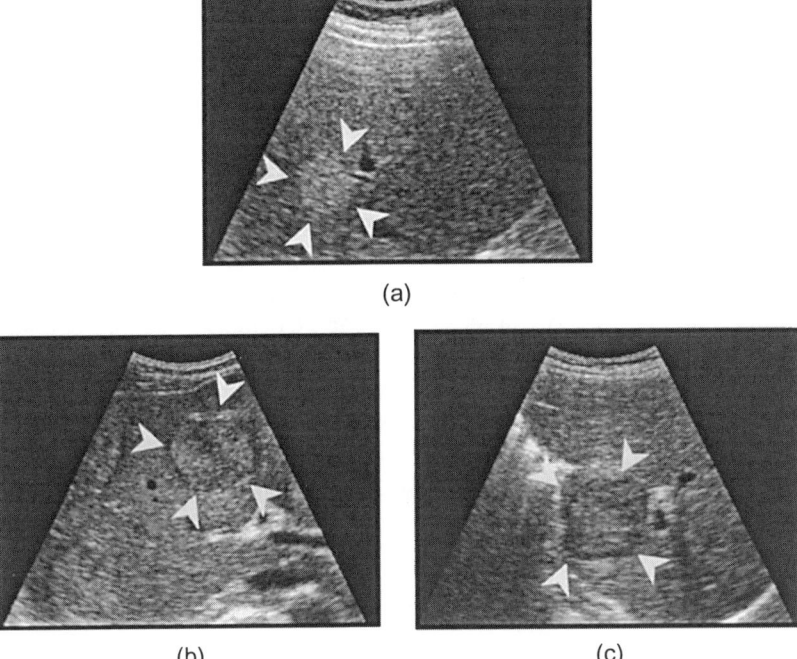

Figure 1. Typical appearance of three types of focal liver lesions. Each lesion is indicated by yellow arrows. (a) Hemangioma. (b) Hepatocellular carcinoma. (c) Metastasis.

In our texture analysis method, we use texture features that quantify the homogeneity of the lesions, because homogeneity is the prominent echo texture feature for the distinction between hemangiomas and HCCs/metastases. Typically, the textural pattern within a hemangioma is hyperechoic and homogeneous (see Figure 1a), whereas that of HCC is often inhomogeneous (see Figure 1b), although a small HCC can be hypoechoic and homogeneous. The internal textural pattern of a metastasis is often inhomogeneous (see Figure 1c), although the sonographic appearance is variable.

In the last several years, there have been significant advances in the area of computerized classification of benign and malignant lesions in various modalities based on NNs: classification of proximal disease in the aorto-iliac artery [9], classification of nodules in chest radiographs [10], differential diagnosis of interstitial lung diseases [11], [12], and classification of mammographic masses [13], [14] and clustered microcalcifications [15]. Success in these approaches is due to their use of unique image features, as well as their effective use of NNs for merging these image features. In these approaches, however, a NN was applied to a set of features that were preselected prior to the training of the NN. Therefore, the selected feature may not be optimal for classification by the NN.

Compared to these approaches, our texture analysis method is unique in the sense that it integrates a process of selection of multiscale texture features and a process of classification by a NN for effective distinction of benign from malignant lesions. To this end, we developed an automated method that selects a set of multiscale texture features which, when combined by a NN, maximizes the performance of the NN in the differentiation of benign from malignant lesions. The following subsections describe the details of our scheme for the classification of focal liver lesions based on multiscale texture analysis.

2   Texture Analysis of Focal Liver Lesions by Neural Networks

A schematic diagram of our scheme for the distinction between benign and malignant lesions is shown in Figure 2. In the first part of the scheme, ultrasound images are postprocessed to minimize the dependence of the images on the ultrasound scanners, which included the following steps: Speckle noise in the images was suppressed by application of a median filter. The ultrasound images might contain a nonuniform background trend due to mismatching of the time-gain settings and the patients' attenuation. We estimated the background trend with a two-dimensional, second-order polynomial surface-fitting technique based on the least-square method [16]. The estimated background was subtracted from the ROI for correction of the difference in global attenuation among sonograms.

Then, regions of interest (ROIs) are manually extracted from within the focal liver lesions. The ROIs were subjected to our new texture analysis method, which divides these ROIs into one of two classes, i.e., benign or malignant. This texture analysis method is described in detail in the following subsections.

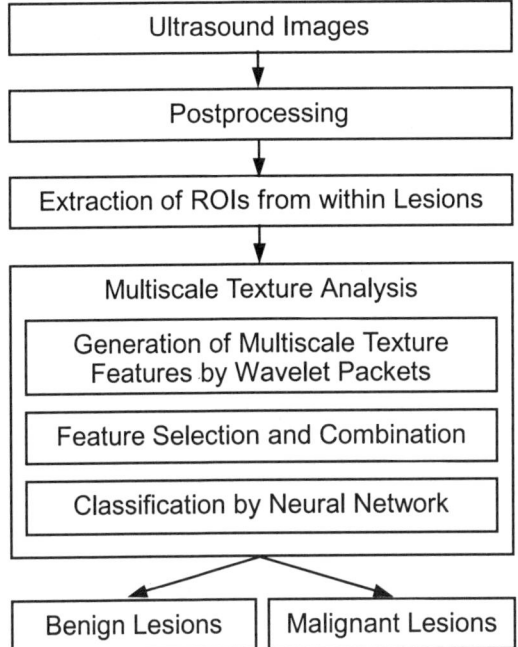

Figure 2. Schematic diagram of our scheme for differentiation between benign and malignant focal liver lesions.

2.1   Wavelet Packets

Wavelet packets generate a multiscale representation of an image, which is obtained by recursively decomposing the image into low- and high-frequency subbands with a set of special filters called *quadrature mirror filters* [17] by using the following filter bank algorithm.

Let H_i and G_i, $i = x, y$, be a one-dimensional low-pass and high-pass filters, respectively. Here, i indicates the orientation to which the filter is applied. These filters form a set of quadrature mirror filters when their impulse responses represented by $h_i(n)$ and $g_i(n)$, respectively, satisfy the following condition:

$$g_i(n) = (-1)^n h_i(1 - n), i = x, y.$$

Various filter coefficients $h_i(n)$ have been proposed corresponding to various types of wavelets. In the wavelet packet decomposition, an image I is decomposed into four different sets of coefficients, H_xH_yI, G_xH_yI, H_xG_yI, and G_xG_yI, corresponding to "low-low," "high-low," "low-high," and "high-high" frequency components, respectively. This operation corresponds to dividing the original image, V^0, into an approximation image, W_0^1, and a detailed image, W_i^1, $i = 1,2,3$, as follows:

$$V^0 = W_0^1 \oplus W_1^1 \oplus W_2^1 \oplus W_3^1. \tag{1}$$

Then the decomposition is repeated recursively on these four components as follows:

$$W_n^j = W_{4n}^{j+1} \oplus W_{4n+1}^{j+1} \oplus W_{4n+2}^{j+1} \oplus W_{4n+3}^{j+1}.$$

Here, j runs from 0 to $J \leq \log_2 N$, where N is the size of the original image, and n runs from 0 to $4^J - 1$. This decomposition can be continued J times, which is the user-defined maximum decomposition level. The index j is the scale index, which indicates that the size of the support of the corresponding wavelet packet is 2^j. The index n is the frequency index, which indicates that the wavelet packet has roughly n oscillations (a frequency $2^{-j} n$). Each subband image, W_n^j, corresponds to a small region specified by (j, n) on a space-frequency plane. Therefore, a particular choice of a subset of the subband images provides a decomposition of the original image over a specific type of tiling on the space-frequency plane. This decomposition yields a quad-tree-structured subband decomposition of the original image, as illustrated schematically in Figure 3. In the figure, the root node (V^0) represents the original image, whereas each of the other nodes in the tree, indicated by W_n^j, represents a subband image of the original image.

2.2   Multiscale Texture Features

Given a texture feature defined on the original image, which we call a *single-scale texture feature*, one can define a set of *multiscale texture features* derived from the feature as a set of feature values calculated on some nodes of the tree in Figure 3. More precisely, let us denote $T^{(\gamma)}$ as a subtree of the complete wavelet packet decomposition, where γ represents a set of indices $\gamma = \{(j, n)\}$ in the subtree, i.e., $T^{(\gamma)} = \{W_n^j;$

$(j, n) \in \gamma_j^{\wedge}$. Let us denote F_n^j as a feature calculated on the node W_n^j. Then a set of multiscale texture features for $T^{(\gamma)}$, denoted by $F^{(\gamma)}$, is defined by a set of features that are calculated from a node included in the subtree, i.e., $F^{(\gamma)} = \{F_n^j; (j, n) \in \gamma_j^{\wedge}\}$. It should be noted that the single-scale texture feature can be seen as a special case of the multiscale texture feature, F_0^0, that consists of a single feature obtained from the original image W_0^0.

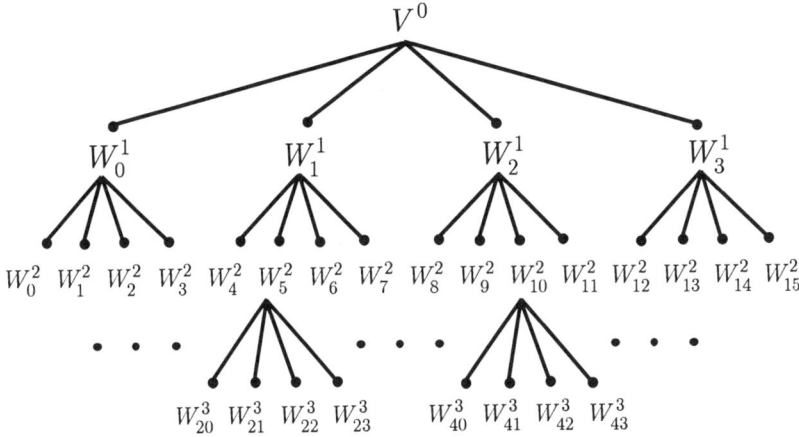

Figure 3. Schematic representation of the wavelet packet decomposition of an image. V^0 represents the original image, and each of the other nodes indicated by Ws represents a subband image of the original image yielded by the wavelet packet decomposition.

2.3   Optimal Selection of Multiscale Texture Features

We used these multiresolution features to differentiate benign from malignant lesions. The novelty of our method is that, for any given classifier that provides a scalar measure of *discrimination performance* (*class separability*), we can automatically generate a set of multiscale features that maximizes the performance of the classifier in distinction between benign and malignant focal liver lesions in the following manner: Assume that we are given a classifier and a set of *testing* images that consists of ROIs each of which is known to be either benign or malignant. First, each ROI in the database is decomposed one level by wavelet packets, which produces four subimages (the four child nodes in the second row in Figure 3), as shown in Equation (1). The multiscale texture features calculated from these four subimages of *all* the ROIs in the testing set are subjected to the classifier, and the discrimination performance is calculated. If the discrimination performance is improved, this means that the set of subimages yields a higher performance than does their parent; therefore, the decomposition is kept. Otherwise, the decomposition is "withdrawn" and only the single-scale feature is used. If the decomposition is kept, each of the four subimages is decomposed further, as shown in the third row in Figure 3, and the features are calculated from the child nodes. The decomposition of each of the subimages is kept *if and only if* the discrimination performance is improved when the classifier employs both the features newly generated by the decomposition *and* the multiscale features generated thus far.

It should be noted that the discrimination is performed using "accumulated" multiscale features. In other words, we use all of the multiscale features generated up to the subband image (node) at which the decomposition at each is performed. This decomposition process is continued, starting with the root node, until a predefined maximum level of decomposition (depth of the tree), J, is reached.

This algorithm is described formally as follows. We denote $P(F^{(\gamma)})$ as the discrimination performance of a set of multiscale texture features $F^{(\gamma)}$. (The definition of P is given at the end of Section 2.4.)

1. Start with the root node by setting a subtree $T^{(\gamma)} = W_0^0$. Set $j = 0$ and $n = 0$.

2. At each non-terminating node W_n^j, construct a tree
   $T_{\text{parent}}^{(\gamma)} = T^{(\gamma)} \cup W_n^j$ and
   $T_{\text{child}}^{(\gamma)} = T^{(\gamma)} \cup W_{4n}^{j+1} \cup W_{4n+1}^{j+1} \cup W_{4n+2}^{j+1} \cup W_{4n+3}^{j+1}$.

3. Calculate the multiscale texture features $F_{\text{parent}}^{(\gamma)}$ and $F_{\text{child}}^{(\gamma)}$ from subtrees $T_{\text{parent}}^{(\gamma)}$ and $T_{\text{child}}^{(\gamma)}$, respectively.

4. Set the tree $T^{(\gamma)} = T_{\text{child}}^{(\gamma)}$ if $P(F_{\text{parent}}^{(\gamma)}) < P(F_{\text{child}}^{(\gamma)})$. Otherwise, set $T^{(\gamma)} = T_{\text{parent}}^{(\gamma)}$.

5. Repeat 2 to 4 from $n = 0$ to 4^j and $j = 0$ to J.

The generation of the multiscale texture features is performed to improve the classification performance at each step. As a result, one can generate a set of multiscale texture features, extracted from a subtree in the wavelet packet decomposition, that maximally separate the ROIs into two classes.

The number of multiscale texture features may become very large when they are generated as described above. Such a large number of features tends to degrade the performance of the classifier because of the curse of dimensionality [6]. (See Section 3.4.3 and the bottom curve in Figure 8.) Therefore, it is desirable to keep only those features that effectively capture the salient differences in the internal sonographic patterns between benign and malignant lesions, and thus actively contribute to the overall classification performance. For this purpose, we used a *backward elimination* method [18]. In this method, we started with the entire set of multiscale features and eliminated a feature whose absence does not decrease the discrimination performance. We continued this process until the discrimination performance started to decrease, or a predefined minimum number of features was reached. An example of a resulting decomposition path is shown in Figure 4b. The kept nodes corresponding to the path from which the multiscale texture features are calculated are shown by one terminal node (subimage) at the 1st level decomposition, and the three terminal nodes at the 2nd level decomposition.

2.4   Combination of Multiscale Features by Neural Network

Because multiple texture features are generated in the above algorithm, an efficient method for combining these features is crucial for achieving a high performance in the distinction between benign and malignant lesions. For this purpose, we employed a NN because NNs were shown to yield a high performance in combining computer-extracted features for effective classification of lesions [10], [19]. We used a simple, two-layer NN trained by backpropagation [8], although other types of NNs can be used without changing the algorithm described in the previous section. The input nodes of NN corresponded to the multiscale texture features. The number of hidden nodes was set equal to one half of the number of input features. All of the input values were normalized to the range from 0 to 1 so that biases caused by unbalanced feature values were avoided. The output of the NN represented a ranked ordering of the likelihood of malignancy, in which 0 and 1 indicate "definitely benign" and "definitely malignant," respectively.

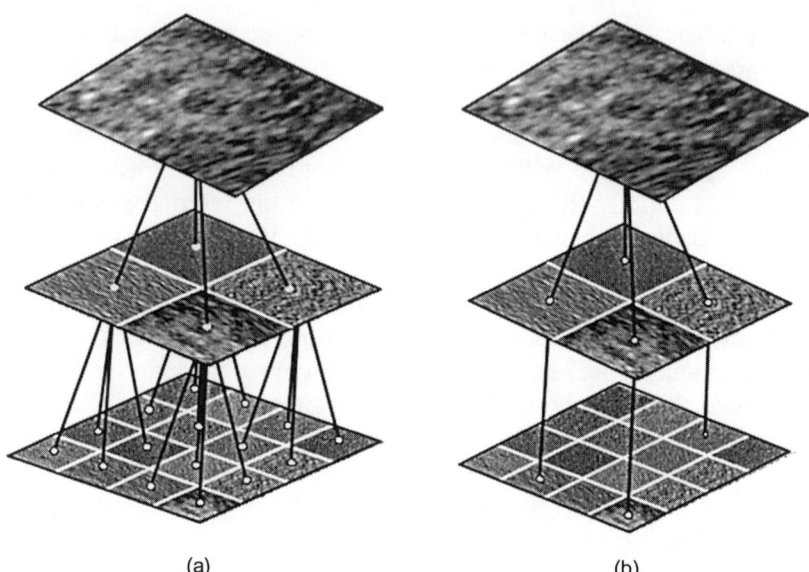

(a)   (b)

Figure 4. (a) Illustration of the wavelet packet decomposition of an ROI extracted from an ultrasound image. The image in the top layer represents an original ROI corresponding to V^0 in Figure 3. Subimages in the second and the third layers are generated by the 1st and the 2nd level wavelet packet decomposition, and they correspond to Ws in Figure 3. (b) Example of the optimally selected subimages, indicated by 4 terminal nodes, by the method described in Section 2.3. The optimal multiscale texture features are the features calculated from these selected subimages.

We trained and tested the NN by using a *jackknife* (or *cross-validation*) method [20], which was shown to be effective in estimation of the performance of NNs [21]. In this method, ROIs in the database were divided randomly into two sets, and one set was used for training, and the other set was used for testing the NN. The jackknife method was applied on a per-case basis, namely, no ROI from the same case as one of the testing ROIs was included in the training set. The advantage of this method is that training and testing cases can be completely separated; therefore, the decomposition criterion is least biased to the cases in the database. It should be noted that "a case" means "a set of ROIs from the same patient" here; therefore, no ROIs from the same patient were included in the training and testing sets simultaneously, thus avoiding a positive bias in evaluation of the NN's performance.

To measure the discrimination performance of the NN, we employed receiver operating characteristic (ROC) analysis [22], [23]. ROC analysis is a statistical technique that assesses performance in any two-group classification task. Given the scalar outputs of a classifier, an ROC curve is generated by use of a maximum-likelihood estimate based on a binormal model, which represents the relationship between the sensitivity (true positive fraction) and the specificity (1 − false positive fraction) of the classifier. The ROC curve yields a value for the area under the ROC curve (A_z) that indicates an unbiased estimation of the performance of the classifier being tested. Generally, a larger A_z value indicates a better overall performance in distinguishing between the two classes. The minimum and the maximum A_z values are 0.5 and 1.0, respectively. In this study, after a NN was trained by a given set of multiscale features $F^{(\eta)}$ and the testing set of ROIs, the output of the NN from individual ROIs in the training set was subjected to the LABROC4 [24] program† to yield an A_z value. This A_z value was used as the discrimination performance $P(F^{(\eta)})$ for the set of multiscale features $F^{(\eta)}$, as described in Section 2.3.

3   Experimental Results

3.1   Database of Focal Liver Lesions

For this study, we developed a digital image database involving benign (hemangiomas) and malignant (HCCs and metastases) focal liver lesions. The database consisted of 44 patients, including 17 hemangioma, 11 HCC, and 16 metastasis cases. The size of the focal liver lesions ranged from 15 mm to 70 mm (mean: 37 mm, standard deviation: 17 mm). The diagnosis of malignant disease was based upon pathologic evaluation. The diagnosis of hemangioma was based upon pathologic evaluation or characteristic CT or MRI appearance.

---

† The latest version of the LABROC4 program can be obtained free of charge from ftp://random.bsd.uchicago.edu/roc .

In all cases, images were obtained either with a GE Logiq 700 (GE Medical Systems, Milwaukee, WI) or a Siemens Sonoline Elegra (Siemens Medical Systems, Ultrasound Group, Issaquah, WA) ultrasound machine, using convex transducers with 3 to 4 MHz, and filmed with multi-format cameras. Filmed images were digitized to 12 bits/pixel with an effective pixel size of 0.1 mm by a high-quality laser scanner (Abe Sekkei, Tokyo, Japan).

A radiologist visually identified individual lesions on sonograms and outlined the boundaries of the lesions. For echo texture analysis, we manually selected, from each lesion, ROIs that were completely circumscribed by the boundary of the lesion. The size of the ROIs was 64 by 64 pixels (i.e., 6.4 by 6.4 mm^2). A total of 193 ROIs, which consisted of 50 hemangiomas, 87 HCCs, and 56 metastases, were used as input to the classification scheme developed in this study. In the following, we refer to the ROIs extracted from hemangioma cases as *benign ROIs*, and those extracted from either HCCs or metastases as *malignant ROIs*.

3.2   Experimental Conditions

The frequency content of the multiscale features provided by the wavelet packets is defined by the *mother wavelet* [25] used for generating the wavelet packets. We examined the effect of various types of orthogonal mother wavelets such as Daubechies, Meyer, and Coiflet, as well as biorthogonal mother wavelets such as the spline wavelet, with various vanishing moments. The best average classification performance was obtained from the wavelet packets generated by the biorthogonal spline with decomposition order of 4. The wavelet packet decomposition was performed up to the 2rd level.

As described in Section 1, homogeneity of the echogenicity within the lesions is the prominent texture feature for the distinction between hemangiomas and HCCs/metastases. We employed three representative types of texture features that quantify homogeneity [26], that is, entropy, root mean square (RMS), and the first moment of the power spectrum (FM), as the single-scale texture features in our multiscale texture analysis. (See Appendix for the definitions of these texture features.) It should be noted, however, that other types of features can be used without changing the algorithm for the classification based on the multiscale texture analysis.

3.3   Distribution of Feature Values

Figure 5 shows the distribution of two of the single-scale features, the RMS and FM, for the ROIs extracted from the three types of lesions. A considerable overlap among these three types of ROIs was observed, and no separation between the benign and malignant ROIs was observed.

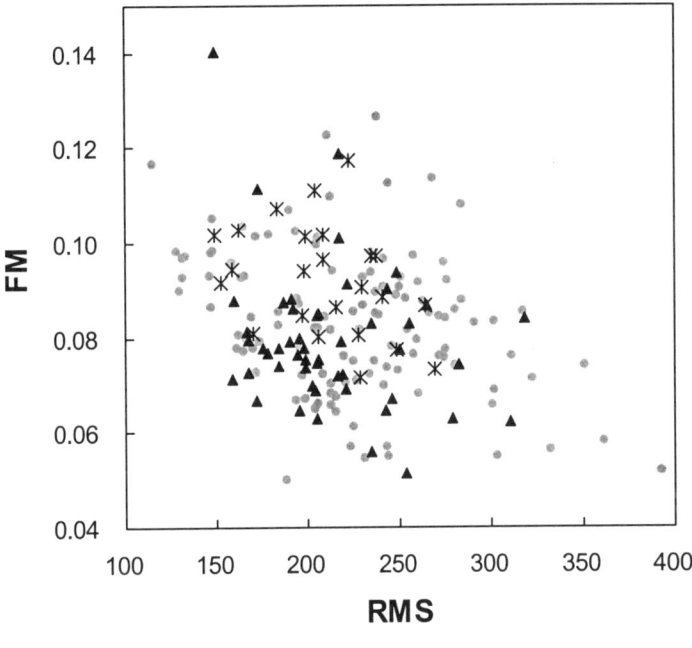

Figure 5. Distribution of two of the single-scale texture features, root mean square (RMS) and the first moment of the power spectrum (FM), for the ROIs extracted from within hemangioma, HCC, and metastatic lesions.

Figure 6 shows the distribution of the multiscale features for the same set of ROIs, in which the horizontal axis corresponds to one of the multiscale features generated by the RMS, and the vertical axis corresponds to one of the multiscale features generated by the FM.

Although there is still a substantial overlap among these ROIs, separation between the benign and malignant ROIs is much improved in this scatter plot. In Figure 6, the benign ROIs tend to have lower RMS and FM values compared to those for the malignant ROIs. This indicates that the classification performance of the multiscale texture features is superior to that of the single-scale features.

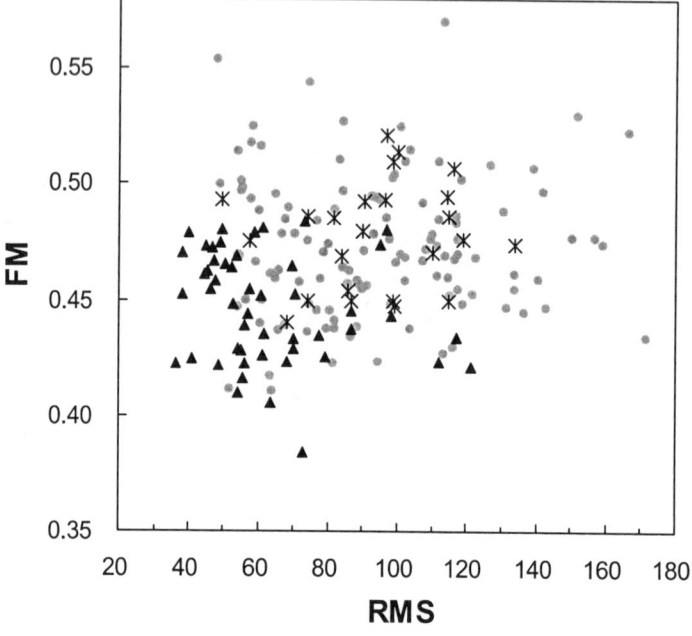

Figure 6. Distribution of the multiscale texture features for the same set of ROIs as those in Figure 5. The horizontal axis corresponds to one of the multiscale features generated by the root mean square (RMS), and the vertical axis corresponds to one of the multiscale features generated by the first moment of the power spectrum (FM).

It should be noted that Figure 6 is only a two-dimensional projection of the multidimensional feature space defined by the multiscale texture features. Therefore, the separation between the benign and malignant ROIs can be much better in the multidimensional feature space than that shown in Figure 6. Because this multidimensional feature space is difficult to visualize, the classification performance of the multiscale texture features was investigated by ROC curves, as described in the next subsection.

3.4   Classification Performance

3.4.1   Performance of Optimally Selected Multiscale Features

When the multiscale texture analysis, in which three types of texture features (i.e., entropy, RMS, and FM) were used, was applied to the analysis of 193 ROIs, the process for the generation of the multiscale features yielded two levels of decomposition; among them, 4 leaf nodes were selected, which yielded 12 multiscale texture features. Our classification method based on these multiscale features yielded a very high A_z value of 0.92 in distinguishing benign from malignant ROIs, as shown in the top curve in Figure 7. An operating point on the curve shows a specificity of 80% at a high sensitivity of 90%.

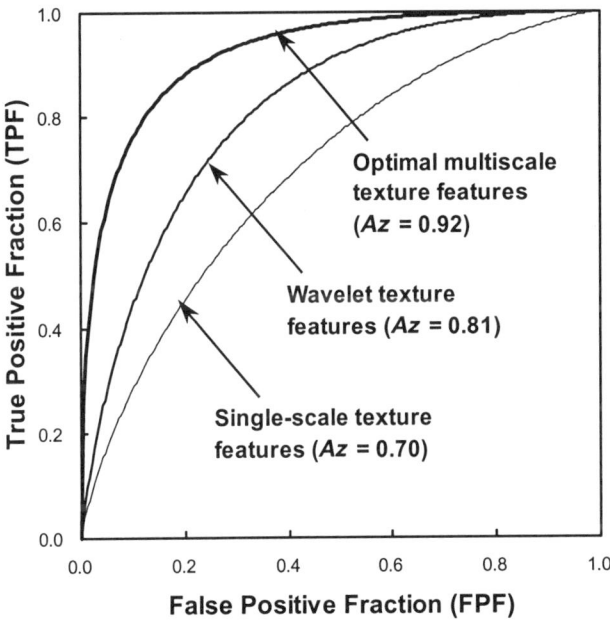

Figure 7.   ROC curves indicating the performance of the texture analysis method in distinguishing between benign and malignant liver lesions. The top curve was obtained by use of the optimally selected multiscale texture features. The middle curve was obtained by uses of the wavelet texture features. The bottom curve was obtained by use of the single-scale texture features.

3.4.2 Performance of Wavelet Texture Features and Single-Scale Texture Features

For comparison, we measured the performance of the *wavelet texture features*, in which the feature values were calculated from the subimages obtained by orthogonal wavelet decomposition. These wavelet texture features were input to the NN, and the NN was trained and tested by the jackknife method. In a similar manner, we combined the three single-scale texture features by NN and measured their performance. The combined wavelet texture features and the combined single-scale features yielded low A_z values of 0.82 and 0.70 in the distinction between benign and malignant ROIs, respectively. The corresponding ROC curves are shown in the middle and bottom curves in Figure 7, which show a point specificity of only 55% and 35%, respectively, at a sensitivity of 90%.

We measured the statistical significance of the differences between the two A_z values for the multiscale and single-scale texture features. For this purpose, we employed a conventional hypothesis test for the difference between the A_z values as described by Metz *et al.* [22]. In this method, the A_z values that were calculated from individual sets of ROIs generated by the jackknife method were subjected to a two-tailed t-test to yield a p-value. The result showed that the p-value was less than 0.001, which indicates that the difference between the two A_z values was statistically significant. In the same manner, the difference between the A_z values for the multiscale and wavelet texture features, as well as the wavelet and the single-scale features, was shown to be statistically significant.

3.4.3 Efficiency of Multiscale Texture Features

To clarify the efficiency of the multiscale texture features, we computed the classification performance obtained from each of the three types of single-scale texture features to that obtained from the multiscale texture features derived from the single-scale feature (Table 1). In the classification of benign and malignant lesions, the A_z values for the single-scale texture features were 0.57, 0.59, and 0.60 for entropy, RMS, and FM, respectively. On the other hand, the A_z values for multiscale texture features derived from entropy, RMS, and FM were 0.83, 0.85, and 0.86, respectively. Clearly, the classification performance was much higher for the multiscale features than that for the single-scale features in each type of texture feature. The difference between the A_z values was statistically significant, as shown by the p-values in the last column in Table 1.

Table 1. Performance of the single-scale and multiscale texture features in distinguishing benign from malignant focal liver lesions. The last row, labeled "All," indicates the performance obtained by combining of the three texture features by NN, whereas the other rows indicate the performance when these features are used individually.

Feature type	Single-scale texture features (Az)	Multiscale texture features (Az)	p-value
Entropy	0.57	0.83	< 0.001
RMS	0.59	0.85	< 0.001
FM	0.60	0.86	< 0.001
All	0.70	0.92	< 0.001

Although the multiscale texture features were effective in the classification of benign and malignant lesions, when all the multiscale features generated by the wavelet packet decomposition were combined by a NN, the performance degraded substantially, as shown by the bottom ROC curve in Figure 8. This is mainly due to the curse of dimensionality, because the number of subimages at the 2nd level wavelet packet decomposition was 16, which generated 48 multiscale texture features. Many of these features were redundant features, which also contributed to the degradation of the performance of the NN.

3.4.4 Performance of Multiscale Texture Features in the Distinction between Different Types of Lesions

We also measured the classification performance of the multiscale texture features in distinguishing between two types of lesions, that is, hemangiomas and HCCs, hemangioma and metastases, and HCCs and metastases. When all three features were used, our method yielded a high A_z value of 0.93 in the classification between hemangiomas and HCCs, and 0.94 in the classification between hemangiomas and metastases. The ROC curves that show the classification performance for the multiscale texture features in these four combinations are shown in Figure 9. This performance is comparable to that in the classification between benign and malignant lesions, which indicates that the multiscale texture features can distinguish the two types of malignant lesions, HCC and metastasis, from hemangioma with high accuracy.

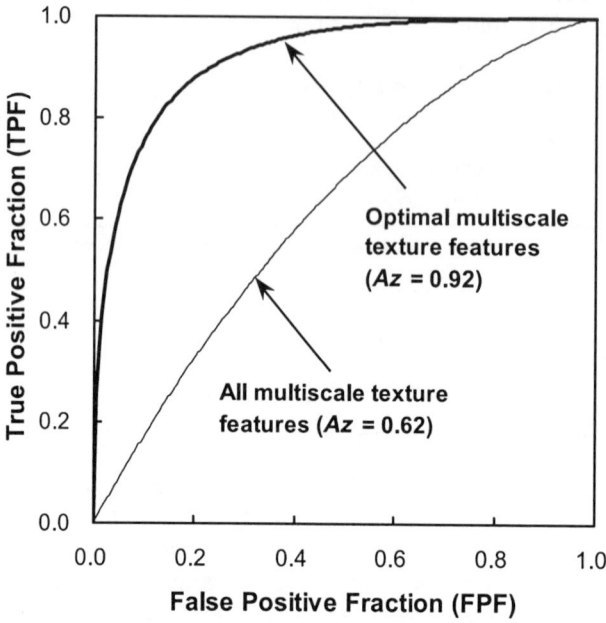

Figure 8. Comparison of the performance of the optimally selected texture features (upper curve) and that of all the multiscale texture features combined. Although the selected set of multiscale texture features yields a high classification performance, when all the multiscale features (up to the 2nd level wavelet packet decomposition) are combined by NN, they show an inferior performance in the same classification task.

However, the accuracy of discrimination between HCCs and metastases is not very high. As indicated by the lowest ROC curve in Figure 9, the performance of the multiscale features in the distinction between HCCs and metastases is much lower (A_z value of 0.85) than that in the distinction between benign and malignant lesions. This is due to the fact that the sonographic appearance of the internal textural patterns of HCCs and metastases often overlaps substantially, whereas there is less overlap between these two lesions and hemangiomas.

These results indicate that our method of multiscale texture analysis based on wavelet packets and NN is very effective in distinguishing benign (hemangiomas) from malignant (HCCs and metastases) lesions.

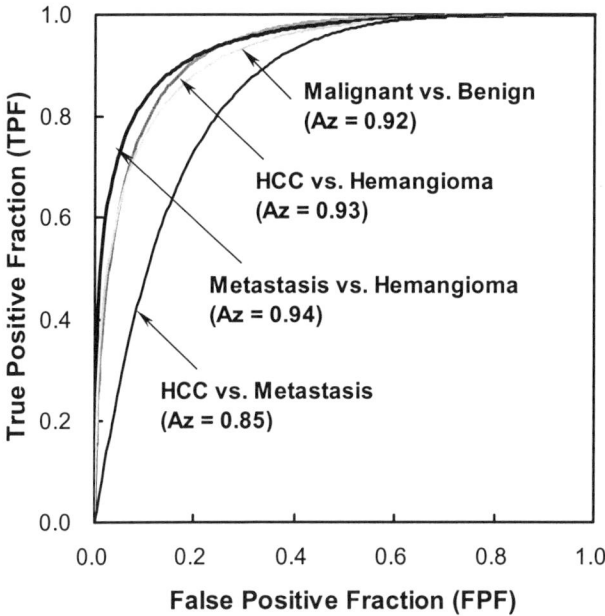

Figure 9. ROC curves indicating the performance of the multiscale texture features in four different classification tasks.

4   Discussion

We developed a texture analysis method that is capable of differentiating the echo texture patterns within focal liver lesions. For this purpose, we developed a novel method of multiscale texture analysis, which yielded a high performance in the classification between benign and malignant lesions. At the high sensitivity of 90%, our method was able to classify 80% of benign lesions correctly as benign. This means that about 80% of patients with hemangiomas can avoid further examination such as biopsy if our method is used for the diagnosis of these patients.

We used a relatively small database in this study, which is one of the limitations on the generalization of the results of this study. Our purpose in this study was to determine whether our multiscale texture analysis of focal liver lesions had the potential to discriminate malignant from benign lesions, and thus, ultimately to help reduce the number of unnecessary examinations. This potential has been shown based upon ROC analysis even though the number of lesions in the database is small. Extension to a larger database will further confirm the usefulness of our method.

The ultimate goal of our classification scheme is to improve the diagnostic accuracy in distinguishing benign from malignant focal liver lesions in ultrasound images, by providing a computer output as a "second opinion." The performance obtained from our multiscale texture analysis is high enough to improve radiologists' diagnostic accuracy, judging from the performance of the other computer-aided diagnosis schemes that were demonstrated to improve the accuracy of radiologists [27]. Observer studies that estimate improvement of radiologists' diagnostic accuracy in distinguishing benign from malignant lesions with and without computer aid are expected to show the benefit of our classification scheme.

5   Conclusions

We have developed a computerized scheme for classification of focal liver lesions based on multiscale texture analysis using wavelet packets and NN. Our new method showed a high performance in distinguishing benign from malignant lesions in ultrasound images, and it outperformed single-scale texture features used for the same differentiation task. We believe that our method can be a powerful tool for developing a computer-aided classification scheme for focal liver lesions in ultrasound images. Such a high-performance scheme will aid in improving the accuracy in distinguishing benign from malignant lesions in sonograms, and thus will reduce the number of patients with benign focal liver disease who require additional imaging, or who need to undergo biopsy or surgery for diagnosis.

Acknowledgments

The authors thank the members of the Kurt Rossmann Laboratories in the Department of Radiology at the University of Chicago for helpful discussions. This work is supported by the Louis Block Fund for Basic Research and Advanced Study, The Division of the Biological

Sciences, and the Pritzker School of Medicine, the University of Chicago.

Appendix

For completeness, we provide in this appendix the definitions of the three texture features [26] that were used in this study. In the following, the image I for which these textures are calculated has dimensions $N \times N$, and has a pixel value $I(x, y)$ at the pixel point (x, y).

Entropy

The non-normalized Shannon entropy is defined by

$$\text{Entropy} = -\sum_{x=1}^{N}\sum_{y=1}^{N} |I(x, y)|^2 \log_2 |I(x, y)|^2 .$$

Root Mean Square (RMS) Variation

The RMS variation represents the magnitude of deviations from the mean value of the image, defined by

$$\text{RMS} = \sqrt{\sum_{x=1}^{N}\sum_{y=1}^{N} (I(x, y) - M)^2} , \quad M = \frac{1}{N^2}\sum_{x=1}^{N}\sum_{y=1}^{N} I(x, y)$$

First Moment of Power Spectrum

The first moment of the power spectrum represents the fineness or coarseness of the texture pattern, defined as follows:

$$\text{FM} = \frac{\int\int \sqrt{\omega^2 + \xi^2}\, F(\omega, \xi)^2\, d\omega d\xi}{\int\int F(\omega, \xi)^2\, d\omega d\xi} ,$$

where $F(\omega, \xi)$ represents the two-dimensional Fourier transform of the image. Generally, the higher the first moment, the more irregular the margin becomes.

References

[1] Chang, T. and Kuo, C.-C.J. (1993), "Texture analysis and classification with tree-structured wavelet transform," *IEEE Trans Image Processing*, vol. 2, pp. 429-441.

[2] Unser, M. (1995), "Texture classification and segmentation using wavelet frames," *IEEE Trans Image Processing*, vol. 4, pp. 1549-1560.

[3] Unser, M. and Eden, M. (1989), "Multiresolution feature extraction and selection for texture segmentation," *IEEE Transactions on Pattern Analysis and Machine Intelligence*, vol. 11, pp. 717-728.

[4] Ros, P.R. (1994), "Malignant liver tumors," in Gore, R.M., Levine, M.S., and Laufer, I (Eds.), *Textbook of Gastrointestinal Radiology, vol. 2.*, Saunders: Philadelphia, pp. 1897-1946.

[5] Jeffrey, R.B. and Ralls, P.W. (1995), *Sonography of the abdomen*, New York: Raven Press.

[6] Coifman, R.R. and Wickerhauser, M.V. (1992), "Entropy-based algorithms for best basis selection," *IEEE Trans Information Theory*, vol. 38, pp. 1713-1716.

[7] Yoshida, H. and Casalino, D. (2000), "Classification of focal liver lesions in ultrasonic images based on echo texture analysis," *Proc SPIE 3982*, pp. 252-256.

[8] Haykin, S. (1999), *Neural networks: a comprehensive foundation*, New Jersey: Prentice Hall.

[9] Wright, I.A. and Gough, N.A. (1999), "Artificial neural network analysis of common femoral artery Doppler shift signals: classification of proximal disease," *Ultrasound Med Biol*, vol. 25, pp. 735-743.

[10] Nakamura, K., Yoshida, H., Engelmann, R., MacMahon, H., Katsuragawa, S., Ishida, T., Ashizawa, K., and Doi, K. (2000), "Computerized analysis of the likelihood of malignancy in solitary

pulmonary nodules by use of artificial neural networks," *Radiology*, vol. 214, pp. 823-830.

[11] Ashizawa, K., MacMahon, H., Ishida, T., Nakamura, K., Vyborny, C.J., Katsuragawa, S., and Doi, K. (1999), "Effect of an artificial neural network on radiologists' performance in the differential diagnosis of interstitial lung disease using chest radiographs," *Am J Roentgenol*, vol. 172, pp. 1311-1315.

[12] Ashizawa, K., Ishida, T., MacMahon, H., Vyborny, C.J., Katsuragawa, S., and Doi, K. (1999), "Artificial neural networks in chest radiography: application to the differential diagnosis of interstitial lung disease," *Acad Radiol*, vol. 6, pp. 2-9.

[13] Huo, Z., Giger, M.L., Vyborny, C.J., Bick, U., Lu, P., Wolverton, D.E., and Schmidt, R.A. (1995), "Analysis of spiculation in the computerized classification of mammographic masses," *Med Phys*, vol. 22, pp. 1569-1579.

[14] Huo, Z., Giger, M.L., Vyborny, C.J., Wolverton, D.E., Schmidt, R.A., and Doi, K. (1998), "Automated computerized classification of malignant and benign masses on digitized mammograms," *Acad Radiol*, vol. 5, pp. 155-168.

[15] Jiang, Y., Nishikawa, R.M., Wolverton, D.E., Metz, C.E., Giger, M.L., Schmidt, R.A., Vyborny, C.J., and Doi, K. (1996), "Malignant and benign clustered microcalcifications: automated feature analysis and classification," *Radiology*, vol. 198, pp. 671-678.

[16] Katsuragawa, S., Doi, K., Nakamori, N., and MacMahon, H. (1990), "Image feature analysis and computer-aided diagnosis in digital radiography: effect of digital parameters on the accuracy of computerized analysis of interstitial disease in digital chest radiographs," *Med Phys*, vol. 17, pp. 72-78.

[17] Verterli, M. and Kovacevic, J. (1995), *Wavelets and Subband Coding*, NJ: Englewood Cliffs.

[18] Fukunaga, K. (1990), *Introduction to Statistical Pattern Recognition*, San Diego: Academic Press.

[19] Wu, Y.C., Doi, K., Giger, M.L., Metz, C.E., and Zhang, W. (1994), "Reduction of false positives in computerized detection of lung nodules in chest radiographs using artificial neural networks, discriminant analysis, and a rule-based scheme,". *J Digit Imaging*, vol. 7, pp. 196-207.

[20] Efron, E. (1982), *The Jackknife, the Bootstrap, and Other Resampling Plans*, Philadelphia: Society for Industrial and Applied Mathematics.

[21] Zhang, W., Yoshida, H., Nishikawa, R.M., and Doi, K. (1998), "Optimally weighted wavelet transform based on supervised training for detection of microcalcifications in digital mammograms," *Medical Physics*, vol. 25, pp. 949-956.

[22] Metz, C.E. (1986), "ROC methodology in radiologic imaging," *Invest Radiol*, vol. 21, pp. 720-733.

[23] Metz, C.E. (1989), "Some practical issues of experimental design and data analysis in radiological ROC studies," *Invest Radiol*, vol. 24, pp. 234-245.

[24] Metz, C.E., Herman, B.A., and Shen, J.H. (1998), "Maximum likelihood estimation of receiver operating characteristic (ROC) curves from continuously-distributed data," *Stat Med*, vol. 17, pp. 1033-1053.

[25] Daubechies, I. (1992), *Ten Lectures on Wavelets*, Philadelphia: Society for Industrial and Applied Mathematics.

[26] Jain, A.K. (1989), *Fundamentals of Digital Image Processing*, New Jersey: Prentice Hall.

[27] Doi, K., MacMahon, H., Giger, M.L., and Hoffmann, K.R. (Eds.) (1999), *Computer-Aided Diagnosis in Medical Imaging*, Excerpta Medica: International Congress Series I182, Elsevier: Amsterdam.

Index